W9-CRB-257

ACS SYMPOSIUM SERIES **373**

Surfactant-Based Mobility Control

Progress in Miscible-Flood Enhanced Oil Recovery

Duane H. Smith, EDITOR

U.S. Department of Energy
Morgantown Energy Technology Center

Developed from a symposium sponsored
by the Division of Colloid and Surface Chemistry
of the American Chemical Society
at the 61st Annual Colloid and Surface Science Symposium
Ann Arbor, Michigan,
June 21–24, 1987

American Chemical Society, Washington, DC 1988

Library of Congress Cataloging-in-Publication Data

Surfactant-based mobility control.
 (ACS Symposium Series; ISSN 0097–6156; 373).

 "Developed from a symposium sponsored by the
Division of Colloid and Surface Chemistry at the 61st
Meeting of the Division of Colloid and Surface Chemistry
of the American Chemical Society, Ann Arbor, Michigan,
June 22–24, 1987."

 Includes bibliographies and indexes.

 1. Secondary recovery of oil—Congresses. 2. Oil field
flooding—Congresses. 3. Surface active agents—
Congresses.

 I. Smith, Duane H., 1943–   . II. American Chemical
Society. Division of Colloid and Surface Chemistry.
III. Series.

TN871.S7678   1988   665.5'3   88–16675
ISBN 0–8412–1491–3

Foreword

The ACS SYMPOSIUM SERIES was founded in 1974 to provide a medium for publishing symposia quickly in book form. The format of the Series parallels that of the continuing ADVANCES IN CHEMISTRY SERIES except that, in order to save time, the papers are not typeset but are reproduced as they are submitted by the authors in camera-ready form. Papers are reviewed under the supervision of the Editors with the assistance of the Series Advisory Board and are selected to maintain the integrity of the symposia; however, verbatim reproductions of previously published papers are not accepted. Both reviews and reports of research are acceptable, because symposia may embrace both types of presentation.

Contents

MECHANISMS AND THEORY OF DISPERSION FLOW

DISPERSION FLOODS IN THE LABORATORY AND FIELD

Preface

> We need only to go back to the Arab oil embargo and the events that immediately followed to get some very good ideas about the consequences for our society when liquid fossil fuels are in short supply. We were not prepared for an energy shortage at that time, and we are not prepared for one now. Then, the public for a few brief moments appreciated that our oil and gas resources were finite and that a consistent national energy policy needed to be developed. That crisis passed, and we are currently faced with an oil glut. The public believes the whole shortage was a hoax. Nevertheless, in the not-too-distant future, we will again be facing the same problems (1).

THESE PITHY REMARKS SUMMARIZE the underlying reasons for the publication of this book.

Despite the massive drop of oil prices in the early 1980s, the crippling or elimination of several major oil companies, and the layoffs of thousands of talented employees, the oil industry has managed to commercialize a major "new" kind of enhanced oil recovery (EOR). This type of EOR is "gas-flooding", in which CO_2 (or less often, some other fluid) is injected into an "old" field at a pressure of about 8 MPa or greater to produce oil that otherwise would not be recovered.

Although it is widely believed that gas-flooding will be commercially successful for many decades, it has been known for more than 30 years that the process suffers from a major shortcoming: The injected fluid never contacts and mixes with most of the oil in a formation; hence, the commercialized process will produce only a fraction (perhaps 20%) of the oil that would be recovered if this mobility control problem could be solved. The only technique that has been shown to overcome the problem in the laboratory is the use of surfactants to create dispersions ("emulsions" and "foams") of the injection and reservoir fluids. Although the use of atmospheric-pressure dispersions (i.e., foams) for analogous processes was patented in 1958, research on the use of dispersions in gas-flooding began only a few years ago, when it became apparent that gas-flooding without surfactant-based mobility control would soon be commercialized.

As the first book on the subject, this volume summarizes the current status of surfactant-based mobility control and a goodly fraction of the research that has been performed on this emergent technology.

Views and conclusions expressed herein are those of the authors. I sincerely thank them for their efforts in presenting their work. We are all indebted to the reviewers. Most of the authors in this book know and understand the history of EOR; many believe that this history is being repeated and are striving to ameliorate the effects of this repetition.

DUANE H. SMITH
Enhanced Oil Recovery Group
Morgantown Energy Technology Center
U.S. Department of Energy
Morgantown, WV 26507–0880

March 25, 1988

Reference

1. Handy, L. L. *J. Petrol. Technol.* **1987,** *39,* 1591.

NEED FOR DISPERSION-BASED MOBILITY CONTROL

Chapter 1

Promise and Problems
of Miscible-Flood Enhanced Oil Recovery

The Need for Surfactant-Based Sweep and Mobility Control

Duane H. Smith

Enhanced Oil Recovery Group, Morgantown Energy Technology Center,
U.S. Department of Energy, Morgantown, WV 26507-0880

There is no known method or collection of methods that can recover
all of the petroleum from an underground reservoir. Even when several different methods are applied sequentially over a period of
decades, on average, they leave about two-thirds of the oil trapped
in the reservoir (1).

Enhanced oil recovery (EOR) is a collective term for various
methods of increasing oil recoveries that have been developed since
about 1970. Up until about 1980, the use of surfactants in EOR was
more or less synonymous with "micellar/polymer" flooding, in which
surfactants are used to decrease the interfacial tension between
"oil" and "water" from ~ 10 dyne/cm to $\leq$ 0.01 dyne/cm.

However, the term "surfactant flooding" is now becoming increasingly associated with ways to improve gas and steam flooding, rather
than with its earlier usage. In the early 1980's disappointment with
micellar/polymer flooding became widespread, as almost all oil being
produced by EOR was due to steam flooding. But steam flooding was
appropriate only for shallow reservoirs containing viscous, "heavy"
oil. Hence, the oil industry turned to "gas" flooding, especially
with CO_2, as the main hope for producing substantially more oil by
EOR. Thus, in the early 1980's the industry began to make very large
investments in gas-flood EOR, and the results of those investments
are just beginning to become known.

Because they have very low viscosities (compared to oil), steam,
CO_2, and other gases suffer from a major problem: these injected
fluids never make contact with much of the reservoir and the oil it
contains, but instead channel more or less directly from injection
to production well, leaving the uncontacted oil unproduced.

By improving "sweep" and "mobility control," surfactant-based
methods offer the most promising ways to alleviate these problems.
This use of surfactants appears to be just on the verge of commercialization for steam flooding. Because miscible CO_2 flooding has
been commercialized more recently, the use of surfactants to improve
gas-flood EOR has not yet been commercialized. Conceivably, however,
the long-term viability of gas flooding could prove to be dependent
on the success of current research efforts in the use of surfactants
to alleviate "bypass" problems.

Gas-Flood Enhanced Oil Recovery

Status and Prospects of Gas Flooding. Gas-flood EOR includes the
injection of CO_2, N_2, stack gases, or mixtures of light hydrocarbons
(e.g., ethane and propane) to produce oil that would otherwise not be
recovered (1-3). It is similar in many aspects to, but usually dif-
ferentiated from, steam flooding, in which the primary function of
the injected fluid is to heat the reservoir. Gas floods are commonly
divided into "miscible floods," in which the injected fluid eventually forms a single phase with the oil, and "immiscible floods" in
which only part of the injected fluid dissolves in the oil. Higher
pressures and the injection of CO_2 or light hydrocarbons into reservoirs that contain light oils favor miscibility, whereas floods with
N_2 or of reservoirs of heavy, high API gravity oil are seldom miscible. Most miscible floods are "multi-contact miscible," that is,
miscibility is not achieved near the injection well at first contact
of the injected fluid with the oil. Instead, the injection fluid
initially tends to extract the lighter hydrocarbons from the oil, and
it is this propagating mixture of changing composition that eventually achieves miscibility with the oil ahead of it. Gas flooding is
usually thought of as a tertiary process, i.e., as a process applied
after primary production and water flooding; but gas flooding may
also be used for secondary, or even primary, production.
    Perhaps the biggest advantage of N_2 flooding is that it poten-
tially can be used anywhere in the world, such as on North Sea plat-
forms, where other injection fluids are not available or their
delivery cost would be prohibitive. The commercial usability of N_2
depends, of course, on the cost of extracting it from the atmosphere.
    Since light hydrocarbons contain considerable combustion ener-
gies, their use as EOR fluids is generally limited to remote locations, from which the cost of their delivery to commercial markets
would be prohibitive. The North Slope of Alaska is a very important
example of a major field in which there is little to do with produced gases except to reinject them (or possibly burn them on-site
for thermal EOR). Since much of the Alaskan oil is heavy, immiscible
processes are of great interest for this area.
    In the contiguous United States, the main emphasis is on misci-
ble CO_2 flooding. Carbon dioxide is relatively inexpensive since it
has no value as a fuel and is available in large amounts from natural
deposits or as a waste product of industrial processes. Table I
shows major natural deposits of CO_2, their approximate amounts, and
their locations with respect to oil fields (1). With (or even without) partial recycling of CO_2, there is no foreseeable shortage of
this injection fluid.
    In 1983, before most current gas-flood projects were started,
gas flooding was the source of 1 percent of the oil produced in the
United States. Steam flooding, a more mature technology, accounted
for 10 percent of United States oil production (4). The importance
of gas flooding was expected to rise considerably over the next few
years (1). This prediction appeared to be reliable because large
investments in field projects and long distance CO_2 pipelines had
already been made, but most had not yet had time to produce significant amounts of incremental oil (1,4). Although some projects have
been postponed or even cancelled, despite the subsequent plunge of
world oil prices, CO_2 miscible flooding has largely lived up to the

TABLE I.   Some Sources and Amounts of
CO_2 for Gas Flooding

Field	Location	Reserves (Tcf)[a]
Sheep Mountain	Southeastern Colorado	1
Bravo Dome Unit	Northeastern New Mexico	> 6
McElmo Dome and Doe Canyon	Southwestern Colorado	10
Jackson Dome	Southern Mississippi	3
LaBarge-Big Piney	Southwestern Wyoming	20

[a] Trillion cubic feet.
NOTE: Data are from ref. 1.

expectation that it would be "the EOR technology" of the second half
of the 1980's. More than 60 CO_2 field projects are planned or in
progress within the United States (5). A recent meeting discussed
17 full-scale field projects, including 15 miscible CO_2 floods, an
enriched gas flood at Prudhoe Bay, and a nitrogen flood (Soc. Petrol.
Engrs. Forum Series, "Monitoring Performance of Full-Scale CO_2 Projects," August 17-21, 1987).

It can be anticipated that all gas-flood projects, as they are
presently being carried out, will leave a large fraction of the
reservoir oil uncontacted by the injected fluids. This bypassed oil
will remain in place, undisplaced by the injected fluid. Thus, in
each current field project, the amount of incremental oil produced by
gas flooding could be substantially increased if the uncontacted oil
could be reached. The improvement of the vertical and areal distribution of injected fluids throughout the reservoir, so that they contact substantially more oil, will require much better methods of
sweep and mobility control.

It has been estimated that CO_2 EOR should recover 15 billion
barrels of oil within the United States alone (5). At a price of
$20/bbl, this oil would be worth $300 billion. Thus, under this
scenario, even a 10-percent production increase from CO_2 flooding due
to the use of surfactants would produce oil valued at $30 billion.

If 10,000 cubic feet (10 Mcf) of CO_2 costing $0.80/Mcf produce
one barrel of oil, the CO_2 cost is $8/bbl. Currently, CO_2 utiliza-
tion ratios of 10 Mcf/bbl are required for profitable operation (5).
Thus, a surfactant-based process for improving sweep could cost as
much as a few dollars per barrel of oil if it doubled the utilization
ratio, and, correspondingly, smaller surfactant-associated costs
would be commercially feasible for smaller improvements in the utilization ratio.

These amounts, although approximate, are important guides to
the economic value of gas flooding with current technology, to the
potential value of research investments that might produce even

incremental improvements, and to the cost limits that must be met by any improvement to attain commercial viability.

Current Field Project Design Technology (1,5). The eventual commercialization of surfactant-based sweep control will require integration of this new technology with existing practices. Because of their huge costs, gas-flood field projects, like all other EOR field projects, demand careful technical design. Basic design parameters include the geology and physical description of the reservoir; amount of residual oil; reservoir porosity, permeability, temperature, and pressure; oil-water relative permeability curves; and the viscosity and so-called "minimum miscibility pressure" of the oil.

The latter quantity is usually determined phenomenologically by measuring displacements of crude oil with carbon dioxide (or other injection fluid) from a long capillary tube (i.e., "slim tube") at a series of successively higher pressures. A plot of displaced oil versus pressure usually has a break at about 95-percent recovery, which is taken as the "minimum miscibility pressure." Often this is not a true miscibility in the correct thermodynamic sense. For example, many crudes contain asphaltenes that precipitate and do not dissolve even after a series of theoretical "multiple contacts" between the crude and the propagating mixture of injection fluid and non-asphaltenic components of the crude.

The almost-total oil recovery obtained in slim tube experiments is an unrealistic measure of the oil recovery that can be obtained from oil reservoirs or even from laboratory floods of parallel cores that have different permeabilities. Roughly, a parallel-core flood might recover half of the fraction of oil obtained from a slim tube experiment, and a field flood might do well to recover half of the fraction of oil obtained in parallel-core floods (6).

These great differences are due to "sweep," "adverse mobility ratios," "conformance," and other fluid mechanical effects (see next section) that are exacerbated by the natural heterogeneities in porous rocks and whose alleviation is the purpose of sweep and mobility control technology.

But despite the great reductions in recovery caused by sweep effects, currently the design of a field project is often dominated by other concerns.

The chemical composition of the oil (especially of the lighter components) may receive considerable attention because this affects not only the miscibility pressure but also the predicted composition of the produced oil as a function of time. This information, in turn, is needed for major process and investment decisions. For example, how long will produced gas be saleable without upgrading, will produced gases be reinjected without separation, or when will equipment be needed to separate produced gases for separate sale and recycle?

In virtually all cases these decisions are made only after lengthy computer simulations. These simulations attempt to optimize the amounts of fluid injected, the injection and production rates, and other operation variables. The timing and equipment sizing for large expenditures such as plants for separation and recompression of CO_2 are based on breakthrough times and production predicted by the simulators.

The simulator may be a four-"component" (CO_2, oil, water, and natural gas) model, or it may a compositional simulator which attempts to calculate the compositions as a function of time and place of the amounts of 10 or more components and pseudo-components in each of the phases.

Despite the many different experimental data that are used as input parameters, the complexity of the simulators, the size of the computers on which they are run, and the large number of simulations that are run, the simulator predictions may be subject to very large uncertainties. Therefore, a field pilot flood is usually performed. At much less financial risk and in years less time than the full-scale flood, the pilot serves as a check on expectations. The reservoir model is "tuned" by changes in the values of the input parameters to make the simulator output fit the pilot data.

For the field-scale projects that have been initiated, calculated optimum CO_2 injection volumes ranged from 20 to 50 percent of the hydrocarbon pore volume. Predicted carbon dioxide utilization factors ranged from 5 to 15 Mcf CO_2/bbl of recovered oil. Projected ultimate enhanced oil recoveries ranged from 5 to 30 percent of the original oil-in-place (Soc. Petrol. Engrs. Forum Series, "Monitoring Performance of Full-Scale CO_2 Projects," August 17-21, 1987).

These numbers represent only the consensus of current expectations, and very large revisions of many predictions may be required after a significant number of actual full-scale production data become available. In short, the American oil industry has invested in a very costly commitment to gas-flood EOR. No one yet knows how successful this experiment will be, or, with complete certainty, what technological pitfalls may be the greatest threat to success.

The Problem of Adverse Mobility Ratio

For at least three decades, long before the recent widespread initiation of field projects, adverse mobility ratio has been recognized as a major technological problem -- perhaps "the major problem" -- of gas-flood EOR (7-9). Adverse mobility ratios produce "viscous fingering," a displacement phenomenon that has been known and studied both experimentally and theoretically for many years (9-14). This phenomenon was a principal reason for the failure of many of the liquified petroleum gas (LPG) floods of the 1950's and early 1960's (15).

Basically, all of these closely related problems occur because gas-flood injection fluids have very small viscosities at the temperatures and pressures at which they are used. For example, the viscosity of CO_2 at 13.8 MPa (2,000 psi) and 38°C (100°F) is about 0.066 cp, whereas the viscosities of reservoir oils are at least an order of magnitude greater (16). This produces a ratio of the mobility of the CO_2 to the mobility of the oil that is much greater than one. (The mobility of a fluid is defined as its relative permeability divided by its viscosity; for the definition of relative permeability, see equations below.)

When the mobility ratio is greater than one, the front between the displaced and displacing fluids is unstable if the porous medium is sufficiently wide (> 10 cm) to allow the formation and propagation of viscous fingers. The lower viscosity fluid channels or "fingers" through the displaced fluid, leaving much of it uncontacted. (See

Figure 1.) The channeling can occur with either miscible or immiscible floods and results in much lower production of the displaced fluid for any given throughput of the injection fluid once the latter reaches the production well (10,12,15). The problem, which is common to water flooding and to all EOR processes, is most severe for gas flooding simply because it is in gas flooding that the injected fluids have the lowest viscosities (most unfavorable mobility ratios).

It is useful to consider a rudimentary introduction to the mobility ratio problem. When a single fluid flows through a permeable medium, the flow is described by Darcy's Law,

$$\bar{V} = - (k/\mu) \nabla(P + pgz) \tag{1}$$

where V is the superficial velocity, k is the permeability, μ is the viscosity, P is the pressure, and pgz is the hydrostatic head (p is the density, g is the gravitational constant, and z is the vertical direction). The operator ∇ is the gradient operator; that is, the vector differentiation is carried out in all three dimensions.

The equation for conservation of mass (incompressible fluid) is

$$\nabla \cdot \bar{V} = 0 \tag{2}$$

and substitution of Equation 1 into Equation 2 gives the Laplace equation

$$\nabla^2(P + pgz) = 0 \tag{3}$$

For two-fluid flow and negligible gravity effects, Equation 1 is replaced by Equations 4a and 4b:

$$\bar{V}_1 = - (k_{r_1} K/\mu_1) \nabla P_1 \tag{4a}$$

$$\bar{V}_2 = - (k_{r_2} K/\mu_2) \nabla P_2 \tag{4b}$$

Here, $_1$ and $_2$ may be taken as the displaced and displacing fluids, respectively, and k_{r_1} and k_{r_2} are their relative permeabilities, which are functions of the volume fractions (i.e., saturations) of the fluids. As shown by Buckley and Leverett, the solution to Equation 4 gives a plot of saturation versus length along the flow direction that has a discontinuity (17). (The saturation of a fluid phase is its volume divided by the sum of the fluid volumes.) This discontinuity is often referred to as a shock front, and the flood is described as a piston-like displacement.

The front is inherently unstable, however, and this is often studied by a linear stability analysis. Infinitesimal perturbations are applied to all of the variables to simulate reservoir heterogeneities, density fluctuations, and other effects. Just as in the Buckley-Leverett solution, the perturbed variables are governed by force and mass balance equations, and they can be solved for a perturbation of any given wave number. These solutions show whether the perturbation dies out or if it grows with time. Any parameter for which the perturbation grows indicates an instability. For water flooding, the rate of growth, B, obeys the proportionality

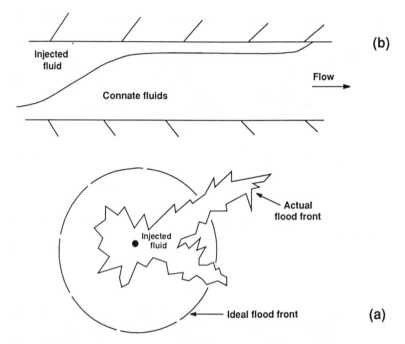

Figure 1.   (a) Example of poor areal sweep caused by fingering from
                  an adverse mobility ratio.
            (b) Example of poor vertical sweep caused by gravity
                  override from an unfavorable density ratio.

$$B \propto (M-1)/(M+1) \tag{5}$$

where M is the all-important mobility ratio, defined as (18)

$$M = (k_{r_2}/\mu_2) \, (\mu_1/k_{r_1}) \tag{6}$$

Whenever $\mu_1/\mu_2 > k_{r_1}/k_{r_2}$ (the usual case since the oil is more viscous), then $M > 1$, $B > 0$, and the flood front is inherently unstable, even in a porous medium that is homogeneous.

Figure 1 shows schematically viscous fingering, which occurs because the injection fluid is less viscous, and gravity override, which occurs because the injection fluid is less dense. Chapter 2 of this book, by Chang and Slattery, describes a linear instability analysis for gas flooding and also introduces a nonlinear treatment of the flow instabilities.

Most treatments of viscous fingering consider only frontal instabilities of two fluids in homogenous media (9-14). Mobility problems are greatly aggravated and their theoretical and experimental study greatly complicated by the facts that (1) the reservoir rock is highly nonuniform at every length scale from submillimeter to kilometer and (2) the number of fluid phases is often three rather than two. These two problems greatly complicate both the design of laboratory floods and the scaling of laboratory results to field dimensions.

The problems of combining flow instabilities with a description of reservoir heterogeneities in a realistic unified treatment is currently of great interest for all types of EOR. Chapter 3 of this book describes the beginnings for new methods of introducing the heterogeneities of a reservoir into simulations of the fluid flow. Treatment of the fully coupled problem, i.e., flow instabilities with three fluids in a field-scale natural reservoir, will require many years of research.

Methods of Gas-Flood Mobility Control

Water-Alternating-Gas (WAG). WAG is the only gas-flood mobility control measure that is regularly used in the field (7-10,19-25). Its chief virtues are familiarity (water flooding and gas flooding are alternated), apparent simplicity (the timing and ratio of water to gas are the only design parameters), and apparent low cost. Often it is used chiefly to maintain pressure while the CO_2 supply is interrupted, or to stretch out injection costs by substituting water for more-expensive CO_2. The WAG method decreases the injectivity and average mobility of CO_2 and other gases.

However, field tests and usage have not demonstrated the beneficial mobility effects claimed for WAG (5,25). To the contrary, a review of 15 CO_2 flooding projects revealed a lower average oil recovery for WAG floods than for floods that used a single injection of CO_2 followed by a water flood (25). The CO_2 retention in many reservoirs is currently between 90 and 100 percent, and poor CO_2 utilization factors substantially increase flooding costs (Soc. Petrol. Engrs. Forum Series, "Monitoring Performance of Full-Scale CO_2 Projects," August 17-21, 1987). These widespread problems may be additional evidence of excessive fingering and poor pattern control in field use of water for mobility control and containment.

Gravity segregation between the water and the less-dense gas can negate the effectiveness of the WAG process. Furthermore, the fronts between gas and the water preceding it are very unstable as described above. Hence, gas fingers through water-saturated zones and may bypass zones of high saturation of oil. The problems of this water-blocking mechanism have been well established by a large number of studies (7,8,20,26-29).

In laboratory measurements, Jackson, Andrews, and Claridge found that WAG ratios in the range from 0.5 to 2.0 were optimum for oil-wet media, but, that for water-wet conditions, the WAG process was detrimental to oil recovery (26). Recent studies indicate that most target reservoirs for CO_2 flooding have both oil-wet and water-wet surfaces (5,25). Hence, some water blocking can be expected, and the use of WAG will exacerbate the problem.

Gas-Soluble Viscosifiers. Conceptually, the simplest and most direct way to ameliorate the adverse mobility ratios encountered in gas flooding would be to dissolve a viscosifying agent in the injection fluid, just as water-soluble polymers are used to improve water flooding. However, it has proven difficult to find viscosifiers for CO_2 (30). The polymers investigated tend to be soluble only at low molecular weights and to increase viscosities by relatively small amounts. The problems of finding viscosifiers for light hydrocarbon solvents appear somewhat less difficult, but the prospects for viscosifying N_2 seem very remote. Recently, polymers have been found that produce somewhat larger viscosity increases for CO_2, but viscosifiers apparently have not yet elicited commercial interest (30).

Gelled and Cross-Linked Polymers. By themselves, water-soluble polymers are unlikely to prove suitable for improving gas-flood mobility control since these agents viscosify the aqueous phase, making the gas-to-aqueous phase mobility ratio even more adverse.

However, one or more polymer-gelling techniques will probably be used for plugging large fractures or other gross flow heterogeneities in reservoirs. In situ polymerization (sometimes called cross linking with an organic agent) has been field tested in a CO_2 flood with favorable results (31). A CO_2-specific, polymer-gelling technique, which utilizes the acidification of water due to dissolution of CO_2 to initiate the chemical reactions of the gelling process, has also been patented recently (32).

Surfactants. Surfactants and other amphiphilic compounds are very versatile materials when mixed with water and/or nonpolar compounds. For this reason, many different ways have been proposed to use surfactants to modify underground flow patterns in various types of EOR. One of these proposals, coprecipitation of a cationic and an anionic surfactant to clog cracks or pores in highly permeable rock zones, might prove useful for gas flooding (K. L. Stellner, J. C. Amente, J. F. Scamehorn, and J. H. Harwell, presented at the symposium, "Use of Surfactants for Mobility Control in CO_2 and N_2 EOR," Ann Arbor, June 22-23, 1987).

However, all methods of current major interest use the ability of surfactants to disperse one (or more) fluid phase(s) in another. Oil recovery processes in which a dispersion might be used to modify flow patterns include water flooding and high-capillary number (i.e.,

"chemical") flooding, as well as the steam- and gas-flooding applications that currently are much more studied.
Since there are various types of fluids, there are different kinds of dispersions that might be encountered in EOR. Fluids may be liquid, gaseous, or in the supercritical state. In EOR, gases are sometimes further classified as condensible (i.e., steam) or as not condensible into a liquid state of essentially the same composition. Certain fluids that contain sufficiently large concentrations of surfactant are termed microemulsions. Hence, depending on the type of oil recovery process and the conditions employed, a dispersion might be a so-called "oil-in-water" emulsion, an emulsion in which one of the fluids is a microemulsion, a foam (i.e., a dispersion of gas in a liquid), or a dispersion in which one of the phases is a supercritical fluid.

In most applications of CO_2 as an oil recovery agent, the CO_2 exists as a supercritical fluid above its critical pressure (7.4 MPa) and temperature (32°C), while its solutions in oil are liquids (5). Hence, the dispersion types of most direct interest are supercritical-fluid-in-a-liquid (for which no specific name yet exists) and emulsions of oleic-in-aqueous liquids (which may be encountered at low CO_2 saturations). However, for historical reasons (described below), all dispersions used in research on gas-flood mobility control are sometimes called "foams," even when they are known to be of another type.

Development of Surfactant-Based Mobility Control

Steps in the Development of Surfactant-Based Mobility Control.
Although surfactant-based sweep and mobility control for gas flooding are still in the research stage, major advances have been made in several areas from which a pattern of past and probable future development can be inferred. In approximate historical order, the steps in this development include the following:

1. Bench-scale demonstration (using floods of porous media) of an effect that, though complex and poorly understood, has the potential to substantially increase oil recoveries if it can be made commercially successful in the field (33,34).
2. Further experiments, including flow through porous media and measurements of fluid properties, to obtain phenomenological descriptions of dispersion flow and to select materials for laboratory study, but without pore-level interpretation of flow experiments, or systematic interpretation of material properties (35).
3. Observation of the mechanisms of lamella formation in single capillaries, etched media, and bead packs, followed by development of pore-level theory for the formation, flow, and disappearance of lamellae ("bubbles") and experimental tests of the theories (36-41).
4. Further refinement of the flow models from Step 3 for adaptation into field-scale simulators and development of one-and three-dimensional simulators for eventual field project design.
5. Phenomenological determinations by high-pressure experiments of dispersion properties required of sweep control surfactant systems and selection of homologous series of surfactants (42-45).

6.   Adaptation of existing experimental methods and theories, and
     development of new ones, to study (usually at reservoir condi-
     tions) the key phase, dispersion, and interfacial properties
     that control dispersion-based mobility processes, as determined
     from pore-level mechanisms and from other studies (D. H. Smith,
     Southwest American Chemical Society Meeting, Houston, Novem-
     ber 19-21, 1986).
7.   Design and selection of surfactants based on measurements of
     surfactant/CO_2/oil/water/electrolyte systems and their disper-
     sions under representative reservoir conditions (especially high
     pressure) and on thermodynamic and other correlations of the
     values of these key parameters with surfactant structures.
8.   Design of field projects using surfactants selected in Step 7
     and a combination of laboratory floods at reservoir conditions,
     computer simulators, and reservoir history matching.
9.   Testing of the laboratory-scale technology in small-scale field
     tests, transfer of the technology to field engineers, and com-
     mercialization of the technology (46-47). (See Chapter 22 for
     a review of field tests.)

Dispersion-Based Flow Control Before 1978

Laboratory Floods
     The use of surfactants to create dispersions that improve
flow behavior in EOR is currently associated quite closely with CO_2
and with steam flooding. However, for two decades research on
dispersion-based flow control focused on the development of low-
pressure, low-temperature foams for other applications, such as
improving water floods and reducing the losses of natural gas from
underground storage. Direct applications to CO_2 flooding did not
appear until 1978 (48,49).
     The use of foams to improve oil recoveries was patented by Bond
and Holbrook in 1958 (33). They increased oil recoveries in laboratory floods from 44 percent to as high as 70 percent by adding a
water-soluble surfactant to the aqueous phase before flooding with
air (33). Shortly thereafter a much more extensive series of
experiments was performed by Fried at the Bureau of Mines (prede-
cessor agency for the U.S. Department of Energy's EOR programs) (34).
Fried's experiments confirmed and extended the beneficial effects
of surfactants that had been claimed in the 1958 patent. In the
20 years following the invention of Bond and Holbrook, studies of
this technique appeared at an average rate of about one per year
(34,35,37,46,48-65).
     In 1964 Bernard and Holm reported a study of the effects of foam
on gas permeabilities (52). They found the seemingly magical property that the permeabilities (to N_2) of sand packs with permeabilities ranging from 100 to 146,000 mD were all reduced to essentially
the same value (7 to 3 mD, respectively) by foams. For a given
pressure gradient, gaseous and aqueous saturations were the same in
the presence of a foam as when no surfactant was used; however, for
a given saturation, the permeability of the gas was greatly reduced
by the foam. Foams reduced gas permeabilities by increasing the
saturation of trapped gas. Foams could be broken down by sufficiently high-pressure gradients, but these gradients were much higher
than would be encountered in reservoir use. Foam stability appeared

to decrease as the permeability of the sand pack increased, and gas permeability reductions were found to increase as the surfactant concentration increased. All of these measurements were made by the steady-state method at a pressure of about 10 atm.

A subsequent study showed that at a given aqueous saturation, the permeability of the aqueous phase was the same whether or not foam was present (53). Foam decreased aqueous permeabilities by increasing the saturation of trapped gas (thus decreasing the saturation of the aqueous phase). The reduction of aqueous permeability was found to last during the passage of many (10 to 25) pore volumes of surfactant-free water.

In a third paper by the Bernard and Holm group, visual studies (in a sand-packed capillary tube, 0.25 mm in diameter) and gas tracer measurements were also used to elucidate flow mechanisms (35). Bubbles were observed to break into smaller bubbles at the exits of constrictions between sand grains (see Capillary Snap-Off, below), and bubbles tended to coalesce in pore spaces as they entered constrictions (see Coalescence, below). It was concluded that liquid moved through the film network between bubbles, that gas moved by a dynamic process of the breakage and formation of films (lamellae) between bubbles, that there were no continuous gas path, and that flow rates were a function of the number and strength of the aqueous films between the bubbles. As in the previous studies (it is important to note), flow measurements were made at low pressures with a steady-state method. Thus, the dispersions studied were true foams (dispersions of a gaseous phase in a liquid phase), and the experimental technique avoided long-lived transient effects, which are produced by nonsteady-state flow and are extremely difficult to interpret.

Mast, in a pioneering 1972 paper, reported visual observations of foam flow in etched glass micromodels (37). His observations showed that some of the conflicting claims about the properties of foam flow in porous media were probably due simply to the dominance of different mechanisms under the various conditions employed by the separate researchers (37). Mast observed most of the various mechanisms of dispersion formation, flow, and breakdown that are now believed to control the sweep control properties of surfactant-based mobility control (37,39-41).

The first studies of CO_2 dispersions at pressures around 10 MPa (1,500 psi) were reported in 1978 (48,49). Before that time virtually all experiments were performed on true foams (i.e., near atmospheric pressure on dispersions of a low-density gas in a continuous liquid phase). For this historical reason, the literature on EOR often refers to high-pressure dispersions as "foams," even though the phase, physical, and flow properties of a dispersion in which both phases have a liquid-like density and compressibility cannot always be assumed similar to those of a true foam (66). For many (but not all) mechanistic studies it may be appropriate to employ a foam, and atmospheric pressure emulsions are appropriate stand-ins for high-pressure emulsions of the same chemical composition. However, atmospheric pressure foams cannot be used when the correct answer depends on replicating all of the physical properties of a dispersion that exists only at high pressure.

Surfactant Selection

An early attempt to correlate the physical properties of surfactant solutions and their foams with oil recoveries was performed
by Deming in 1964 (50). Deming concluded that high foaming ability
favored high displacement efficiency, but that high foam stability
were not required for high displacement efficiency. Bernard and Holm
found that oil substantially decreased the abilities of most surfactants to reduce aqueous permeabilities, but that some surfactants
remained effective even in the presence of oil (52,53).

Early researchers turned to foamability or "foaminess" measurements to screen surfactants for flow experiments (51). In one variation of this test, a long, vertical glass cylinder with a frit at the
bottom was filled with the test solution, and gas was forced through
the frit. The height of foam formed in the column was then measured,
or, the foam was collected and the amount of liquid in the foam
determined (51). In his screening of some 200 materials, Raza measured interfacial tensions of aqueous solutions with respect to air
and to oil, and the foamability and foam stability in the absence and
presence of oil. The latter experiment consisted simply of shaking
the solution in a test tube and measuring the volume of foam at various times (60).

The development of improved methods of surfactant design
required progress in several other areas: (1) understanding the
mechanisms of dispersion flow in porous media, to determine which
physical properties should be measured, and how their values would
affect sweep control; (2) measurements of these properties that are
valid at the conditions under which the surfactants will be used;
and (3) understanding of how the values of these parameters depend
on phase behavior, molecular structure, and other thermodynamic
variables.

The next section introduces recent progress in the area of dispersion formation and flow.

Mechanisms of Dispersion (Lamella) Formation and Flow

Dispersion Formation, Subdivision, and Coalescence. The ability to
create and control dispersions at distances far from the injection
well will be critical to the field-use of dispersion-based mobility
control. The early studies of Bernard and Holm, followed by more
recent work by Hirasaki, Falls, and co-workers, and others showed
that the flow properties of surfactant-induced dispersions depend
on the presence and composition of oil, volume ratio of the dispersed and continuous phases, capillary pressure, and capillary
number (35,37,39-41,52-54,68). However, it is the size of the droplets or bubbles that dominates dispersion flow (39,68). Moreover,
early debates on the ratio of droplet (or bubble) size to pore size
have been resolved by ample evidence showing that, under commonly
employed conditions, droplets are larger than pores (39). Only for
very large capillary numbers (i.e., for interfacial tensions of ca.
1 mdyne/cm or less) do droplets become smaller than pores (68).

Films of wetting fluid that extend across pores and may cause
dispersion formation are called lamellae. (See Figures 2 and 3.)
Several mechanisms have been identified that collectively determine
the number of lamallae and the distribution of droplet sizes of a
dispersion in a porous medium. For noncondensible fluids they

include (1) "leave-behind," (2) capillary snap-off, (3) division, (4) coalescence, and (5) diffusion of the dispersed phase between droplets.

"Leave-Behind"
    The "leave-behind" mechanism of lamella formation, which was observed by Mast and named by Radke and Ransohoff, is illustrated by Figure 2 (37,40). It is the dominant mechanism of lamella generation below a critical flow velocity in homogenous bead packs and may play a similar role in natural porous media (40). As the nonwetting phase invades a region previously filled by the wetting phase, two local nonwetting flow fronts may approach the same pore space from different directions, squeezing the wetting fluid in the pore into a lamella between them. The two fronts of the nonwetting phase need not arrive simultaneously at the pore site. Instead, they can arrive at different times and form the lamella when the local capillary pressure becomes sufficient (40). Unlike other "forward" mechanisms of lamella generation and division, the lamellae created by the leave-behind mechanism do not subdivide the nonwetting phase into discontinuous droplets, and, thus, they do not form a dispersion. For this reason, lamellae created by leave-behind offer considerably less resistance to flow. A second important feature of this mechanism is that each site can form only a single lamella, unless the pore body is refilled by the wetting fluid once the first lamella ruptures or leaves the site.

Thread Breakup
    A long, stationary droplet or "thread" of one fluid in another can break up into a string of smaller droplets, and this breakup mechanism has been further treated for slowly moving bubbles by Flumerfelt and co-workers (69,70). In tubes, thread breakup produces bubbles whose lengths are on the order of four times the diameter of the unbroken thread. The incorporation of this mechanism into a constricted tube model for dispersion flow in porous media is described in the chapter by Prieditis and Flumerfelt.

Capillary Snap-Off
    Capillary snap-off (Chapter 14, by Ratulowski and Chang), was described by Roof and is illustrated by Figure 3 (36). It is the same mechanism by which capillary forces trap oil in water flooding. In the exploitation of this mechanism for gas-flood mobility control, a surfactant solution wets the porous medium, forming a thin film between it and the second fluid phase (which in miscible CO_2 flooding could be a supercritical fluid or a liquid of any CO_2/oil ratio). If the mean curvature in the fluid-fluid interface is greater in the pore throat than in the rest of the capillary, then the capillary pressure is also greater than elsewhere ($P \sim 1/r$, where r is the mean radius). In the usual case that the pressure gradient in the nonwetting phase is negligible, the pressure in the wetting phase is less in the restriction than in the rest of the wetting film. Hence, the wetting phase flows into the pore throat, forming a thicker film there. Once this film becomes sufficiently thick, the interfacial tension makes it unstable; and the free energy can then be lowered by formation of a film, or lamella, across the pore throat. Continued flow of the discontinuous phase pushes the lamella from the

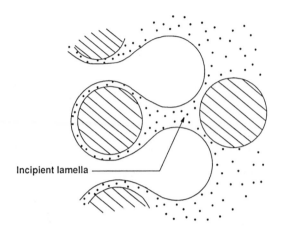

Incipient lamella

Figure 2.   Schematic of lamella formation by the "leave-behind"
            mechanism. (Reproduced with permission from Ref. 40.
            Copyright 1986 SPE-AIME.)

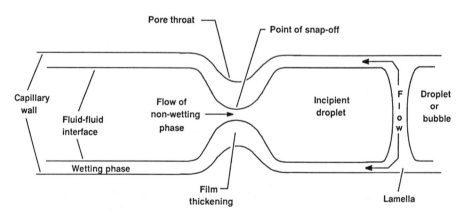

Figure 3.   Illustration of the formation of dispersions and the
            lamellae that separate them by the mechanism of capillary
            snap-off.

pore throat and down the capillary, allowing the process to repeat
itself. In steady-state flow this produces a line of equal-sized
droplets separated by lamellae of equal thickness. The pressure
gradient is proportional to the linear density of lamellae and is,
to a first approximation, independent of lamella thickness. Hence,
until lamellae become too thin to be stable, large reductions in the
mobility of the nonwetting phase are produced by conditions that
favor the formation of many droplets and the lamellae that separate
them.

A key factor in the commercialization of surfactant-based
mobility control will be the ability to create and control dispersions at distances far from the injection well (71). Capillary
snap-off is often considered to be the most important mechanism for
dispersion formation, because it is the only mechanism that can form
dispersions directly when none are present (39,40). The only alternative to snap-off is either leave-behind, or else injection of a
dispersion, followed by adequate rates of thread breakup and division to maintain the injected lamellae.

Division

When a dispersion is already present in a porous medium, division is another mechanism by which coarse droplets can become
smaller, and more droplets and lamellae can be formed (37,39). The
division mechanism also requires that the lamellae be mobile. Figure 4 illustrates how division can occur at a fork in a capillary.
If the pressure gradient is too small for the lamellae to move, no
event occurs. If the gradient is large enough to move lamellae
through one of the exits of the fork but not the other, droplets
leave by that exit after entering the Y without any change of size.
However, if the pressure gradient is sufficiently large to mobilize
lamellae in both exits, an entering lamella splits into two, one
lamella for each of the exit pores. Thus, the droplet of the nonwetting phase divides into two smaller droplets, and each entering
lamella divides into two thinner lamellae. The divisions increase
the local pressure gradient, unless the new lamellae are too thin
to be stable.

Countering snap-off and division, there are two mechanisms by
which droplets of noncondensible fluids can become larger and dispersions can coarsen and disappear.

Coalescence

When the lamella between two droplets thins and breaks, the
droplets on either side coalesce into a single, larger droplet
(41,72). Continuation of this "backward" process eventually leads
to the disappearance of the dispersion, if it is not balanced by the
"forward" mechanisms of snap-off and division. Lamellae are thermodynamically metastable, and there are many mechanisms by which static
and moving thin films can rupture. These mechanisms also depend on
the molecular packing in the film and, thus, on the surfactant structure and locations of the dispersed and dispersing phases in the
phase diagram. The stability and rupture of thin films is described
in greater detail in Chapter 7.

Diffusion Between Droplets
      Molecular diffusion of components of the dispersed phase from
smaller droplets to larger ones makes larger droplets grow at the
expense of smaller ones and tends to make dispersions disappear.
This diffusion occurs because the pressure is greater in the smaller
droplets. The pressure difference is proportional to the interfacial tension and to the difference between the inverse radii
($\Delta P/\gamma \sim 1/R_2 - 1/R_1$). Diffusion is thought to be a relatively slow
process, but it increases when pressure increases or other changes
increase the solubility of the components of the dispersed phase in
the dispersing fluid.

Effects of Capillary Number, Capillary Pressure, and the Porous
Medium. Since the mechanisms of leave-behind, snap-off, lamella
division and coalescence have been observed in several types of
porous media, it may be supposed that they all play roles in the
various combinations of oil-bearing rocks and types of dispersion-
based mobility control (35,37,39-41). However, the relative importance of these mechanisms depends on the porous medium and other
physico-chemical conditions. Hence, it is important to understand
quantitatively how the various mechanisms depend on capillary number,
capillary pressure, interfacial properties, and other parameters.
The quantitative description of these mechanisms is required for the
construction of meaningful flow simulators that can be scaled up
from the dimensions of laboratory experiments to the dimensions of
field use, and for scientifically based surfactant design.

Capillary Number
      Capillary snap-off occurs over a very wide range of flow rates
when the nonwetting phase flows from a lower permeability region into
a region where the permeability is suddenly higher (39,40). This is
a promising phenomenon in dispersion-based mobility control, because
it creates dispersions and their accompanying flow restrictions where
they are most needed to correct for heterogeneities in the porous
medium.
      However, using homogenous glass bead packs and defining the
capillary number as

$$C_r = \mu ULR/Kk_{nw} \, \gamma_{nw} \tag{7}$$

Ransohoff and Radke found that for $C_r \lesssim 10$ lamella generation
occurred only by the leave-behind mechanism. The lamellae moderately
increased the resistance to flow, raising it by about a factor of
five (40). (Here μ is the viscosity of the nonwetting phase, U is
the total superficial velocity, R is the bead radius, K is the absolute permeability, k_{nw} is the relative permeability, L is length,
and γ_{nw} is the interfacial tension.) The use of this definition of
the capillary number, instead of the usual $C_a = \mu U/\gamma_{nw}$, with μ the
viscosity of the wetting phase, was justified in the theoretical
analysis (40).
      For capillary numbers greater than the critical value, snap-off
occurred even in homogenous bead packs (40). The resulting dispersions caused much greater resistance to flow than the resistance produced by leave-behind lamellae (which do not disperse the nonwetting
phase).

Capillary Pressure and the Porous Medium
    There are three important ways by which capillary pressure
affects the dynamics of dispersion formation and disappearance: the
lower and upper limits on the range of capillary numbers over which
capillary snap-off can occur in homogenous media, and an upper limit
on the capillary number above which lamellae are unstable and droplets quickly coalesce.
    A lower limit on the capillary number required for snap-off
arises from a static analysis of nonwetting flow into the constriction. Roof's analysis of snap-off in symmetric constrictions shows
that there is a strictly geometrical requirement for the nonwetting
phase to enter the constriction. Below this limit snap-off cannot
occur. For example, in a circular constriction the relation

$$R_b > 2R_c R_g / (R_g - R_c) \qquad (8)$$

must hold for the wetting phase to flow into the constriction to
begin snap-off (36). (See Figure 3.) (Here R_b is the radius of
the pore, R_c is the radius of the pore throat, and R_g is the bead
radius.) For different pore geometries the factor 2^g must be replaced
by other dimensionless quantities, but these quantities also depend
only on the geometry of the pore structure (73).
    An upper limit on the capillary number required for snap-off
arises from the dynamics of wetting fluid flow into the constriction.
The capillary number must be below the upper limit for a long enough
time that sufficient wetting fluid can flow back into the constriction to form a lamella (40). If the volume of wetting fluid is too
small, the lamella cannot form.
    A third, related limit on the capillary pressure is created by
the existence of an upper critical capillary pressure above which the
life times of thin films become exceedingly short. Values of this
critical capillary number were measured by Khistov and co-workers for
single films and bulk foams (72). The importance of this phenomenon
for dispersions in porous media was confirmed by Khatib and col-
leagues (41). Figure 5 shows the latter authors' plot of the capillary pressures required for capillary entry by the nonwetting fluid
and for lamella stability versus permeability of the porous medium.
As discussed above, the upper limiting capillary pressure for lamella
formation must be considered, as well as the limiting capillary pressure for lamella stability.
    Capillary pressure effects appear to explain the very important
1964 discovery of Bernard and Holm that dispersions could make the
mobility of the nonwetting phase essentially independent of the absolute permeability of the porous medium (52). (See above.) Indeed,
the theoretical analysis of Khatib, et al., which was corroborated
by experiments, gave dispersed-phase mobilities at the upper limiting
capillary pressure (for coalescence) that were nearly constant for
absolute permeabilities ranging from 7D to ca. 1,000D (41).

Simulators and Core Floods for Dispersion-Based Sweep Control

Simulators. Now that substantial progress has been made in describing the formation, flow, and collapse of lamellae in single capillaries, the construction of detailed mechanistic descriptions of flow
in porous media has begun. The ultimate engineering objective of

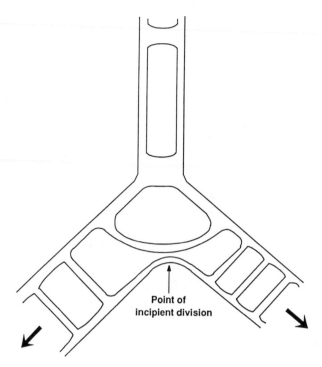

Figure 4.   Illustration of the formation of smaller droplets from a
             larger one by the division mechanism.

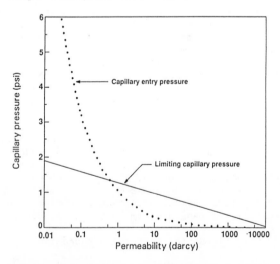

Figure 5.   Relationship between capillary entry pressure, limiting
             capillary pressure, and permeability of the medium.
             (Reproduced with permission from Ref. 41. Copyright
             1986 SPE-AIME.)

this work is a simulator that is sufficiently complex to capture the physico-chemical effects that are of major importance in field use of the technology, while remaining sufficiently manageable to be useful for designing field projects. Although work in this direction is very new, it is already evident that a variety of models of different complexities and utilizing different approaches will emerge. Some of these will be better adapted to planning and interpretion on the laboratory scale, whereas others, with a somewhat different mix of complexities, will be designed for direct field use.

After the single-capillary model discussed in the previous section, the next most complex "porous medium" is a bundle of unconnected capillaries of different radii and/or cross-sectional shapes. Porous media of this type have been much studied for other types of problems, but they appear too simplistic for dispersion flow (65).

However, two-dimensional networks appear to capture almost all of the important physics and chemistry of the problem. (Their dimensionality, two dimensions instead of the three of a real porous medium, is fundamentally incorrect.) Figure 6 illustrates a square-lattice network in which all tubes have the same length and connectivity but different radii. Important parameters for a network include the population distribution of radii, the physical distribution of those radii in the medium, and the connectivity (number of tubes that meet at a node).

In actual use for mobility control studies, the network might first be filled with oil and surfactant solution to give a porous medium with well-defined distributions of the fluids in the medium. This step can be performed according to well-developed procedures from network and percolation theory for nondispersion flow. The novel feature in the model, however, would be the presence of equations from single-capillary theory to describe the formation of lamellae at nodes where tubes of different radii meet and their subsequent flow, splitting at other pore throats, and destruction by film drainage. The result should be equations that meaningfully describe the droplet size population and flow rates as a function of pressure (both absolute and differential across the medium).

One of the very attractive features of this approach for research purposes is that the model can incorporate measurable physical parameters (e.g., pore-size populations, disjoining pressures of the wetting film) for both the fluid phases and their dispersions. A field-scale simulator based on this approach would probably require simpler equations that captured the relevant phenomena without explicitly addressing many of them. Chapter 15, by Prieditis and Flumerfelt, models two-phase flow in a network of interconnected channels that consist of constricted tube segments. Work on the creation of a model that contains capillary snap-off in a network similar to that of Figure 6 has very recently been started at the University of Texas (R. Schechter, personal communication, October 26, 1987).

Another attractive feature of two-dimensional network models is the fact that they can be carefully tested with laboratory experiments on well-defined porous media that closely replicate the media in the model. Thus, at this level the ambiguities that arise from the necessarily simplified description of a natural porous medium can be avoided. Experimental work on flow in two-dimensional etched media is represented by Chapter 12 by Shirley. Chapter 13, by Elsik

and Miller, describes visual studies of flow in a porous medium of
packed glass beads.
    The population balance simulator has been developed for three-
dimensional porous media. It is based on the integrated experimental
and theoretical studies of the Shell group (38,39,41,74,75). As
described above, experiments have shown that dispersion mobility is
dominated by droplet size and that droplet sizes in turn are sensitive to flow through porous media. Hence, the Shell model seeks to
incorporate all mechanisms of formation, division, destruction, and
transport of lamellae to obtain the steady-state distribution of
droplet sizes for the dispersed phase when the various "forward" and
"backward" mechanisms become balanced. For incorporation in a reservoir simulator, the resulting equations are coupled to the flow
equations found in a conventional simulator by means of the mobility
in Darcy's Law. A simplified one-dimensional transient solution to
the "bubble population balance" equations for capillary snap-off was
presented and experimentally verified earlier. Patzek's chapter
(Chapter 16) generalizes and extends this method to obtain the population balance averaged over the volume of mobile and stationary
dispersions. The resulting equations are reduced by a series expansion to a simplified form for direct incorporation into reservoir
simulators.

Core Floods. At present the strong coupling between droplet size and
flow has major experimental consequences: (1) flow experiments must
be performed under steady-state conditions (since otherwise the
results may be controlled by long-lived, uninterpretable transients);
(2) in situ droplet sizes cannot be obtained from measurements on an
injected or produced dispersion (because these can change at core
faces and inside the core); and (3) care must be taken that pressure
drops measured across porous media are not dominated by end effects.
Likewise, since abrupt droplet size changes can occur inside a porous
medium, if the flow appears to be independent of the injected
droplet-size distribution, it is likely that a new distribution is
quickly forming inside the medium (38).
    While mechanistic simulators, based on the population balance
and other methods, are being developed, it is appropriate to test the
abilities of conventional simulators to match data from laboratory
mobility control experiments. The chapter by Claridge, Lescure, and
Wang describes mobility control experiments (which use atmospheric
pressure emulsions scaled to match miscible-CO_2 field conditions) and
attempts to match the data with a widely used field simulator that
does not contain specific mechanisms for surfactant-based mobility
control. Chapter 21, by French, also describes experiments on emulsion flow, including experiments at elevated temperatures.
    For mechanistic studies, ambient pressure experiments on emul-
sions and foams often offer significant experimental advantages over
high-pressure experiments. However, high-pressure measurements are
also needed since the phase behavior, physical properties of the
fluids, and dispersion flow may all depend on pressure. Experiments
under laboratory conditions that closely match reservoir conditions
are particularly important in the design of projects for specific
fields. Chapter 19, by Lee and Heller, describes steady-state flow
experiments on CO_2 systems at pressures typical of those used in
miscible flooding. The following chapter, by Patton and Holbrook,

also contains results from high-pressure flooding experiments with CO_2 and N_2 in the presence of surfactants. Chapter 17, by Wellington and Vinegar, describes the use of computerized tomography (CT) to image the flow of mobility-control dispersions of CO_2 at typical reservoir pressures and temperatures.

Interfacial Properties, Dispersion and Phase Behavior, and Surfactant Design

Determination of Important Parameters in Surfactant Design. Recent work (Chapters 8 and 9) demonstrates the utility of correlating test results with surfactant structures. But as the complexities of pore level mechanisms, dispersion properties, and fluid behavior become better understood, it is also becoming increasingly clear that a variety of physical property measurements will be required for advanced surfactant design. Many of these measurements will be needed at pressures (ca. 10 MPa) that are characteristic of gas-flood conditions.

As discussed above, for capillary snap-off to operate as desired, the intended noncontinuous phase must be nonwetting, and the capillary number and capillary pressure must fall within certain limits. The fluid-fluid interfacial tension is the only parameter in the capillary number and capillary pressure that is subject to effective control by the process designer. Hence, the capillary number and pressure establish limits for acceptable values of this tension.

The fluid-fluid tension and the wettability requirement in turn set limits on the tension between the porous medium and each of the fluids. These fluid-solid interfacial tensions are affected by the isotherm for surfactant adsorption.

In summary, reservoir characteristics establish most of the parameters that control capillary snap-off, and interfacial tensions are the only controllable snap-off parameters. The dependencies of interfacial tensions on phase behavior and surfactant structure define many of the objectives of the surfactant designer.

Although large gaps exist in our knowledge, this section discusses various aspects of the design sequence from capillary number to surfactant structure. First fluid-fluid tensions are briefly considered. Surfactant adsorption is then described, leading to a discussion of how adsorption can affect wettability.

Since a sequence of dispersion structures in bulk dispersions has been correlated with flooding results, the dependence of dispersion structure on phase behavior is also briefly reviewed. This leads to a discussion of phase behavior and its dependence on surfactant structure and other thermodynamic parameters.

Interfacial Properties

Fluid-Fluid Tensions
As defined by Radke and Ransohoff (Equation 7), the "snap-off" capillary number, C_r, contains the effective grain radius, R; the permeability, K; and the relative permeability of the nonwetting phase, k_{nw} (40). In field applications, the values of all of these parameters are set by the reservoir. Also contained in C_r are the total superficial velocity, U, and the distance between injection and production wells, L. Within narrow limits, L can be changed by

choices of the number and location of wells. But the reservoir engi-
neer has little control over U, except very close to injection wells.
Hence, only the fluid-fluid interfacial tension, γ_{nw}, through its
dependence on surfactant structure, is subject to a significant
degree of control.
    The fluid-fluid interfacial tension also appears as the con-
trollable design parameter in the capillary pressure and its effects
on dispersion-based mobility control. As described by Equation 9,
the capillary pressure, P_c, is directly proportional to the fluid-
fluid tension, γ_{nw}:

$$P_c = \gamma_{nw} \; (\phi/K)^{0.5} \; J \qquad\qquad (9)$$

(Here J denotes the Leverett J function, and ϕ is the porosity.)
Khatib and co-workers used Equation 9 in their demonstration of the
dependence of coalescence and gas mobility on the limiting capillary
number (41).
    Much is known about the surface tensions between surfactant/
water mixtures and air at 1 atm. However (except for thermodynamic
equations), hardly anything is known about tensions between aqueous
solutions saturated with CO_2 at ~ 10 MPa and their conjugate CO_2-rich
phases. Although interfacial tension measurements at such pressures
are very uncommon, values of the capillary number and their dependence on surfactant and hydrocarbon structures cannot be determined
without such data.

Surfactant Adsorption
    In surfactant-based mobility control, surfactant adsorption is
important in at least four ways:
- Surfactant that is adsorbed by the reservoir may be effectively
  lost, decreasing the effectiveness of the process and/or
  increasing the cost of the project by increasing surfactant
  cost.
- Chromatographic effects caused by surfactant adsorption tend to
  separate the various compounds of a surfactant mixture. Since
  commercial surfactants contain many different members of a
  homologous series, or even compounds that are chemically very
  different, chromatographic separation could alter the physical
  properties and effectiveness of the surfactant mixture.
- Surfactant adsorption and its effect on chromatographic flow can
  sometimes be exploited to control the distance from the injection well at which different surfactants mix. Thus, if the
  transport rates of two different chemicals are known, the slower
  moving one can be injected first, followed by the faster moving
  one at an appropriate later time. The distance from the well at
  which the two mix will be determined by the delay time between
  the injections and the difference in the transport velocities.
- Surfactant adsorption can change the wettability of the porous
  medium from hydrophilic to hydrophobic and even back again.
  Since capillary snap-off disperses the nonwetting phase in the
  wetting fluid, wettability changes may drastically alter the
  effectiveness of the process.
    The surfactant adsorption isotherm depends on surfactant struc-
ture, temperature, mineral content of the solid surface, and other
parameters. As illustrated by Figure 7, the isotherm for the

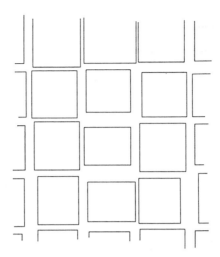

Figure 6.   Two-dimensional, square-lattice network with connectivity
             four and a distribution of capillary radii.

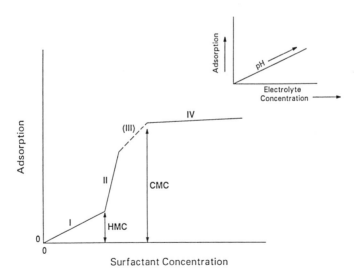

Figure 7.   Isotherm for surfactant adsorption as a function of
             surfactant concentration; salinity and pH have similar
             effects.

dependence of surfactant adsorption on concentration of surfactant,
hydrogen ion, or other species is often divided into four regions:
(See Chapter 10.)

- In Region I the concentration of surfactant in the bulk solution
  is very low, and typically the solution obeys Henry's Law. The
  concentration of surfactant adsorbed on the surface of the solid
  is also very low. As shown in Figure 8, the adsorbed molecules
  are believed to "lie down," maximizing their contact with the
  solid.

- The boundary between Regions I and II is appropriately known as
  the admicelle or hemimicelle concentration, depending on the
  structure postulated for the patches of adsorbed surfactant in
  Region II. As shown by Figure 8, the patches may be analogous
  to micelles, i.e., the patches may be bilayers, with polar
  groups on the outside of the admicelle contacting both the solid
  and the bulk solution. Or, the patches may be hemimicelles,
  i.e., monolayers adsorbed with the polar groups adjacent to the
  solid and hydrocarbon chains exposed to the bulk solution.

- Moreover, the structure of the adsorbed surfactant in Region II
  may depend on which of two conjugate bulk phases is in contact
  with it. According to one hypothesis, the bilayer structure
  could obtain when the solid is in contact with an aqueous phase.
  However, replacement of the aqueous phase with the conjugate
  oleic phase (while leaving the surfactant activity unchanged)
  might convert the admicelles into hemimicelles. In this
  hypothesis the outer surfactant layer of the bilayer would dissolve in the oleic phase and, perhaps, partially readsorb in
  the gaps between patches, with the polar groups in contact with
  the solid (J. H. Harwell, University of Oklahoma, personal communication, November 12, 1987). Such a change should lower the
  interfacial free energy.

- Region III is now thought to be a transition between Regions II
  and IV, in which the patches of adsorbed surfactant start to
  interact with their neighbors, as opposed to a true, separate
  region.

- The beginning of Region IV may or may not coincide with the
  critical micelle concentration (CMC). When it does, once
  micelles begin to form the thermodynamic activity of the surfactant rises only very slowly with further increases of surfactant concentration. Hence, the isotherm becomes essentially
  level for an extended range of surfactant concentration. The
  adsorbed surfactant is thought to form a bilayer, with polar
  groups towards both the solid and the solution.

Wettability

Because capillary snap-off produces a dispersion of the nonwetting fluid in the fluid that wets the porous medium, if the
"wrong" fluid wets the solid, the "wrong" type of dispersion will
be produced. As described by the Young-Dupre equation,

$$\gamma_{ns} - \gamma_{ws} = \gamma_{nw} \cos \theta \qquad (10)$$

the wettability, or contact angle θ, depends on the difference
between the nonwetting fluid-solid interfacial tension (γ_{ns}) and
the wetting fluid-solid interfacial tension (γ_{ws}). For the wetting

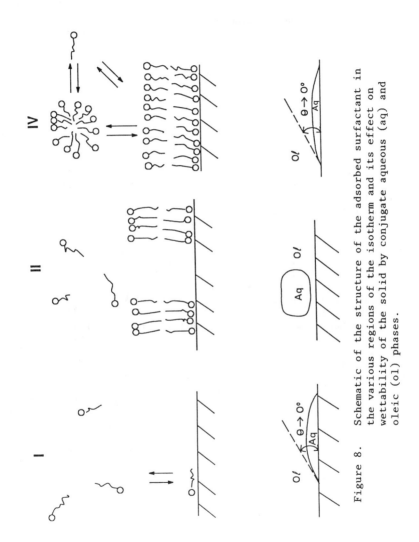

Figure 8.   Schematic of the structure of the adsorbed surfactant in the various regions of the isotherm and its effect on wettability of the solid by conjugate aqueous (aq) and oleic (ol) phases.

condition $\cos \theta = 1$, the desired value of $\gamma_{ns} - \gamma_{ws}$ is γ_{nw}, whose preferred value is determined by the desired values of C_r and P_c.

The wettability of the porous medium was found to have a significant effect on foam flow as early as 1966 (D. C. Bond and G. G. Bernard, AIChE 58th Annual Meeting, Dallas, February 7-10, 1966). Later, Kanda and Schechter showed that a foam produced a large reduction of permeability only if the aqueous phase wet the porous medium (64). Thus, various flow studies confirm the importance of wettability.

The Amott index is widely used as a measure of the wettability of porous media, for which contact angles cannot be directly measured (76). Figure 9 shows the Amott "wettability" index of Berea sandstone as a function of surfactant adsorbed for a quaternary amine surfactant (J. Comberiati and D. H. Smith, Morgantown Energy Technology Center, unpublished data). (The adsorption was changed by varying the pH at constant surfactant concentration.) Since, by definition, the value of the Amott index varies from - 1 to + 1, it is evident that surfactant adsorption produced a significant change in this wettability parameter. Other studies (see Chapter 11) have shown directly that different amounts of adsorption of a single surfactant can cause large changes in the contact angle (77).

Figure 8 shows schematically how the contact angle depends on the adsorption. In the absence of adsorption, the solid surface is wet by aqueous phases that spread over the surface. In Region I of the adsorption isotherm, the surface coverage by the surfactant is too small to significantly affect the wettability.

However, in Region II a significant fraction of the solid can be covered by the exposed oleophilic hydrocarbon chains of the surfactant. Depending on the exact location within Region II, the contact angle may be somewhat or greatly changed from hydrophilic towards hydrophobic. In addition, as described above, replacement of a bulk aqueous phase with its conjugate oleic phase may sometimes change the structure of the adsorbed surfactant, so that hydrocarbon chains instead of polar groups are presented to the bulk solution. Such a change would also convert the wettability from aqueous to oleic.

In Region IV the surface is completely covered by a surfactant bilayer, with the solution again in contact with polar groups. Hence, the wettability again returns toward the hydrophilic condition exhibited at low surfactant concentration.

The admicelle and critical micelle concentrations depend on the surfactant structure, temperature, pH, and many other variables. Thus, depending on the mineral and also these variables, the system may be in any of the various regions of the isotherm. Hence, the porous medium may have greatly different wettabilities, and the chance of forming the desired type of dispersion may also change.

Surfactant adsorption is addressed in Chapter 10 by Lapata, Harwell, and Scamehorn, which describes the different adsorption regions. Following their contribution is a chapter by Louvisse and González, in which surfactant adsorption and its effect on contact angles are measured.

Dispersion and Phase Behavior. The selection of surfactants for high-pressure gas-flood mobility control effectively began in 1978 when Bernard and Holm received a patent on the use of alkyl polyethoxy sulfates $(C_n H_{2n+1}(OC_2H_4)_m SO_4 M$ as mobility control agents for

a

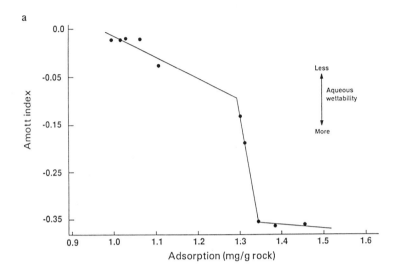

b

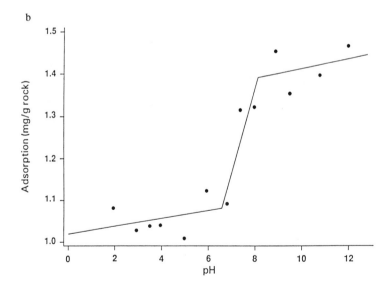

Figure 9.   (a)   Effect of adsorbed surfactant on the Amott index
                    (a measure of wettability).
             (b)   Effect of pH on surfactant adsorption (and thus with
                    Figure 7a on wettability).

CO_2 floods at pressures above 10 MPa ($\underline{49}$). Only flooding experiments were used to choose surfactants in this first extension of
low-pressure foam flooding to high-pressure, CO_2-miscible flooding.
    Subsequently, Wellington and colleagues have used high-pressure,
sight-cell tests to choose surfactants for CO_2 flooding ($\underline{42-44}$). The
oil-producing abilities of the different types of dispersions
observed in the cell tests were measured in flooding experiments.
The cell tests consisted of three steps: (1) In the first step, the
cell was charged with approximately equal volumes of CO_2 and an aqueous solution of the test surfactant in reservoir brine. The desired
behavior was formation of an emulsion-like dispersion of the CO_2-rich
phase in the aqueous phase. (2) In the second step, a small amount
of reservoir oil was added. Desirable surfactants formed three-phase
dispersions in which both the CO_2-rich and oil-rich phases were dispersed in the aqueous phase. (The crude oil was not miscible with
CO_2.) (3) In the third step of the test, the amount of oil in the
cell was increased until it was somewhat larger than the volumes of
CO_2 and of aqueous phase. Although relatively few surfactants passed
this third step, the desired dispersion structure was believed to be
droplets of the CO_2-rich phase dispersed in the continuous oleic
phase, with films of aqueous surfactant solution encasing the dispersed droplets ($\underline{42,43}$, S. L. Wellington, Shell Development Company,
personal communication, November 13, 1987). "Foaminess" tests performed under these conditions correlated with the results of flooding
experiments. Both nonionic alkoxylated surfactants and their anionic
sulfonated derivatives were tested by these methods ($\underline{42,43}$).
    Dispersion types in two-phase systems have been shown to corre-
late with the region of the phase diagram ($\underline{i.e.}$, the miscibility gap)
in which the system lies ($\underline{78-81}$). On the basis of these correlations
and of the phase behavior reported in Chapter 4, it is expected that
for many systems, increases of pressure, temperature, oil concentration, or salinity will cause dispersions of the CO_2-rich phase in the
aqueous phase to invert.
    However, the structures of three-fluid dispersions have not yet
been well characterized. Nevertheless, a starting point for investi-
gating these structures is provided by existing knowledge of the
phase behavior and critical point locations of systems that form
three liquids, combined with correlations between this phase behavior
and the structures of two-phase dispersions ($\underline{78-81}$). The dispersion
behavior postulated by Wellington ($\underline{42}$) suggests near-tricritical
phase behavior in which the phase of greatest surfactant concentration was thermodynamically the "middle" phase. We may further speculate that, with the second addition of oil, the system passed through
a wetting transition, in which the middle phase wet the interface
between the other two fluids only at higher oil concentrations.
Such wetting transitions have recently been observed in less-complex
systems (D. H. Smith, Morgantown Energy Technology Center, unpublished data, December 3, 1987).
    The three dispersion types described by Wellington $\underline{et\ al.}$ are
important mechanistically, in view of the apparent importance of
capillary snap-off. Extant descriptions of the snap-off mechanism
explicitly treat the first type of dispersion, and they should be
able to accommodate the second dispersion type by addition of a
second fluid that does not wet the porous medium. However, if the
aqueous phase of the first two dispersion types wets the porous

medium, the oleic continuous phase in the third type of dispersion cannot also wet the porous medium unless it is able to change the medium's wettability. The latter possibility suggests that the systems in Wellington's core flood experiments could have been in Region II of the isotherm for surfactant adsorption. An alternative hypothesis is that the surface of the solid phase always remained hydrophilic, and thus, the third type of dispersion formed in the right-cell tests but not in the flooding experiments.

Phase Behavior and Surfactant Design. As described above, dispersion-based mobility control requires capillary snap-off to form the "correct" type of dispersion; dispersion type depends on which fluid wets the porous medium; and surfactant adsorption can change wettability. This section outlines some of the reasons why this chain of dependencies leads, in turn, to the need for detailed phase studies. The importance of phase diagrams for the development of surfactant-based mobility control is suggested by the complex phase behavior of systems that have been studied for high-capillary number EOR (78-82), and this importance is confirmed by high-pressure studies reported elsewhere in this book (Chapters 4 and 5).

Enhanced oil recovery always deals with two or more fluids. By implication these fluids are conjugate phases in equilibrium with each other, although Chapter 6 shows that nonequilibrium mixing can sometimes be important when surfactants are used. When one considers the role of the critical micelle concentration (CMC) in CO_2 mobility control, it is the CMC of the aqueous phase saturated with CO_2 that is important. As illustrated in Figure 10, this CMC may be much lower than the CMC of CO_2-free surfactant solutions (R. S. Schechter, University of Texas, personal communication, October 26, 1987).

Furthermore, the compositions and other properties of the conjugate phases are also needed. Figure 10 shows two systems that can form "normal" micelles in the aqueous phase and "inverted" micelles in the conjugate, oleic phase. As surfactant is added to the system with the lower partial phase diagram in Figure 10, the tieline for normal micelles is encountered before the surfactant concentration at which inverted micelles first form is reached. Hence, a study confined to measuring the CMC and wettabilities of CO_2-saturated aqueous phases might not be misleading, although it would be incomplete.

As surfactant is added to the second system of Figure 10, however, the tieline that terminates at the CMC for formation of inverted micelles in the oleic phase is reached before the normal CMC is encountered. Once the former tieline is reached, further additions of surfactant merely increase the concentration of inverted micelles. Only with much larger additions of surfactant will normal micelles begin to form. If only aqueous phases were studied, this could lead to the belief that normal micelles and the associated wettability would be encountered in a flow experiment, when, in fact, inverted micelles and different wettabilities would actually occur.

Figure 11, which is based on the phase behavior of surfactant/ oil/water systems, illustrates just a few of the many different patterns of phase behavior that may be encountered. On the left is a simple, "well-behaved" system, such as is implicitly assumed in most mobility control studies on "foams." Barring unforeseen wettability problems, the system can be expected to form a "CO_2-in-water foam."

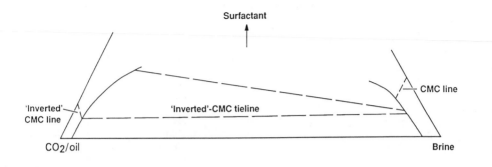

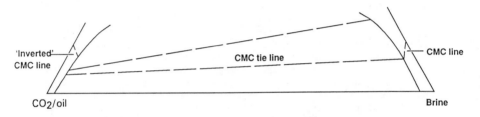

Figure 10.   Illustration of why surfactant design requires the study of both conjugate phases.

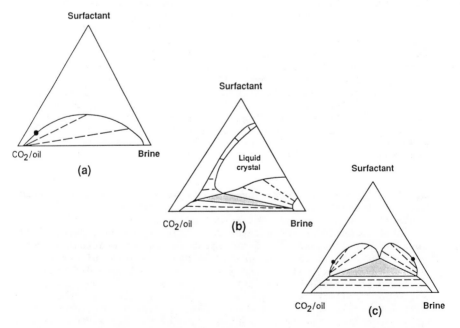

Figure 11.   Examples of various types of phase diagrams which would produce much different mobility control effects.

The middle diagram of Figure 11 illustrates phase behavior that may be more desirable for some applications. In this case the presence of a lamellar liquid crystal region in the phase diagram, although it will not be revealed by experiments confined to low surfactant concentrations, may be exploitable for the formation of very stable dispersions (82).

Finally, the phase diagram of Figure 11c illustrates the formation of three liquid phases in which the middle phase is a microemulsion. Since such systems typically have ultra-low interfacial tensions, they can form fine dispersions (droplet radius less than pore radius) with only a minimum of shear (68).

Early researchers sought to choose appropriate surfactants for mobility control from the hundreds or thousands that might be used, but very little of the technology base that they needed had yet been created. Since then, work on micellar/polymer flooding has established several phase properties that must be met by almost any EOR surfactant, regardless of the application. This list of properties includes a Krafft temperature that is below the reservoir temperature, even if the connate brine contains a high concentration of divalent ions (i.e., hardness tolerance), and a lower consolute solution temperature (cloud point) that is above the reservoir temperature.

The development of a rational strategy of surfactant design requires some way of estimating the dependence of phase parameters on surfactant structure and reservoir characteristics (e.g., salinity). Chapter 9, by Borchardt, describes a method for correlating phase and physical properties of mobility control surfactants with their molecular structures.

High-pressure studies of the phase behavior of systems that contain amphiphiles, water, and supercritical solvents (e.g., ethane and CO_2) are very recent, and most of the active research groups are represented in this volume. Chapter 4, by DiAndreth and Paulaitis, shows that nonionic amphiphiles can exhibit the type of phase behavior illustrated in Figure 11c. The progression from two liquids to three liquids to two liquids (plus gas in all cases) that is observed with increasing temperature or salinity for amphiphile/oil/water systems is mirrored by a similar progression with increasing pressure on amphiphile/CO_2/water systems. Hence, it may prove possible to exploit for CO_2 EOR the rich patterns of phase behavior and physical properties that up to now have been investigated only for improving water flooding.

The chapter by Fulton and Smith (Chapter 5) shows that ionic surfactants can form microemulsions with ethane and water under conditions that might be encountered in miscible floods with light hydrocarbons. These microemulsions correspond to the single-phase regions of the model diagrams in Figure 11.

Summary

We have seen that a very large amount of oil is believed recoverable by gas flooding (more than 15 billion barrels in the United States alone) and that large financial commitments already have been made to produce that target oil. Over 60 field projects are planned and many of these are already in operation and beginning to produce incremental oil.

However, we have also considered that the low viscosities of
"gas" injection fluids, combined with density differences, cause
these fluids to finger through and override the target oil, leaving
a substantial fraction of it uncontacted by the injected fluid and,
thus, unproduced. These detrimental sweep effects present the greatest single target for improvement of gas-flood EOR, and their
alleviation is currently the object of a wide spectrum of applied
research.

Of the processes that might be used to improve sweep and
mobility control in gas flooding, processes that exploit the ability
of surfactants to form dispersions are by far the most promising.

The beginnings of dispersion-based mobility control can be
traced back to 30 years ago. But application of these ideas to gas
flooding began only about 10 years ago, and extensive research began
only with the commercialization of miscible CO_2 flooding in the
early 1980's.

Early flow-visualization studies revealed the various mechanisms
for creation and destruction of lamellae and dispersions during flow
through porous media. Starting with these studies, rapid progress
has recently been made in mathematically describing pore-level
mechanisms. These descriptions form the basis for two divergent,
but complementary branches of research, both of which are needed
for soundly engineered field use.

One of these branches is the evolution of current pore-level
flow models into computerized simulators for testing with laboratory
floods in artificial and natural porous media, followed by the development and the use of field-scale simulators for designing field
tests.

The other application of pore-level mechanisms exploits their
dependence on dispersion type, wettability, capillary number, and
capillary pressure to design surfactants that will optimize these
parameters. Measurements of phase behavior, interfacial tensions,
surfactant adsorption, wettability, and related parameters will be
needed to fit the various requirements of different reservoirs, each
of which has a unique combination of mineralogy, pore structure,
temperature, pressure, oil and brine composition, etc.

These two branches, simulator development and materials selection, can then come together in well-engineered designs for field
use of surfactant-based mobility control in gas flooding.

Literature Cited

1. Enhanced Oil Recovery; National Petroleum Council: Washington,
    DC, 1984.
2. Stalkup, F. I., Jr. Miscible Displacement; Society of Petroleum
    Engineers: Dallas, 1983.
3. Blackwell, R. J. In Enhanced Oil Recovery; F. J. Fayers, Ed.;
    Elsevier: New York, 1981.
4. Krouskaa, V. A. Proc. SPE/DOE 5th Symposium, Enhanced Oil
    Recovery, 1986, SPE 14951.
5. Holm, L. W. J. Petrol. Technol. 1987, 39, 1337.
6. Casteel, J. F.; Djabbarah, N. F. Proc. 60th Ann. Tech. Conf.
    Soc. Petrol. Engrs., 1985, SPE 14392.
7. Caudle, B. H.; Dyes, A. B. Trans. Am. Inst. Min. Metall.
    Petrol. Engrs. 1958, 213, 281.

8. Saffman, P. G.; Taylor, G. I. Proc. Royal Soc., London, 1958, A245, 312.
9. Blackwell, R. J.; Terry, W. M.; Rayne, J. R.; Lindley, D. C.; Henderson, J. R. Trans. Am. Inst. Min. Metall. Petrol. Engrs. 1960, 219, 293.
10. Chuoke, R. L.; VanMeurs, P.; Van Der Poel, C. Trans. Am. Inst. Min. Metall. Petrol. Engrs. 1959, 216, 188.
11. Habermann, B. Trans. Am. Inst. Min. Metall. Petrol. Engrs. 1960, 219, 264.
12. Perrine, R. L. Trans. Soc. Petrol. Engrs. AIME 1963, 228, 205.
13. Perkins, T. K.; Johnston, O. C.; Hoffman, R. N. Soc. Petrol. Engrs. J. 1965, 5, 301.
14. Heller, J. P. J. Appl. Phys. 1966, 37, 1566.
15. Craig, F. F. J. Petrol. Technol. 1970, 22, 529.
16. L'air Liquide, Division Scientifique. Gas Encyclopaedia; N. Marshall, Trans; Elsevier: New York, 1976.
17. Scheiddeger, A. F. The Physics of Flow Through Porous Media; University of Toronto Press: Toronto, 1974.
18. Neogi, P. In Microemulsions: Structure and Dynamics; S. E. Friberg and P. Brothorel, Eds., CRC Press: Boca Raton, 1987.
19. Fitch, R. A.; Griffith, J. D. J. Petrol. Technol. 1964, 16, 1289.
20. Shelton, J. L.; Scheinder, F. N. Soc. Petrol. Engrs. J. 1975, 15, 217.
21. Raimondi, P.; Torcaso, M. A. Soc. Petrol. Engrs. J. 1964, 4, 49.
22. McKee, B. Proc. 4th Ann. DOE Symp., 1978, p. c-4/1.
23. San Filippo, G. P.; Guckert, L. G. Proc. ERDA Enhanced Oil, Gas Recovery, Improved Drilling Methods, 1977, p. c-3/1.
24. Conner, W. D. Proc. ERDA Enhanced Oil, Gas Recovery, Improved Drilling Methods, 1977, p. c-2/1.
25. Huang, E. T.; Holm, L. W. Proc. 61st Ann. Tech. Conf. Soc. Petrol. Engrs., 1986, SPE 15491.
26. Jackson, D. D.; Andrews, G. L.; Claridge, E. L. Proc. 60th Ann. Tech. Conf. Soc. Petrol. Engrs., 1985, SPE 14303.
27. Graue, D. J.; Blevins, T. R. Proc. 5th Symp., Improved Methods Oil Recovery, 1978, SPE 7090.
28. Jenkins, M. K. Proc. SPE/DOE 4th Symp., Enhanced Oil Recovery, 1984, SPE 12632.
29. Tiffin, D. F.; Yellig, W. F. Soc. Petrol. Engrs. J. 1983, 23, 447.
30. Heller, J. P.; Dandge, D. K.; Card, R. J. Soc. Petrol. Engrs. J. 1985, 25, 679.
31. Woods, P.; Schramko, K.; Turner, D.; Dalrymple, D.; Vinson, E. Proc. 5th SPE/DOE Symp., Enhanced Oil Recovery, 1986, SPE/DOE 14958.
32. Bruning, D. D. U.S. Patent 4 569 393, 1986.
33. Bond, D. C.; Holbrook, O. C. U.S. Patent 2 866 507, 1958.
34. Fried, A. N. The Foam-Drive Process for Increasing the Recovery of Oil, Bureau of Mines Rept. 5866, 1961.
35. Holm, L. W. Soc. Petrol. Engrs. J. 1968, 8, 359.
36. Roof, J. G. Soc. Petrol. Engrs. J. 1970, 10, 85.
37. Mast, R. F. Proc. 47th Ann. Fall Mtg, Soc. Petrol. Engrs., 1972, SPE 3997.

38. Hirasaki, G. J.; Lawson, J. B.   Soc. Petrol. Engrs. J. 1985,
     25, 176.
39. Falls, A. H.; Gauglitz, D. A.; Hirasaki, G. J.; Miller, D. D.;
     Patzek, T. W.; Ratulowski, J.   Proc. SPE/DOE 5th Symp.,
     Enhanced Oil Recovery, 1986, SPE/DOE 14961.
40. Radke, C. J.; Ransohoff, T. C.   Proc. 61st Ann. Tech. Conf.
     Soc. Petrol. Engrs., 1986, SPE 15441.
41. Khatib, Z. I.; Hirasaki, G. J.; Falls, A. H.   Proc. 61st Ann.
     Tech. Conf. Soc. Petrol. Engrs., 1986, SPE 15442.
42. Wellington, S. L.   U.S. Patent 4 380 266, 1983.
43. Wellington, S. L.; Reisberg, J.; Lutz, E. F.; Bright, D. B.
     U.S. Patent 4 502 538, 1985.
44. Borchardt, J. K.; Bright, D. B.; Dickson, M. K.;
     Wellington, S. L.   Proc. 60th Ann. Tech. Conf., Soc. Petrol.
     Engrs., 1985, SPE 14394.
45. Wellington, S. L.; Vinegar, H. J.   Proc. 61st Ann. Tech. Conf.
     Soc. Petrol. Engrs., 1985, SPE 14393.
46. Holm, L. W.   J. Petrol. Technol. 1970, 22, 1499.
47. Heller, J. P.; Boone, D. A.; Watts, R. J.   Proc. Soc. Petrol.
     Engrs., East. Reg. Mtg., 1985, SPE 14519.
48. Fischer, P. W.; Holm, L. W.; Pye, D. S.   U.S. Patent 4 088 190,
     1978.
49. Bernard, G. G.; Holm, L. W.   U.S. Patent 4 113 011, 1978.
50. Deming, J. R.   M.S. Thesis, The Pennsylvania State University,
     University Park, Pennsylvania, 1964.
51. Bernard, G. G.   Producers Monthly 1963, 27, 18.
52. Bernard, G. G.; Holm, L. W.   Soc. Petrol. Engrs. J. 1964, 4,
     267.
53. Bernard, G. G.; Holm, L. W.; Jacobs, W. L.   Soc. Petrol. Engrs.
     J. 1965, 5, 295.
54. Marsden, S. S.; Khan, S. A.   Soc. Petrol. Engrs. J. 1966, 6, 17.
55. Jacobs, W. L.; Bernard, G. G.   U.S. Patent 3 330 346, 1967.
56. Bernard, G. G.; Holm, L. W.   U.S. Patent 3 342 256, 1967.
57. Raza, S. H.; Marsden, S. S.   Soc. Petrol. Engrs. J. 1967, 7,
     359.
58. David, A.; Marsden, S. S.   Proc. 4th Ann. Fall Mtg., Soc.
     Petrol. Engrs., 1969, SPE 2544.
59. Bernard, G. G.; Holm, L. W.   Soc. Petrol. Engrs. J. 1970, 10, 9.
60. Raza, S. H.   Soc. Petrol. Engrs. J. 1970, 10, 328.
61. Albrecht, R. A.; Marsden, S. S.   Soc. Petrol. Engrs. J. 1970,
     10, 51.
62. Bernard, G. G.   U.S. Patent 3 529 668, 1970.
63. Minssieux, L.   J. Petrol. Technol. 1974, 26, 100.
64. Kanda, M.; Schechter, R. S.   Proc. 51st Ann. Fall Tech. Conf.,
     Soc. Petrol. Engrs., 1976, SPE 6200.
65. Aizad, T.; Okandan, E.   Proc. SPE-AIME Intl. Symp. Oilfield
     Geotherm. Chem., 1977, SPE 6599.
66. Heller, J. P.; Lien, C. L.; Kuntamukkula, M. S.   Soc. Petrol.
     Engrs. J. 1985, 25, 603.
67. Falls, A. H.; Musters, J. J.; Ratulowski, J.   Soc. Petrol.
     Engrs., Unsolicited Manuscript, 1986, SPE 16048.
68. Reed, R. L.; Carpenter, C. W.   British Patent 2 078 281, 1982.
69. Tomotika, S.   Proc. Roy. Soc. London 1935, A150, 322.
70. Lee, W. K.; Flumerfelt, R. W.   J. Multiphase Flow 1981, 7, 363.

71. Irani, C. A.; Solomon, C., Jr. Proc. 5th SPE/DOE Symp., Enhanced Oil Recovery, 1986, SPE/DOE 14962.
72. Khristov, K. I.; Ekserova, D. R.; Kruglyakov, P. M. Colloid J. USSR 1981, 43, 80.
73. Mayer, R. P.; Stowe, R. A. J. Colloid Interface Sci. 1965, 20, 893.
74. Vinegar, H. J. J. Petrol. Technol. 1986, 38, 257.
75. Wellington, S. L.; Vinegar, H. J. J. Petrol. Technol. 1987, 39, 885.
76. Amott, E. Trans. Am. Inst. Min. Metall. Petrol. Engrs., 1959, 216, 156.
77. Shergold, H. L.; Crawley, R. S. Colloids Surfaces 1981, 3, 253.
78. Smith, D. H. J. Colloid Interface Sci. 1984, 102, 435.
79. Smith, D. H. J. Colloid Interface Sci. 1985, 108, 471.
80. Smith, D. H. In Microemulsion Systems; H. L. Rosano and M. Clausse, Eds.; Marcel Dekker: New York, 1987.
81. Smith, D. H. Proc. SPE/DOE Fifth Symp. Enhanced Oil Recovery, 1986, SPE/DOE 14914.
82. Friberg, S.; Larsson, K. In Advances in Liquid Crystals; G. H. Brown, Ed.; Academic Press: New York, 1976, 2, 173.

RECEIVED April 7, 1988

Chapter 2

Stability of Miscible Displacements

Shih-Hsien Chang and John C. Slattery

Department of Chemical Engineering, Northwestern University,
Evanston, IL 60208

Previous linear and nonlinear stability analyses for
miscible displacement are reviewed.

When one phase is displaced by another in a porous medium,
instabilities may develop that allow the displacing phase to finger
through the displaced phase, bypassing major portions of it. For
this reason, sweep control measures are very important to the
success of commercial displacement operations.

We will focus our attention here upon miscible displacements
(1-4). In a miscible displacement, there is no phase boundary
between the displaced and displacing fluids, only a mixing zone in
which composition is a function of position. A general review of
viscous fingering is given by Homsy (5).

The inherently unstable nature of miscible displacements with
unfavorable mobility ratios (the viscosity of the displacing fluid
is less than the viscosity of the displaced fluid) and unfavorable
density ratios (for a downward vertical displacement, the density of
the displacing fluid is greater than the density of the displaced
fluid) has been well documented (6-18). For a downward vertical
displacement with a favorable mobility ratio and an unfavorable
density ratio, Hill (6) proposed an approximate theory that
neglected the effects of dispersion to predict a critical velocity
above which the displacements were stable; with an unfavorable
mobility ratio and a favorable density ratio, the displacements were
unstable above this critical velocity. It has been demonstrated
both experimentally (18, 19) and theoretically (20) that, for
neutral density ratios, instabilities attributable to unfavorable
mobility ratios can be eliminated by gradually changing from a more
viscous to a less viscous displacing fluid. Similar observations
including the effect of the density ratios have not been reported.
[There are many related studies that are outside the context of this
review. For example, Gorell and Homsy (21) analyze an immiscible
displacement in which the viscosity of the displacing phase is
graded. Hickernell and Yortsos (22) examined the stability
characteristics of miscible flow with spatially varying mobility,
but they neglected the effects of diffusion and dispersion.]

0097–6156/88/0373–0038$06.00/0

In order to understand this instability problem, the first step is to construct a linear stability analysis. This is used to define the parameter limits within which instabilities can be triggered by infinitesimal perturbations to the system. Within these parameter limits, a nonlinear stability analysis should be used to study the development of these instabilities. In this way, one can determine the parameter limits within which instabilities may be of practical concern.

We would like to emphasize that parameter limits within which instabilities can be triggered by macroscopic perturbations to the system may be considerably broader. Within a certain range of parameters, a flow that may be stable to infinitesimal perturbations may be unstable to macroscopic perturbations. This distinction is important, since heterogeneities may represent macroscopic perturbations to an otherwise homogeneous system.

Local Volume Averaging

As we do for all mass transfer problems, we must satisfy the differential equation of continuity for each species as well as the differential momentum balance. Since we are dealing with a porous medium having a complex and normally unknown geometry, we choose to work in terms of the local volume averaged forms of these relations. Reviews of local volume averaging are available elsewhere (23-25).

Let us define V to be the volume of the region enclosed by the averaging surface S. We will denote by $R^{(f)}$ the region occupied by the fluid enclosed by S; $V^{(f)}$ is the volume of $R^{(f)}$. Assume that B_f is some quantity associated with the fluid. We will have occasion to speak of at least two averages (26): the superficial average of B_f (the mean value of B_f in the region enclosed by S)

$$\overline{B_f} \equiv \frac{1}{V} \int_{R^{(f)}} B_f \; dV \tag{1}$$

and the intrinsic average of B_f (the mean value of B_f in $R^{(f)}$)

$$<B_f> \equiv \frac{1}{V^{(f)}} \int_{R^{(f)}} B_f \; dV \tag{2}$$

Here dV denotes that a volume integration is to be performed.

We will limit our attention here to the creeping flow of a Newtonian fluid in a rigid, isotropic, permeable medium having a uniform porosity ψ, and we will assume that the total mass density ρ^* of the fluid is nearly a constant within the averaging region. Under these conditions, the local volume average of the equation of continuity may be written as (23)

$$\psi \frac{\partial <\rho^*>}{\partial t^*} + \text{div} \; (<\rho^*>\overline{v^*}) = 0 \tag{3}$$

where t^* is time, v^* is the (mass-averaged) velocity of the fluid phase, and the superscript $*$ denotes a dimensional variable. The local volume average of Cauchy's first law normally reduces to Darcy's law (23)

$$\nabla<P^*> - <\rho^*>b^* + \frac{\mu^*}{k^*} \; \overline{v^*} = 0 \tag{4}$$

in which P^* is the thermodynamic pressure, b^* gravity, μ^* the viscosity of the fluid, and k^* the permeability.

In a miscible displacement, one follows the transport of one of the species, say species A. We will consider two limiting cases: (1) a binary system of species A and B with a constant diffusion coefficient $\mathcal{D}^*$ and

(2) a N component mixture consisting of a solvent and N - 1 components (including species A) that are present in only trace amounts. Fick's first law may again be used to describe the mass flux of species A; the diffusion coefficient $\mathcal{D}^*$ will be nearly constant and equal to that for a dilute binary solution of A in the solvent (23).

Let us begin with the first of these cases, binary diffusion of species A and B with a constant diffusion coefficient. If there are no chemical reactions and neglecting any adsorption on the pore walls, the local volume-averaged equation of continuity for a particular species A may be written as [the more general case is treated by Slattery (23)]

$$\frac{\partial \left(\psi \langle \rho^* \rangle \langle \omega_{(A)} \rangle \right)}{\partial t^*} + \text{div} \left(\langle \rho^* \rangle \langle \omega_{(A)} \rangle \overline{\mathbf{v}^*} \right) = - \text{div} \ \mathbf{j}_{\{A\}}^{\{e\}^*} \tag{5}$$

where the effective mass flux vector

$$\mathbf{j}_{\{A\}}^{\{e\}^*} = - \psi \langle \rho^* \rangle \mathbf{D}_{\{AB\}}^{\{e\}^*} \cdot \nabla \langle \omega_{(A)} \rangle \tag{6}$$

the effective dispersion tensor

$$\mathbf{D}_{\{AB\}}^{\{e\}^*} \equiv \mathcal{D}^* (1 + D_{(A1)}) \mathbf{I} - \frac{\ell_0^* \ |\nabla \langle \omega_{(A)} \rangle| \ D_{(A2)}}{\overline{\mathbf{v}^*} \cdot \nabla \langle \omega_{(A)} \rangle} \ \overline{\mathbf{v}^*} \ \overline{\mathbf{v}^*} \tag{7}$$

and, for $j = 1, 2,$

$$D_{(Aj)} = D_{(Aj)} \left(N_{Pe}, \ \frac{\overline{\mathbf{v}^*} \cdot \nabla \langle \omega_{(A)} \rangle}{|\overline{\mathbf{v}^*}| \ |\nabla \langle \omega_{(A)} \rangle|}, \ \frac{\ell_0^* \ |\nabla \langle \omega_{(A)} \rangle|}{\langle \omega_{(A)} \rangle}, \ \psi \right) \tag{8}$$

Here ℓ_0^* is a characteristic dimension of the local pores,

$$N_{Pe} \equiv \frac{\ell_0^* \ |\overline{\mathbf{v}^*}|}{\mathcal{D}^*} \tag{9}$$

a local Peclet number,

$$\omega_{(A)} \equiv \frac{\rho_{(A)}^*}{\rho^*} \tag{10}$$

the mass fraction of species A, and $\rho_{(A)}^*$ is the mass density of species A. For a binary mixture, it can be shown that Equations 6-8 must satisfy the additional constraint

$$\mathbf{j}_{\{A\}}^{\{e\}^*} + \mathbf{j}_{\{B\}}^{\{e\}^*} = 0 \tag{11}$$

This is satisfied, so long as we require

$$\mathbf{D}_{\{AB\}}^{\{e\}^*} = \mathbf{D}_{\{BA\}}^{\{e\}^*} \tag{12}$$

For a N component solution consisting of a solvent and N - 1 components present in only trace amounts, Equations 6-8 again apply for each trace species A. But in this case, there is no constraint similar to Equation 12, since the solvent would not obey Equations 6-8 [because it would not follow Fick's first law].

Equations 6-8 represent a class of empirical relationships, one member of which is the traditional model (27-32):

$$D^{(e)*}_{\{AB\}} = (D^*_d + a^*_\ell \ |<\mathbf{v}^*>|)\,\mathbf{I} + (a^*_\ell - a^*_t)\,\frac{<\mathbf{v}^*><\mathbf{v}^*>}{|<\mathbf{v}^*>|} \tag{13}$$

This traditional model contains three parameters: D^*_d is called the effective diffusion coefficient in a porous medium, a^*_ℓ the longitudinal dispersivity, and a^*_t the transverse dispersivity. The coefficients D^*_d and a^*_ℓ can be measured in one-dimensional experiments, but a^*_t must be measured in experiments in which a two-dimensional concentration profile develops.

In a detailed comparison, Chang and Slattery (Transport in Porous Media, currently being reviewed) show that one of the simpler members of this class (Equations 6-8)

$$D^{(e)*}_{\{AB\}} \equiv A^*_{(A)}\,\mathbf{I} - \frac{|\nabla<\omega_{(A)}>|\ B^*_{(A)}}{\mathbf{v}^* \cdot \ \nabla<\omega_{(A)}>}\,\frac{}{\mathbf{v}^*}\,\frac{}{\mathbf{v}^*} \tag{14}$$

compares favorably with the traditional description of dispersion, Equation 13. The advantage of this simplified model is that it contains only two empirical parameters, $A^*_{(A)}$ and $B^*_{(A)}$, both of which can be determined in one-dimensional experiments. Equation 14 is consistent with saying that the longitudinal dispersion coefficient is a linear function of N_{Pe} as suggested by Perkins and Johnston (33).

For the common one-dimensional experiment in which the signs of the concentration gradient and of velocity are different,

$$A^*_{(A)} = D^*_d \tag{15}$$

$$B^*_{(A)} = a^*_\ell / \psi \tag{16}$$

Previous Linear Stability Analyses

Perrine (34, 35) determined the marginal stability criteria for displacement in a semi-infinite reservoir of finite thickness and unbounded width, oriented at some given angle with respect to gravity. The longitudinal and transverse dispersion coefficients were assumed to be constants. [Lee et al. (36) assumed that the longitudinal and transverse dispersion coefficients were constants in writing their first equation, but subsequently considered them to be functions of velocity in the perturbation analysis.]

Most studies (22, 37-42) have limited themselves to porous media that were unbounded in the direction of flow. We have two objections to this. First, how are the fluids to be introduced to a region far removed from the entrance to the system without the formation of any instabilities? Second, Chang and Slattery (20) as well as our analysis suggest that, if a displacement is unstable, the instabilities can develop at the entrance to the system.

With the exceptions of Schowalter (38) and of Hickernell and Yortsos (22), all previous linear stability analyses (34-37, 39-42) have used the local volume-averaged equation of continuity for an incompressible fluid, although they assumed that density was a function of concentration and therefore position and time. This is

not generally correct, the following analysis indicates that, when
density is a weak linear function of concentration, the local volume
averaged equation of continuity for an incompressible fluid can be
used. While Schowalter (38) and Hickernell and Yortsos (22)
correctly posed this portion of their problems, Schowalter (38)
limited himself to an unbounded porous medium as discussed above;
Hickernell and Yortsos (22) did not consider the effects of
diffusion and dispersion in their analysis.

Recent Linear Stability Analyses

The system we consider here is a semi-infinite porous medium having
a uniform porosity ψ, and a uniform permeability k^*, a finite
thickness L^* (the smaller lateral dimension) in the z_3^* direction,
and a finite width hL^* in the z_2^* direction. The gravity acts in the
z_1^* direction. The magnitude of the fluid velocity has a uniform
value v_0^* over the injection face. We will consider two cases in
which the injection fluid is well mixed and diffusion upstream of
the injection face can be neglected. Either the composition at the
injection face is $\omega_{(A)\infty}$, or it increases linearly with time from
$\omega_{(A)0}$ until it reaches $\omega_{(A)\infty}$ and remains fixed at that value. Both
the viscosity and the density of the fluid are recognized to be
functions of the composition of the fluid.

As an illustration, we use the dispersion model employed by
Chang and Slattery (20). As we will show, the results obtained by
using the traditional model, Equation 13, are qualitatively similar.

We will find it convenient to introduce as dimensionless
variables

$$t \equiv \frac{t^* v_0^*}{L^*} \qquad\qquad \mathcal{P} \equiv \frac{k^*}{L^* \mu_0^* v_0^*} (<P^*> + \rho_0^* \phi^*)$$

$$\mathbf{v} \equiv \frac{\mathbf{v}^*}{v_0^*} \qquad\qquad \mu \equiv \frac{\mu^*}{\mu_0^*}$$

$$w \equiv \frac{<\omega_{(A)}> - \omega_{(A)0}}{\omega_{(A)\infty} - \omega_{(A)0}} \qquad\qquad \rho \equiv \frac{<\rho^*>}{\rho_0^*}$$

$$z_j \equiv \frac{z_j^*}{L^*} \qquad (j = 1, 2, 3) \qquad\qquad \mathbf{b} \equiv \frac{\mathbf{b}^*}{g^*} \qquad\qquad (17)$$

Here μ_0^* is the viscosity of the displaced fluid, ρ_0^* is the total
mass density of the displaced fluid, g^* the acceleration of gravity,
and

$$\mathbf{b}^* = - \nabla \phi^* \qquad\qquad (18)$$

where ϕ^* is potential energy per unit mass.

The dimensionless intrinsic average of the total mass density
having been expanded in a Taylor series

$$\rho = 1 + \frac{\rho_1^* \Delta\omega}{\rho_0^*} w + \cdots \qquad\qquad (19)$$

Equations 3-5 become in terms of these dimensionless variables

$$\operatorname{div} \mathbf{v} + \left(\frac{\rho_1^* \Delta\omega}{\rho_0^*} \right) \left(\cdots \right) + \cdots = 0 \qquad\qquad (20)$$

$$\nabla \mathcal{P} - Gw\mathbf{b} + \mu\mathbf{v} - \cdots = 0 \tag{21}$$

$$\frac{\partial w}{\partial t} + \frac{1}{\psi} \nabla w \cdot \mathbf{v} - A \, \mathrm{div}(\nabla w) + B \, \mathrm{div}(|\nabla w|\mathbf{v})$$

$$+ \left(\frac{\rho_1^* \Delta\omega}{\rho_0^*} \right) \left(\cdots \right) + \cdots = 0 \tag{22}$$

in which

$$A \equiv \frac{A_{(A)}^*}{v_0^* L^*} \tag{23}$$

$$B \equiv \frac{B_{(A)}^*}{L^*} \tag{24}$$

$$G \equiv \frac{\rho_1^* \Delta\omega g^* k^*}{\mu_0^* v_0^*} \tag{25}$$

$$\Delta\omega \equiv \omega_{(A)\infty} - \omega_{(A)0} \tag{26}$$

The quantity $A_{(A)}^*$ is the effective diffusion coefficient for the porous media and A is the dimensionless effective diffusion coefficient; B characterizes the contribution of convection to dispersion; G characterizes the ratio of the excess gravitational forces (attributable to the density difference between the displacing fluid and the displaced fluid) to the viscous forces. In writing Equation 22, we have recognized Equation 14. We will confine our attention to the limit that density is a weak linear function of concentration

$$\frac{\rho_1^* \Delta\omega}{\rho_0^*} \to 0 \tag{27}$$

in which case Equations 20-22 reduce to

$$\mathrm{div} \, \mathbf{v} = 0 \tag{28}$$

$$\nabla \mathcal{P} - Gw\mathbf{b} + \mu\mathbf{v} = 0 \tag{29}$$

$$\frac{\partial w}{\partial t} + \frac{1}{\psi} \nabla w \cdot \mathbf{v} - A \, \mathrm{div}(\nabla w) + B \, \mathrm{div}(|\nabla w|\mathbf{v}) = 0 \tag{30}$$

Note that Equation 27 implies nothing about G.
　　Since the porous medium is initially filled with a fluid having a uniform composition $\omega_{(A)0}$, we will require

$$\text{at} \quad t = 0 \quad \text{for} \quad z_1 > 0 : \quad w = 0 \tag{31}$$

and

$$\text{for} \quad t \geq 0 \quad \text{as} \quad z_1 \to \infty : \quad w \to 0 \tag{32}$$

Two cases are considered at the injection boundary. In the first case, the composition at the injection face is $\omega_{(A)\infty}$, which means

$$\text{for } t > 0 \text{ at } z_1 = 0 : w = 1 \tag{33}$$

In the second case, the composition at the injection face increases linearly with time from $\omega_{(A)0}$ until it reaches $\omega_{(A)\infty}$, which implies

$$\text{for } 0 < rt \leq 1 \text{ at } z_1 = 0 : w = rt$$

$$\text{for } rt > 1 \text{ at } z_1 = 0 : w = 1 \tag{34}$$

In what follows, we seek a perturbation solution that is correct to the first order:

$$\mathbf{v} = \mathbf{v}^{(0)} + \varepsilon \mathbf{v}^{(1)} + \cdots$$

$$\mathcal{P} = \mathcal{P}^{(0)} + \varepsilon \mathcal{P}^{(1)} + \cdots$$

$$w = w^{(0)} + \varepsilon w^{(1)} + \cdots \tag{35}$$

The stable solution is denoted by the superscript (0); the first perturbation by the superscript (1); the parameter ε characterizes a random perturbation in the physical problem that is left unspecified. If the first perturbation decays with time, the displacement is stable to infinitesimal perturbations; if it grows with time, the displacement is unstable to such perturbations. In Figure 1, we plot stability curves corresponding to various values of G, $r = 1$, and $h^{-1} = 1$ for the case where the mobility ratio is favorable (the viscosity of the displacing fluid is larger than that of the displaced fluid) and the density ratio is unfavorable (the density of the displacing fluid is larger than that of the displaced fluid during a downward vertical displacement). The region above each curve designates those values of A and B corresponding to stable displacements. This suggests that the smaller G is, the more likely that a given displacement will be stable.

If either the mobility ratio or the density ratio is unfavorable, instabilities can form at the injection boundary as the result of infinitesimal perturbations. But if the concentration is changed sufficiently slowly with time at the entrance to the system, the displacement can be stabilized, even if both the mobility ratio and the density ratio are unfavorable. At least for neutral density ratios, this was anticipated by the experimental observations of Slobod and Lestz ([19]) and of Kyle and Perrine ([18]).

We found that the displacement is stable, when both the mobility ratio and the density ratio are favorable.

When the mobility ratio is unfavorable and the density ratio is favorable, we can define a critical velocity v_c^* below which a displacement is stable. In the case that the mobility ratio is favorable and the density ratio is unfavorable, the displacement is stable if the velocity exceeds v_c^*. These were anticipated by the experimental observations of Hill (6).

When G equals to zero, Equations 28-30 are identical with those given by Chang and Slattery (20, Equations 15-17). In this case, density is independent of concentration, and we do not limit ourselves to the vertical system. If the mobility ratio is unfavorable, the displacement is more likely to be stable as the aspect ratio h^{-1} (ratio of thickness to width, which is assumed to be less than one) is increased. Commonly the laboratory tests supporting a field trial use nearly the same fluids, porous media, and displacement rates as the field trial they are intended to support. For the laboratory test, the aspect ratio may be the order of one; for the field trial, it may be two orders of magnitude smaller. This means that a laboratory test could indicate that a displacement was stable, while an unstable displacement may be observed in the field.

For the case that density is independent of concentration, we have also examined the case in which the traditional model Equation 13 is used to describe dispersion. Figure 2 shows the resulting stability curves corresponding to various values of a_t, r = 1, and h^{-1} = 1 for the case of an unfavorable mobility ratio. The region above each curve designates those values of D_d and a_1 corresponding to stable displacements. This suggests that the larger a_t is, the more likely that a given displacement will be stable. Here

$$D_d \equiv \frac{D_d^*}{v_0^* L^*} \tag{36}$$

$$a_1 \equiv \frac{a_t^*}{L^* \psi} \tag{37}$$

$$a_t \equiv \frac{a_t^*}{L^* \psi} \tag{38}$$

Figures 3 and 4 show stability curves corresponding to various values of r for the traditional and simplified models, respectively. Both figures all imply that, if the concentration is changed sufficiently slowly with time at the entrance to the system, the displacement can be stabilized, even if the mobility ratio is unfavorable.

From Equations 15, 16, 23, 24, 36, and 37, we have

$$D_d = A \tag{39}$$

$$a_1 = B \tag{40}$$

For reservoir conditions, a_t is typically smaller than 0.001 (33, 36). Comparing Figures 2 and 3 with 4, we can see that for reservoir conditions the stability curves based upon the simplified model are nearly identical with those based upon the traditional model.

Previous Nonlinear Stability Analyses

Let us assume that the parameter limits within which an unstable

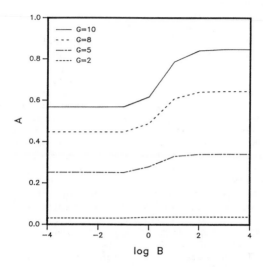

Figure 1.   Stability curve as a function of G for M = 0.0253, D = 1.17, r = 1, and h^{-1} = 1.

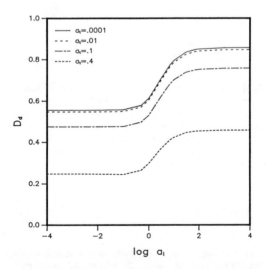

Figure 2.   Stability curve as a function of a_t for M = 68.9, r = 1, and h^{-1} = 1.

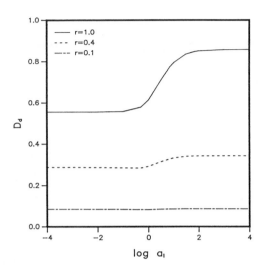

Figure 3. Stability curve as a function of r for M = 68.9, a_t = 0.0001, and h^{-1} = 1.

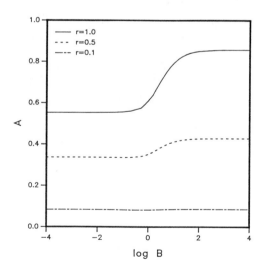

Figure 4. Stability curve as a function of r for M = 68.9, and h^{-1} = 1.

displacement can be expected to be triggered by infinitesimal perturbations have been established. The next step would be to carry out numerical simulations of the full nonlinear problem within these parameter limits, in order to determine how instabilities develop as functions of time.

Unfortunately, most studies have not taken this approach. Numerical simulations of displacement have not been preceded by linear stability analyses to define the parameter limits within which an unstable displacement could be expected. [In their nonlinear stability analysis, Perrine and Gay (43) allowed their dispersion coefficients to be functions of velocity. In his linear stability analysis, Perrine (34, 35) assumed that his dispersion coefficients were constants.]

Christie and Bond (44) began with a linear stability analysis, but they did not construct a stability curve to define the parameter domain in which instabilities could be expected. In their nonlinear analysis, instabilities were initiated by random perturbations in the initial concentration distribution at the entrance (macroscopic perturbations). [Peters and Kasap (45) used the same method to initiate instabilities.]

On the basis of their computations, most workers (43, 46-49) have concluded that viscous fingers would not develop during the simulation of a displacement through a homogeneous porous medium. Perrine and Gay (43) used a higher-order perturbation analysis and found that instabilities would not develop in their simulations, unless the permeability was permitted to be a function of position briefly at the beginning of the computation. Others (46-49), using various forms of numerical simulations, have found that instabilities were not triggered by truncation and round-off errors (infinitesimal perturbations). It was necessary to introduce some permeability variation in the system (macroscopic perturbations).

In apparent contrast with this experience, several laboratory-scale experiments with miscible systems (9-11, 13, 14) as well as the study of Giordano and Salter (49) with partially miscible systems clearly showed the development of instabilities in homogeneous porous media. One explanation of these reports is that real porous media are never truly homogeneous.

On the other hand, Settari et al. (50) used a finite-element analysis in examining the consequences of both velocity-dependent and constant dispersion coefficients during a two-dimensional displacement. They found that fingers in the concentration distribution developed when the permeability was homogeneous, so long as the dispersion coefficients were sufficiently small. This was apparently the first successful use of truncation and round-off errors to play the roles of physical perturbations in initiating instabilities. Russell (51) later had a similar experience.

Recently, Giordano et al. (52) used a finite difference simulator together with experiments to analyze the physical processes that control the initiation and propagation of fingers in porous media. They found that numerical round-off errors (variations in the 14th place) were sufficient to trigger a single finger (Section IV, first paragraph, page 8).

Our tentative conclusion is that Settari et al. (50), Russell (51), and Giordano et al. (52) were correct in concluding that instabilities can develop during the simulations of displacements in homogeneous porous media and that truncation and round-off errors can play the roles of physical perturbations in initiating instabilities during simulated displacements. We do know that in another context truncation and round-off errors have been used successfully to trigger instabilities (53) in a numerical simulation of an unsteady-state flow.

We think that there may be two reasons why others (43, 46-49) have come to a contrary conclusion.

First, since most of these studies were not preceded by a linear stability analysis to define the parameter limits within which an unstable displacement could be expected, we are free to speculate that in some cases the displacements were actually stable to infinitesimally small perturbations (but not necessarily stable to the macroscopic perturbations).

Second, we know that numerical errors (such as a grid that is too coarse in a finite difference computation) can mimic some of the effects of dispersion. It may be that such "numerical dispersion" is responsible for unexpectedly stable displacements in some cases. This may be supported by the experience of Settari et al. (50) and of Russell (51), who found that their displacements were stable, unless the effects of dispersion were sufficiently small.

Acknowledgments

The authors are grateful for the support that they have received for this work from the Amoco Production Company.

Legend of Symbols

a_ℓ^* :	longitudinal dispersivity in Equation 13
a_t^* :	transverse dispersivity in Equation 13
a_1 :	defined by Equation 37
a_t :	defined by Equation 38
$A_{(A)}^*$:	effective diffusion coefficient, defined by Equation 14
b^* :	gravity
$B_{(A)}^*$:	parameter characterizing the effect of convection upon dispersion, defined by Equation 14
B_f :	some quantity associated with the fluid
D_d^* :	effective diffusion coefficient in Equation 13
$D_{(A\mathcal{J})}$:	parameters upon which these functions depend indicated by Equation 8 ($\mathcal{J} = 1, 2$)
$D_{\{AB\}}^{\{e\}*}$:	effective dispersion tensor, defined by Equation 7
$\mathcal{D}^*$:	diffusion coefficient
g^* :	acceleration of gravity
G :	defined by Equation 25
h :	reciprocal of the aspect ratio, which is the ratio of thickness to width and assumed to be less than one
I :	identity tensor that leaves vectors unchanged
$j_{\{A\}}^{\{e\}*}$:	effective mass flux vector with respect to $\mathbf{v}^*$, represented by Equation 6
k^* :	permeability of the porous structure to the fluid
ℓ_0^* :	characteristic dimension of the local pores
L^* :	thickness of the reservoir
N_{pe} :	local Peclet number, defined by Equation 9

P^* :   thermodynamic pressure
$\mathcal{P}$:   defined by Equation 17
r: parameter in Equation 34
$R^{(f)}$:   region occupied by the fluid enclosed by S
S: averaging surface
t^* :   time
$\mathbf{v}^*$:   (mass-averaged) velocity of the fluid
v_c^* :   critical velocity
v_0^* :   uniform magnitude of fluid velocity over the injection face
$V^{(f)}$:   volume of $R^{(f)}$ enclosed by S
$\mathcal{V}$:   volume of the region enclosed by S
w: defined by Equation 17
z_j^* :   rectangular Cartesian coordinates (j = 1, 2, 3)

Greek Letters

$\Delta\omega$:   defined by Equation 26
ε :   perturbation parameter
μ^* :   viscosity of the fluid
μ_0^* :   viscosity of the displaced fluid
ρ^* :   total mass density of the fluid
ρ_0^* :   total mass density of the displaced fluid
ρ_1^* :   defined by Equation 19
$\rho_{(A)}^*$:   mass density of species A
ϕ^* :   potential energy, defined by Equation 18
ψ: porosity
$\omega_{(A)}$:   mass fraction of species A, defined by Equation 10
$\omega_{(A)\,0}$: initial mass fraction of species A in the displaced fluid
$\omega_{(A)\,\infty}$: final mass fraction of species A in the injection fluid

Other

$\ldots^{(0)}$:   superscript denoting the stable solution
$\ldots^{(1)}$:   superscript denoting the first perturbation variable
$\ldots^*$:   superscript denoting the dimensional variable
div: divergence operation
∇:   gradient operator
dV: indicating that a volume integration is to be performed
$\overline{\ldots}$:   indicating a superficial averaged variable defined by Equation 1
$\langle\ldots\rangle$:   indicating an intrinsic averaged variable defined by Equation 2

Literature Cited

1. Clark, N. J.; Shearin, H. M.; Schultz, W. P.; Garms, K.; Moore, J. L. J. Pet. Technol. June 1958, 11.
2. Craig, F. F., Jr.; Owens, W. W. J. Pet. Technol. April 1960, 11.
3. Stalkup, F. I., Jr. J. Pet.Technol. 1983, 35, 815.
4. Stalkup, F. I., Jr. Miscible Displacement; Society of Petroleum Engineers: Dallas; Monograph Series Volume 8.
5. Homsy, G. M. Ann. Rev. Fluid Mech. 1987, 19, 271.

6. Hill, S. Chem. Eng. Sci. 1952, 1, 247.
7. Offeringa, J.; van der Poel, C. Trans. AIME 1954, 201, 310.
8. Craig, F. F., Jr.; Sanderlin, J. L.; Moore, D. M.; Geffen, T. M. Trans. AIME 1957, 210, 275.
9. Blackwell, R. J.; Rayne, J. R.; Terry, W. M. Trans. AIME 1959, 216, 1.
10. Habermann, B. Trans. AIME 1960, 219, 264.
11. Brigham, W. E.; Reed, P. W.; Dew, J. N. Soc. Pet. Eng. J. 1961, 1, 1.
12. Gardner, G. H. F.; Downie, J.; Kendall, H. A. Soc. Pet. Eng. J. 1962, 2, 95.
13. Slobod, R. L.; Thomas, R. A. Soc. Pet. Eng. J. 1963, 3, 9.
14. Benham, A. L.; Olson, R. W. Soc. Pet. Eng. J. 1963, 3, 138.
15. Crane, F. E.; Kendall, H. A.; Gardner, G. H. F. Soc. Pet. Eng. J. 1963, 3, 277.
16. Slobod, R. L.; Howlett, W. E. Soc. Pet. Eng. J. 1964, 4, 1.
17. Dumore, J. M. Soc. Pet. Eng. J. 1964, 4, 356.
18. Kyle, C. R.; Perrine, R. L. Soc. Pet. Eng. J. 1965, 5, 189.
19. Slobod, R. L.; Lestz, S. J. Producers Monthly August 1960, 13.
20. Chang, S. -H.; Slattery, J. C. Transport in Porous Media 1986, 1, 179.
21. Gorell, S. B.; Homsy, G. M. SIAM J. Appl. Math. 1983, 43, 79.
22. Hickernell, F. J.; Yortsos, Y. C. Stud. Appl. Math. 1986, 74, 93.
23. Slattery, J. C. Momentum, Energy, and Mass Transfer in Continua; McGraw-Hill, New York, 1972; second edition, Robert E. Krieger, Malabar, FL 32950, 1981.
24. Alemán, M. A.; Ramamohan, T. R. ; Slattery, J. C. SPE 13265, Society of Petroleum Engineers, P. O. Box 833836, Richardson, TX 75083-3836, 1984.
25. Jiang, T. -S., Kim, M. H.; Kremesec, V. J., Jr.; Slattery, J. C. Chem. Eng. Commun. 1987, 50, 1.
26. Slattery, J. C. AIChE J. 1967, 13, 1066.
27. Nikolaevskii, V. N. PMM, J. Appl. Math. Mech. (Engl. Transl.) 1959, 23, 1492.
28. Bear, J. J. Geophys. Res. 1961, 66, 1185.
29. Scheidegger, A. E. J. Geophys. Res. 1961, 66, 3273.
30. de Josselin de Jong, G.; Bossen, M. J. J. Geophys. Res. 1961, 66, 3623.
31. Peaceman, D. W. Soc. Pet. Eng. J. 1966, 6, 213.
32. Bear, J. Dynamics of Fluids in Porous Media; American Elsevier Publishing Company, New York, 1972.
33. Perkins, T. K.; Johnston, O. C. Soc. Pet. Eng. J. 1963, 3, 70.
34. Perrine, R. L. Soc. Pet. Eng. J. 1961, 1, 9.
35. Perrine, R. L. Soc. Pet. Eng. J. 1961, 1, 17.
36. Lee, S. T.; Li, K. M. G.; Culham, W. E. SPE/DOE 12631, Society of Petroleum Engineers, P.O. Box 833836, Richardson, TX 75083-3836, 1984.
37. Wooding, R. A. Z. Angew. Math. Phys. 1962, 13, 255.
38. Schowalter, W. R. AIChE J. 1965, 11, 99.
39. Heller, J. P. J. Appl. Phys. 1966, 37, 1566.
40. Gardner, J. W.; Ypma, J. G. J. Soc. Pet. Eng. J. 1984, 24, 508.
41. Peters, E. J.; Broman, W. H., Jr.; Broman, J. A. SPE 13167, Society of Petroleum Engineers, P.O. Box 833836, Richardson, TX 75083-3836, 1984.

42. Tan, C. T.; Homsy, G. M. Phys. Fluids 1986, 29, 3549.
43. Perrine, R. L.; Gay, G. M. Soc. Pet. Eng. J. 1966, 6, 228.
44. Christie, M. A.; Bond, D. J. SPE/DOE 14896, Society of
    Petroleum Engineers, P.O. Box 833836, Richardson, TX 75083-
    3836, 1986.
45. Peters, E. J.; Kasap, E. SPE 15597, Society of Petroleum
    Engineers, P.O. Box 833836, Richardson, TX 75083-3836, 1986.
46. Peaceman, D. W.; Rachford, H. H., Jr. Soc. Pet. Eng. J. 1962,
    2, 327.
47. Claridge, E. L. Soc. Pet. Eng. J. 1972, 12, 352.
48. Young, L. C. Soc. Pet. Eng. J. 1981, 21, 115.
49. Giordano, R. M.; Salter, S. J. SPE 13165, Society of Petroleum
    Engineers, P.O. Box 833836, Richardson, TX 75083-3836, 1984.
50. Settari, A.; Price, H. S.; Dupont, T. Soc. Pet. Eng. J. 1977,
    17, 228.
51. Russell, T. F. SPE 10500, Society of Petroleum Engineers, P.O.
    Box 833836, Richardson, TX 75083-3836, 1982.
52. Giordano, R. M.; Salter, S. J.; Mohanty, K. K. SPE 14365,
    Society of Petroleum Engineers, P.O. Box 833836, Richardson, TX
    75083-3836, 1985.
53. Neitzel, G. P.; Davis, S. H. J. Fluid Mech. 1981, 102, 329.

RECEIVED January 29, 1988

Chapter 3

Reservoir Description: Key to Success in Chemical-Enhanced Oil Recovery

Larry W. Lake

Department of Petroleum Engineering, University of Texas, Austin, TX 78712

Reservoir definition impacts on chemical enhanced oil recovery processes more than any other type of Enhanced Oil Recovery (EOR). This is because these processes are exceptionally prone to deterioration through mixing and through bypassing. Both mixing and bypassing relate directly to the spatial distribution of permeabilities which is the major topic of this paper.

Spatial permeability distributions are so complex that we normally describe them through statistical distributions. Our inquiries into geologically-realistic permeabilities indicate that the mean and variance of such distributions are scale-independent but the range of spatial correlation is not. Through flow modeling, we are able to relate the properties of the distribution to parameters characterizing gross mixing and to the geological make-up of the environment. Such a linkage holds promise for more intimate impact of geology into reservoir description and thereby to the prediction and design of chemical EOR processes.

Chemical enhanced oil recovery processes are now in a period of retrenchment caused mainly by less than successful performance of several field tests. Nearly all of these tests have, to at least some extent, been affected by shortcomings in the reservoir description which compound process complexities and uncertainties. Surprises in the reservoir description should be regarded as normal in EOR applications; however, we can do a far better job of anticipating by adopting an integrated approach to reservoir simulation based on geology, statistics and fluid flow modeling known as conditional simulation.

<u>A Characterization Procedure</u>. A plausible procedure for accomplishing a geological/engineering prediction might be the following:

1. Generate a statistical description (means, trends, variances and correlations) of the reservoir flow field. Doing this requires a rather massive amount of data; primary sources are well data, outcrop analogues, seismic profiling and "type" functions based on stratification types and depositional environment. In an ideal case, there should be such a statistical description for every input variable for the reservoir simulator.

2. Generate a synthetic flow field consisting of one or more stochastic variables. There are a number of techniques to convert raw statistics to a collection of point values which consist of a single "realization" of the flow field.

3. Condition the flow field to be consistent with deterministic information from isolated point or line observations. Points at which values are actually measured should not be treated as random. Such conditioning will allow information from well tests, specific core observations, and seismic data to be entered into the procedure. Also, the flow field in step two must be globally conditioned to make sure the statistics derived therefrom are the same as those desired.

4. Lay down a simulation grid on the conditioned stochastic flow field. All practical simulation procedures require assigning a discrete static volume (grid block) to each point in the flow field. The number (and hence the size) of these volumes are usually determined by the available computation facilities, although in some cases geologic phenomena will indicate an "best" grid size and geometry.

5. In even the most favorable case the minimum granularity of the simulation (smallest possible grid size) will be larger than the smallest discernible geologic event. If such events are important to the fluid flow, and we make this presumption here, they must be added to the simulation by defining appropriate averaged or lumped "pseudo" properties. The latter should reflect the detail within the block as well as the block geometry and size. Typical properties which must be averaged in this fashion are porosity, absolute permeability, relative permeability, capillary pressure, and dispersivity. We deal with permeability and dispersivity in more depth below.

6. Perform a fluid flow simulation using the tabular input from step five. Just as the flow field itself, the results of this simulation are a single realization. Repeating steps 2-6 will lead to a distribution of results from which one may estimate the risk inherent in the flow process in the subject environment.

The above procedure requires a level of technology and computer commitment which is currently quite laborious; however, several aspects are becoming better defined with active research. The procedure is clearly desirable from two standpoints: it provides a direct mechanism for incorporating detailed geologic observation (steps 1 and 2) and it provides a means for assessing the effects of the geologic uncertainties on the prediction (step 6).

This paper touches on all aspect of the above procedure, though it is steps 1 and 5 (in reverse order) that occupy the most attention.

Permeability Adjustment

Shales, very fine grain clastic material, are probably the most significant barrier to flow in natural media. They occur in two forms: segregated and dispersed. Dispersed shales affect fluid transmissivity significantly, particularly if chemical changes occur, but their effect is usually incorporated in the intrinsic permeability of the medium itself. Segregated shales are more complex. If they are of great lateral extent, their presence can be easily detected and incorporated into simulation; these are "deterministic" shales. If their extent is smaller than the well spacing, their position (and indeed sometimes even their presence) is unknown; these are "stochastic" shales. Our first example deals with stochastic shales.

Figure 1 shows a synthetic cross-section containing randomly distributed, stochastic shales. In this profile, the shale positions are random and uncorrelated, the shale width distribution from cores, and the length distribution from outcrop studies according to environment type. These sources are a blend of direct observation, geologic inference and statistics. The procedure that develops such a profile adds shales to the cross-section (with correction for overlap) until a pre-specified and known ratio of sand to shale is reached. This constitutes the global conditioning step in the shale generation. In an actual application the shales would be adjusted to conform to the actual observations at observations at the wells from where the thickness statistics were derived (local conditioning); this conditioning was not done in Figure 1.

As elaborate as the presentation in Figure 1 is, it is of little use to simulation unless the shale distribution can be translated into effective block properties. To do this we explicitly simulated single-phase flow in media with several specific shale distributions. Figure 2 shows the results of this work expressed as a ratio of effective permeability to intrinsic "sand" permeability versus aspect ratio of the system. To be sure, the specific shale arrangement strongly affects the permeability reduction; however, certain patterns emerge in the limits of large or small aspect ratios. For large aspect ratios, as are invariably encountered for horizontal permeability in numerical simulation, the permeability reduction is slight and independent of aspect ratio, being mainly attributable to the reduction in cross-sectional area caused by the shales. For small aspect ratios, as in vertical permeability, the reduction is severe and varies as the first power of aspect ratio. In this limit, the reduction is because of the increased tortuosity of the flow paths. Both extremes can be accurately calculated from the specific distribution using analytic formulas; hence, we can now associate effective permeabilities with each of the grid blocks in Figure 1.

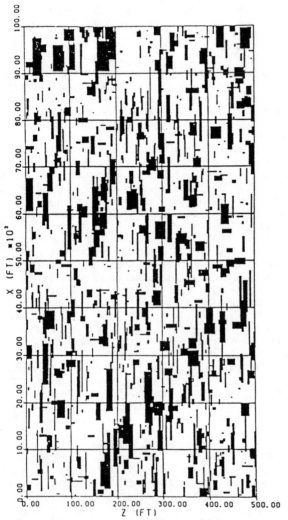

Figure 1. A single realization of a synthetic reservoir cross-section containing sand (light) and shales (dark). The lengths, thicknesses and center coordinates of the shales are independent, random events taken from known distribution functions. Shales are accumulated until a pre-specified global degree of shale area/total area (f_s = 0.24 in this case) is attained. The light boxes correspond to grid blocks. (Reproduced from Ref. 2.)

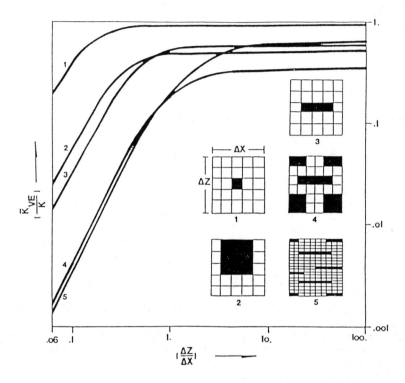

Figure 2. Permeability reduction because of the presence of non-communicating shales. When the aspect ratio ($\Delta z/\Delta x$) is small, as in the case of vertical permeability, the reduction is pronounced. When it is large, as for horizontal permeability, the reduction is small. Both extremes can be dealt with analytically. (Reproduced from Ref. 2.)

Dispersivity Estimates

Permeability deals with average properties of the flow paths in a simulation block; dispersivity deals with the variability of the flow paths. Our second illustration of conditional simulation deals with estimating dispersivity.

Simple Theory. The concentration C of a miscible agent of unit injected concentration displacing a fluid of zero concentration in a one-dimensional, semi-infinite medium is given by the familiar error function solution:

$$C = \frac{1}{2} \left[1 - erf \left(\frac{x - t}{2\sqrt{t/N_{Pe}}} \right) \right] \qquad (1)$$

where N_{Pe} is the Peclet number, a dimensionless ratio of convective to dispersive transport

$$N_{Pe} = \frac{uL}{\phi K_\ell} \qquad (2)$$

See the Nomenclature for the definition of other symbols. The term K_ℓ is the longitudinal (for parallel to bulk fluid flow) dispersion coefficient. A great deal of experimental evidence, mainly in laboratory corefloods, suggests that K_ℓ takes the form

$$K_\ell = \frac{D_o}{\phi F} + \alpha_\ell v^\beta \qquad (3)$$

where D_o is t he effective binary diffusion coefficient between the displacing and displaced fluid. α_ℓ is the longitudinal dispersivity, a characteristic length of the medium and the prime subject of this section. The exponent β is determined by experiment to be about 1-1.2; in what follows we take it to be exactly one. If diffusion is neglected, then, Equation 3 serves as a definition for the diffusivity. Note that the Peclet number now becomes

$$N_{Pe} = \frac{L}{\alpha_\ell} \qquad (4)$$

because $v = u/\phi$. In much of what follows we will use the inverse Peclet number (α_ℓ/L) rather than N_{Pe}.

Mixing Scales. A stochastic simulation will generate $C_s(t,x,y)$ for a unit inlet concentration in a two-dimensional flow field. The goal of this generation is to deduce the behavior of the dispersivity and relate these back to the statistical properties of the field; however, there are two ways to do this each manifesting a particular scale of mixing.

Let the operator $E(\cdot)$ denote averaging in the y-direction.

Then $E(C_s)$ is not a function of y. We can denote one scale of mixing by minimizing the function

$$S_{ME} = \int_0^\infty (C(t,x;\alpha_{ME}) - E(C_s))^2 dx \qquad (5)$$

where the subscript ME denotes megascopic and the semicolon in the argument for C separates variables from parameters. In actual practice we use a more sophisticated analytic solution than Equation 1 in the minimization Equation 5 but the idea is the same. Also, the integration in Equation 5 must be truncated to reflect the finite nature of the system, an effect which will be apparent below, and C_s must be corrected for truncation error. The minimization in Equation 5 is through least-squares regression. The process yields a megascopic dispersivity which is, in general, a function of time or mean distance traveled.

If we perform the minimization in (5) on C_s rather than $E(C_s)$ we obtain a dispersivity α_ℓ which is a function of t and y only, the x-dependency having been eliminated by the integration in (5). For this case we define the macroscopic dispersivity as $\alpha_{MA} = E(\alpha_\ell)$. Macroscopic dispersivity is a measure of the local or point mixing taking place in the medium.

Megascopic Dispersivity. The megascopic scale is the full-aquifer dispersivity whose value determines the volumetric sweep in numerical simulation blocks. Figure 3 shows the behavior of α_{ME} (expressed as inverse Peclet number) as a function of time for miscible displacements in a two-dimensional stochastic permeability field. The parameter V_{DP} is the Dykstra-Parsons coefficient, a dimensionless measure of the spread of the permeability distribution to which the flow field was conditioned. $V_{DP} = 0$ corresponds to a homogeneous medium and $V_{DP} = 1$ is infinitely heterogeneous; see the Appendix for more discussion of V_{DP}. In Figure 3 and elsewhere the results have been corrected for the presence of numerical truncation error in these explicit finite-difference simulations.

As intuitively expected, the higher heterogeneity runs have larger dispersion, but α_{ME} grows with time in general. In fact for small t we see linear growth passing over to near constancy at larger time. Clearly, there are other factors in the permeability distribution besides its spread that account for this behavior. One of these factors is the presence of correlation in the permeability field.

Figure 4 shows a similar plot of α_{ME} versus t except now V_{DP} is being held constant while the dimensionless correlation length λ_D is changed for each run. The correlation length expresses the distance over which the covariance of the permeability distribution falls to near zero. Figure 5 shows the autocorrelation functions (covariance/variance) for the permeability fields used to construct Figure 4. In general, runs with small λ_D (λ_D is the correlation length divided by the system length) have relatively constant and small α_{ME}; runs with large λ_D have α_{ME} that is larger and grows significantly with t.

Another way to view correlation length is that it is the

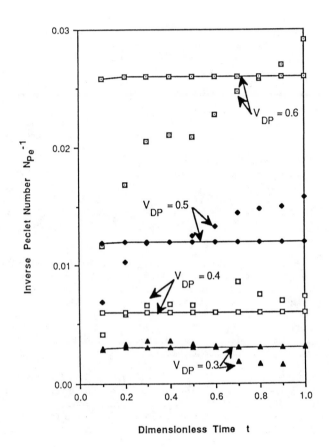

Dimensionless Time   t

Figure 3.   Growth of megascopic dispersivity (points) with throughput or dimensionless travel distance for four different degrees of heterogeneity. Dispersivity generally grows with distance and is larger at larger heterogeneities (as measured with the Dykstra-Parson coefficient V_{DP}). Dispersivities are from a single-phase, diffusion-free simulation in a stochastic permeability generated by the Heller procedure. Solid curves are from Taylor's equation. (Reproduced from Ref. 4.)

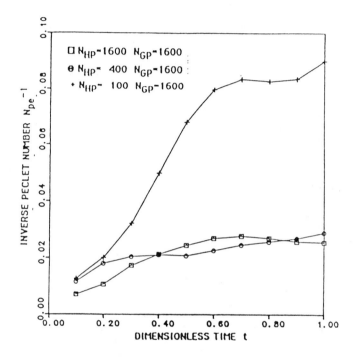

Figure 4. Growth of megascopic dispersivity (points) with throughput or dimensionless travel distance for three different dimensionless correlation lengths. Cases with small correlation lengths generally have constant dispersivities; those with large correlation lengths generally grow, reaching a plateau at large time. Heterogeneity is constant for all cases. Abnormal growth near effluent end (x_D = 1) here and in Figure 3 is an artifact from the averaging of the outflow boundary condition in the finite difference solution. Diffusion is zero. (Reproduced from Ref. 4.)

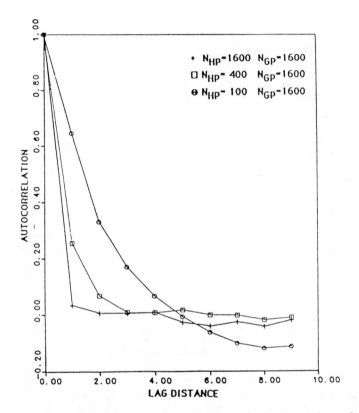

Figure 5. Permeability autocorrelograms for the three realizations whose dispersivity is in Figure 4. We can change the dimensionless correlation length by varying the ratio of Heller points to grid blocks $\lambda_D = [N_{HP}]^{-\frac{1}{2}}$. The horizontal axis is in grid block units. The curves are only approximately described by an exponential model. In each case, the permeability distribution itself is log-normal. (Reproduced from Ref 4.)

distance over which permeability values are roughly the same on average. Thus, it is not surprising that α_{ME} grows with time when λ_D is large because the medium is becoming more layered, a situation for which it is well-known that megascopic dispersivity, as defined here, grows linearly with time. Figures 6 and 7 confirm that this is so. Recall when comparing these two figures that they both have the same degree of heterogeneity V_{DP} and both were generated through stochastic means--the permeability field was not conditioned to be layered. Yet, layering of the concentration distribution $C_s(x,y,t)$ is evident in Figure 7 and absent in Figure 6.

Taylor's Theory. To explain the behavior of α_{ME} in Figures 3 and 4 we borrow from G. I. Taylor's theory of continuous movements, one of the pioneering works in the representation of turbulence. Casting Taylor's equation over into our notation and restricting the theory to an exponentially-declining correlation function, as seems appropriate from Figure 5, yields the following for the dependence of α_{ME} upon t:

$$\frac{\alpha_{ME}}{L} = C_v^2 \lambda_D (1 - e^{-t/\lambda_D}) \qquad (6)$$

This equation is shown on Figure 3 as the solid lines. The coefficient of variation C_v and the correlation length λ_D were all derived from the properties of the underlying permeability field--there was no history match. (Strictly speaking, the statistical properties should come from the velocity field, but they appear to be the same within the accuracy of determining C_v and λ_D.)

We accept the quality of agreement between Equation 6 and the simulated dispersivities in Figures 3 and 4, and will use it to draw further conclusions below. However, the agreement is not perfect. Some of the reasons for this are the absence of a permeability correlation function that is precisely exponential (Figure 5), and the lack of multiple realizations in the simulated results. Equation 6 actually expresses the expectation of the dispersivity; thus, it should agree best with the mean of a large number of simulations each with the same statistical properties but with differing flow fields. In fact, the agreement improves by taking such means. In actual practice there will also be uncertainties in estimating V_{DP} (see Appendix).

Equation 6 has two limiting cases which provide a basis for classifying displacements and permeable media. If the correlation length is small Equation 6 becomes

$$\frac{\alpha_{ME}}{L} = C_v^2 \lambda_D \qquad (7)$$

which predicts that dispersivity is independent of t (that is, an intrinsic property of the medium) and proportional to the coefficient of variation squared and the dimensionless correlation length. We should call such a medium uniform. When the correlation length is large Equation 6 becomes

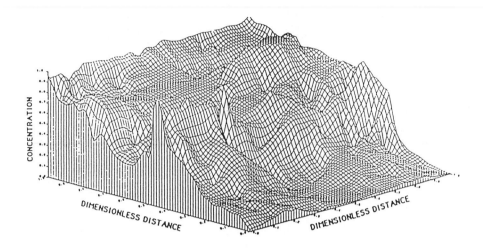

Figure 6.   Isometric map of injectant concentration for a case
with small correlation lengths and no diffusion.   The longitudinal direction (parallel to flow) is x; the   transverse direction (perpendicular to flow) is y. Concentration is quite irregular because there are no preferred paths. (Reproduced from
Ref. 4.)

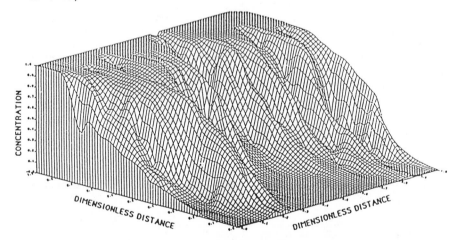

Figure 7.   Isometric map of injectant concentration for a case
with large correlation lengths and no diffusion.   The longitudinal direction (parallel to flow) is x; the transverse direction (perpendicular to flow) is y.   Concentration gives the
appearance of layering even though there are none present.
(Reproduced from Ref. 4.)

$$\frac{\alpha_{ME}}{L} = C_v^2 \, t \tag{8}$$

a result which says that dispersivity grows in proportion to time or travel distance. Since this consequence occurs when the displacement is channeling, media with λ_D large are layered. In general, the entire Equation 6 applies; however it is curious that layering predominates at relatively small correlation lengths.

Molecular diffusion adds an additional correlation scale to the behavior of α_{ME}. As Figure 8 shows, increasing the diffusion hastens the attainment of a constant dispersivity that is always greater than what would be present because of diffusion alone. The result of diffusion is to lessen the transverse concentration gradients and thereby reduce channeling. The increased "apparent" dispersivity is because of the combined effects of isotropic diffusion and the velocity fluctuations. The values of diffusion shown in Figure 8, however, are generally quite a bit larger than what is evidenced is actual displacements.

Figure 9 summarizes the observations on megascopic dispersivity. All displacements go through regimes of channeling (non-Fickian), transition and diffusive (Fickian) mixing. This behavior is characterized by Taylor's theory in Equation (6) with the appropriate limiting cases as in Equations 7 and 8. The extent to which, if any, dominate a particular displacement depends on the heterogeneity and correlation structure of the permeability field, a subject discussed in the last part of this paper. However, the behavior suggests two important conclusions about how such mixing can be handled in numerical simulators, scaling, and, indeed, in the very nature of permeability heterogeneity.

Lumping Mixing Effects. The two limiting cases, Equations 7 and (8), can be easily captured in heuristic models--that is transport relations whose effect is to mimic the actual behavior. In the case of diffusion mixing, Equation 7, we simply input the appropriate dispersivity to the simulator. In the case of channeling, as in Equation 8, such fingering/bypassing models as the Koval, Todd-Longstaff, and Young models are available. However, none of these will mimic the general behavior in Equation 6 of Figure 9. To be sure it is possible to use Equation (6) directly as a heuristic model, but this is impractical since time-dependent parameters would be required. Deducing a practical but efficient way to model the general megascopic mixing case is a very fruitful avenue for future research.

In the same vein, the general behavior of Figure 9 is not easy to scale from field to laboratory conditions. To see this imagine that we have a laboratory and a field displacement which follow exactly the same unknown α_{ME} curve. (This does not actually happen, but the additional complexity is superimposed on these remarks.) From the laboratory experiment we know α_{ME} at some small time t_1 and we adjust the heuristic models, Equations 7 and 8, to these values. Now let use use the adjusted heuristic models to predict mixing at a much longer time t_2. Since the models do not in general model the true behavior, α_{ME} and the mixing zone size will not be predicted correctly at t_2. Using Equation 7 will under

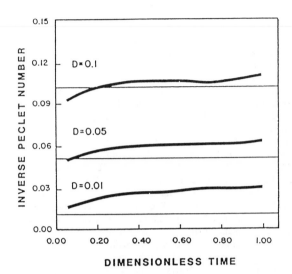

Figure 8. Growth of megascopic dispersivity with distance traveled for cases with varying amounts of diffusion. The dimensionless (input) diffusion coefficient D is (D_0/vL) which is show in the above as dark horizontal lines. Dispersivity approaches pure diffusion for large D wherein all transverse concentration gradients are zero. Dispersivities being larger than D is caused by velocity variations (as in Figures 3 and 4) and by the combined effect of such variations with transverse diffusion (Taylor's diffusion). (Reproduced from Ref. 5.)

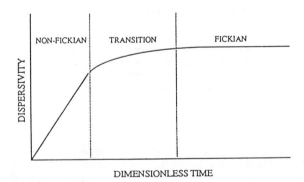

Figure 9. Schematic representation of the growth of megascopic dispersivity with distance traveled. All displacement pass through three regimes (non-Fickian, transition and Fickian) which may be large or small depending on the correlation length and D. Even displacement which are strongly Fickian do not lose the evidence of their non-Fickian beginning. (Reproduced from Ref. 1.)

predict the mixing zone size and Equation 8 will over predict. Indeed, experience with using Equation 8 suggests that the "effective" coefficient of variation must be reduced during scaling.

Finally, let use consider the behavior of actual field-measured dispersivities. Figure 10 shows a remarkable plot of u_{ME} versus measurement distance from over 60 datum taken from both laboratory and field measurements. Despite the considerable scatter there is a clear trend of increasing dispersivity with distance. If we assume that the measurement distance is the same as the distance traveled or the horizontal axis in Figures 3 and 4 then we immediately notice that there is no suggestion of a leveling off even at very large measurement distances. Two contradictory dictory conclusions are possible based on this: measurement distance and travel distance are not synonymous, or the correlation structure of real media do not lead to constant dispersivities at large travel distances. Considering the latter, one could always argue that the correlation distance is actually larger than the horizontal scale of Figure 10, but this seems unlikely given the very large range encompassed. An alternate possibility is that permeability is distributed as a fractal whence we expect no leveling out regardless of the measurement length. Indeed, the field data in Figure 10 can be fit with a fractal dimension of 1.12. In the last part of this paper we discuss our attempts to directly measure correlation in naturally--occurring media.

Figure 10 makes apparent the futility of estimating large-scale dispersivities from laboratory experiments: megascopic dispersivities are different between the two by several factors of ten even in the same media and under the same flow conditions. What is more likely to be comparable is the field-scale macroscopic dispersivity and the laboratory-scale megascopic dispersivity, but even these comparisons are tenuous because the laboratory experiments generally are more homogeneous, have smaller length scales and are more influenced by diffusion. We discuss macroscopic dispersivity below, but first we conclude this section by giving results on megascopic dispersivity on the laboratory scale.

Figure 11 shows the results of the simulations, plotted as a normalized dispersion coefficient versus a dimensionless rate, compared to experimental data from sandstone cores. The acceptable agreement indicates three conclusions about the nature of dispersion in laboratory experiments:

1. Diffusion and mechanical mixing dominate at low and high rates, respectively. This is consistent with Equation 2.

2. In the mechanical mixing regime (large velocities), dispersion coefficients become proportional to velocity to the first power. Thus, we are justified in taking $\beta = 1$ in Equation 2 since a rate-independent u_{ME} agrees with Equation 7.

3. Correlation lengths are as little as five grain diameters in laboratory experiments. This length is very much smaller than even the smallest experimental dimensions and we, therefore, expect time-invariant dispersivities even in the absence of diffusion. It is possible, in principle, to calculate C_v and λ_D from this observation and Equation 7.

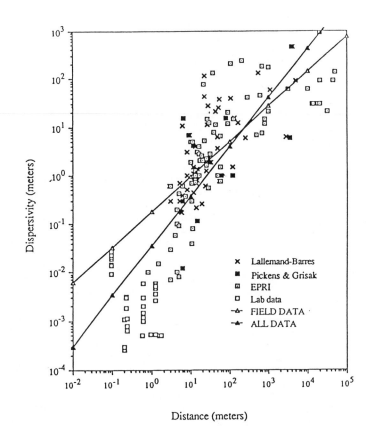

Distance (meters)

Figure 10. Summary of experimentally measured dispersivities ranging from small (laboratory) to large (field) scale. Solid curves are best fit to all data (slope = 1.13) and to field data only (slope = 0.755). Data includes experiments in a large variety of permeable medium, tracers and experimenters which accounts for much of the scatter, but the trend with increasing dispersivity with distance is evident. (Reproduced from Ref. 1.)

Macroscopic Dispersivity. α_{MA} is of interest in many enhanced oil recovery displacements (such as development of miscibility or generation of optimal salinities) where local mixing controls the success of the displacement. We can also extract α_{MA} from the randomly heterogeneous runs discussed above but our conclusions are much less precise because we have no theoretical model like Equation 1 for local mixing in two-dimensional flow. But we can compare α_{MA} to α_{ME} under identical circumstances.

Figures 12 and 13 make this comparison for two divergent cases. We see that in general like Equation 1 α_{MA} grows with time in the same manner as α_{ME}. Its dependence on V_{DP} may be somewhat less. Most importantly, α_{MA} is always smaller than α_{ME}, but the difference decreases as correlation length becomes small. The reduction in the difference with small correlation length is consistent with Equation 7 for in this limit of a uniform medium sampling one point would be, on average, equivalent to sampling all points at the same cross-section. The difference cannot be made to vanish entirely, however, as α_{ME} always retains its initial channeling component, a feature which is absent in α_{MA}. We are very much in need of a consistent way to estimate α_{MA} from medium properties--an analogue to Equation 6--for in this way we could devise a laboratory medium that would truly represent a characteristic degree of mixing in large-scale displacements.

Geologic Descriptions

The most significant barrier to wide-spread use of conditional simulation is the lack of insight into the statistical properties of real permeable media. While we can and are making great strides into understanding flow in generic media (as was done above) there can be no predictive power unless the stochastic field is realistically representable for specific cases. Stochastic assignments and conditional simulation are probably the only fruitful line of inquiry in making predictions, but tools are still lacking when confronted with realistic media. In this final section we describe efforts to statistically describe a large-scale outcrop.

The subject outcrop was an ancient eolian sand near Glen Canyon dam in northern Arizona. Eolian sands were originally deposited by wind-blown deposition, but their character manifests itself in several scales of heterogeneity as illustrated in Figure 14. We observed the following stratification types: intradune (sand-free zones between fingers of sand deposition), grainfall (avalanching caused by oversteepening of dune faces) and wind-ripples (reworking of sand grains into undulating patterns caused by low velocity winds). Even with this relatively few number of types the outcrop displayed formidable complexity through the ordering, geometry, orientation and repetition of the patterns.

We sampled the outcrop for permeability using a minipermea-meter, a device designed to make a large number of nondestructive permeability measurements in a short time. This device was invaluable inasmuch as very large data sets are required to make statistically meaningful statements. Nearly as important was the sampling schemes. We used three: sampling according to stratifica-tion type, sampling along a line (traverses) and sampling along

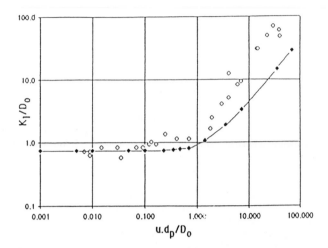

Figure 11. Comparison of calculated macroscopic dispersivities (solid curve) to experimentally measured megascopic (laboratory scale) dispersivities. d_p is the grain diameter for the packing in the experimental measurements and the correlation length for the calculated curve. Better agreement could be obtained by letting d_p be about five times the correlation length in the calculated curve. (Reproduced from Ref. 4.)

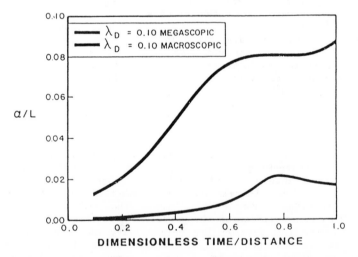

Figure 12. Comparison of megascopic (upper curve) and macroscopic (lower curve) dispersivities from simulations with large correlation length and no diffusion. In all cases, the macroscopic dispersivity is smaller and grows slower. (Reproduced from Ref. 5.)

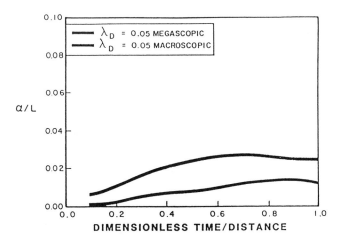

Figure 13. Comparison of megascopic (upper curve) and macroscopic (lower curve) dispersivities from simulations with small correlation length and no diffusion. When correlation length decreases the two dispersivities approach each (compare with Figure 12) an observation consistent with less channeling when media are uncorrelated. (Reproduced from Ref. 5.)

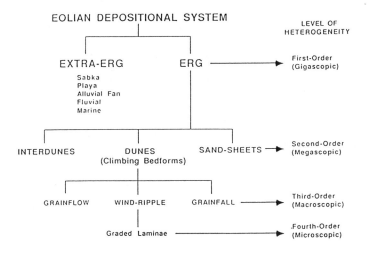

Figure 14. The heirarchy of heterogeneity scales for an eolian sand. Permeability groups by the third-order heterogeneity classifications. Our measurements deal primarily with wind ripple, grainflow and intradune groups. (Reproduced from Ref. 5.)

concentric grids. Each yielded particular insights, but it is only
the latter that we discuss here.

Figure 15 shows the five concentric grids superimposed on one
wall of the outcrop. The grids were intended to all be centered on
a single point, but the irregular surface prevented us from doing
this exactly. We intended to make the largest grid completely cover
the outcrop face and the smallest grid to be such that adjacent
samples would be of the same separation distance as in laboratory
cores. The largest three grids sampled several beds and the
smallest two are entirely within a single bed. Each grid
contained more than 50 samples on a regular two-dimensional
pattern.

Figure 16 shows the coefficient of variation C_v for each grid
plotted as a function of the grid area. C_v is the same for the
three largest grids, dips to a minimum at the fourth grid and then
returns to a large value at the smallest grid. Statistically, the
fourth grid is different from all the others which are themselves
the same. We can explain the dip in the C_v curve of this data by
observing that the fourth grid is the first one entirely within one
bed. Indeed, histograms of grids one through three show evidence of
two or more populations. The increase in the fifth grid is
because of very small-scale laminae which are present all over the
outcrop, but noticeable only in the statistics of the fifth grid.
The mean permeability was the same for all grids.

Whatever the geologic causes, there are several purely statistical inferences to be drawn from Figure 16 which bear directly on
the issue of reservoir simulation. The size of grid four may be a
natural choice for the grid block size in a deterministic simulation
model. Such a selection would minimize the variation between blocks
and may, in fact, make stochastic assignments of secondary importance (thus, reducing the differences between realizations). The
variation of the fifth scale would be incorporated as pseudo
functions or megascopic dispersivity into individual blocks.

However, a representative elementary volume (REV) --the volume
below which permeable media properties uniformly approach single-
value limits--is not evident in the data. To justify this statement
imagine that the individual data in each grid are close to the mean
for a large number of measurements of several entire grids. As the
size of these grids shrinks we see that the variability of their
means goes through a minimum. But if there exists a REV, then the
variability should approach zero at the size of the REV. Of
course there still might exist a REV of size smaller than the
fifth grid, but this size is not too different from the pore
dimensions where grain-to-grain fluctuations begin to be important.

We mentioned the possibility of permeability being distributed
as a fractal in conjunction with Figure 10. A fractal distribution
would show C_v decreasing monotonically at a rate prescribed by its
factal dimension. This is not the case with the eolian outcrop
which clearly shows two scales of heterogeneity. Given its
correspondence with the geologic features, the two-scale interpretation is the only possible consistent interpretation.

In discussing the behavior of dispersivity we emphasized the
role of the autocorrelation structure of permeability. The concentric grid sampling allows a very efficient investigation of both
the magnitude and orientation of the correlation. Figure 17 shows

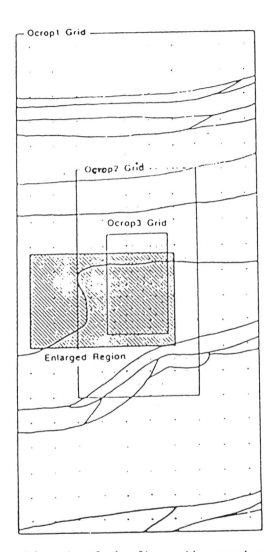

Figure 15.   Schematic of the five grids superimposed on the
outcrop face of an eolian sand (Page sandstone, Coconino county
Arizona).   The three largest grids contains more than one bed.
Size of the largest grid is about 100 by 180 ft. and the
smallest about 15 inches square. (Reproduced from Ref. 5.)

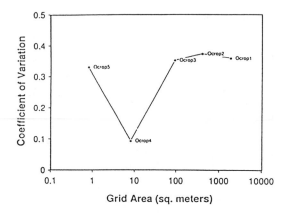

Figure 16.   Coefficient of variation for the five grids in
Figure 14.   The   statistically   significant   drop   for OCROP4
reflects properties being measured entirely within one bed.
Inter-septile range shows the same effect although the mean
permeability did not vary greatly.   (Reproduced from Ref. 5.)

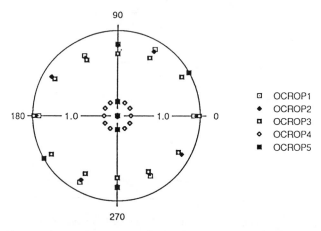

Figure 17.   Ellipses of variogram range (correlation length)
for the grids in Figure 13.   Three largest ellipses are essentially the same, fourth outcrop is uncorrelated and smallest
outcrop has a correlation orientation parallel to local
laminae.   (Reproduced from Ref. 5.)

the correlation ellipses for the correlation lengths (expressed as variogram ranges) of the five grids. To compress all of the data on to one plot shows, Figure 17 is on log-polar (units: log-ft.) coordinates.

Once again grid four is pathologic; it and grid five have different properties from the largest grids. The major correlation length for the largest grids is about 50 feet and orientated at an angle of about 30° above horizontal. The major correlation length for the small scales is about 2 feet and oriented nearly vertically. The latter scale does not appear in grid four (which appears entirely uncorrelated) because its sampling interval is about the same length as this.

Such complexity requires considerable sophistication in generating stochastic fields for conditional simulation. What is required is a generator which can deal with multiple and imbedded correlation scales each with a different variability in three dimensions. The fields must also be anisotropic and have orientation which is also scale-dependent. Generation of fields with this complexity being beyond any currently available generator, this provides another avenue for future research.

The above observations require several qualifications. Several of the variograms used to infer Figure 17 do not show strict spherical behavior. On the largest scales they are somewhat better described by hole models, but this is entirely consistent with the observation of sampling of multiple beds. More importantly, the data in both Figure 16 and 17 represent only a single realization on the outcrop. We have no way of knowning if a repetitive experiment on the same outcrop at a different location would give the same results.

With regards to the last point, the confidence of an interpretation is greatly increased (and the number of data required greatly diminished) when the statistics is backed up with the geologic observations. For example, we would be much more hesitant to conclude about the existence of two scales of heterogeneity in the eolian outcrop if such were not consistent with the classifications of Figure 15. Generally speaking, the geology is consistent with the statistics (the bed orientations on the outcrop do match the orientation angle in Figure 17) and once this is verified for several diverse outcrops we can have confidence in generating random fields directly from the geologic descriptions. Indeed, it seems reasonable that a taxonomy of reservoir types based on statistical descriptions is possible.

Concluding Remarks

Conditional simulation provides a potential means for efficiently merging geologic observation and engineering. What is required to bring this to common practice are (1) highly time-efficient computer models which can effect multiple realizations without excessive burden, (2) statistical generation procedures that can handle the complexities of geologically realistic media, (3) geologic descriptions which extract quantitative data, and (4) representations or models which can bring about a translation between geological and statistical model with accuracy and confidence.

Appendix

Bias and Precision of the Dykstra-Parsons Coefficient.

The Dykstra-Parsons coefficient is a normalized measure of the spread of a permeability distribution that is bounded between zero and one. The formal definition is

$$V_{DP} = 1 - \frac{k_\sigma}{k}$$

where k and k_σ are the medium permeability and the permeability at one standard deviation below the median, respectively. Both quantities are taken from best-fit straight lines to the stochastic data plotted on log-probability paper. Inaccuracies caused by this practice are small when the permeability data are log-normally distributed as assumed is the case in the work here. In what follows we discuss the bias and precision in estimating V_{DP}. For log normal distributions C_V and V_{DP} are related as

$$C_V^2 = \exp[\ln(1 - V_{DP})]^2 - 1$$

Imagine that there exists a population of permeabilities consisting of a very large number of datum. The population has a known V_{DP}. Let us draw a certain number N from this population and calculate an estimated V_{DP} from this sample. Repeating the procedure several times will generate a probability distribution function (p.d.f.) for the estimate V_{DP}. This function that illustrates the difficulties in estimating V_{DP}.

Suppose each repetition gives exactly the same value for $\hat{V}_{DP}$, but this value is different from V_{DP}. The estimator is biased and the magnitude of the bias b_V is the expected value of V_{DP} minus V_{DP} divided by the true value. The estimator is precise, however.

On the other hand, suppose the p.d.f. has a finite spread, but the mean of the distribution is V_{DP}. The estimator is unbiased, but imprecise and the measure of the imprecision is the standard error s_V of the p.d.f. In actuality, the estimate for V_{DP} is neither unbiased nor precise, but both can be estimated.

Figure 18 plots the bias as a function of the number of datum N in the samples and the true value of V_{DP}. As expected, the bias approaches zero with increasing sample size, and is largest for small sample sizes. The bias is also largest when V_{DP} is about 0.7; however, in no case is the bias large. We conclude that V_{DP} is essentially unbiased except perhaps at very small sample sizes.

Unfortunately, the estimator is fairly imprecise, Figure 19. Here we are plotting N versus V_{DP} for a given standard error. Once again, precision improves as the sample size becomes larger, but even for fairly large sample sizes the error is not negligible.

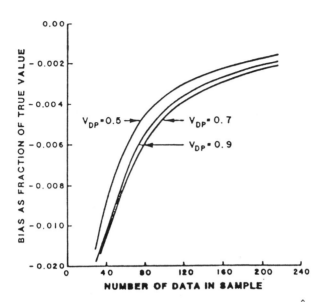

Figure 18. Bias b_y of Dykstra-Parsons coefficient $\hat{V}_{DP}$ estimator of heterogeneity as functions of number of samples and numerical value of V_{DP}. Curves were theoretically derived from a single-population, log-normal distribution. (Reproduced from Ref. 5.)

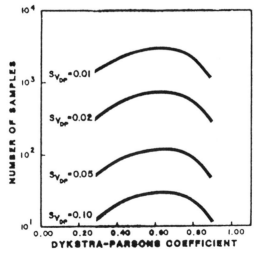

Figure 19. Number of samples required for a given level of precision for the Dykstra-Parsons coefficient V_{DP}. s_V is the standard error of the estimate at the given value of V_{DP} and sample number. Curves were derived from theoretical considerations simulations based on a single-population, log-normal distribution. (Reproduced from Ref. 5.)

Acknowledgments

This work was supported by the U. S. Department of Energy through grants, DE-AS19-82BC10744 and DE-AC19-85BC10849, and the Center for Enhanced Oil Recovery Research at The University of Texas at Austin.

Nomenclature

b_v	Bias
C^v	Concentration, analytic solution
C	Concentration, numerical solution
C^s	Coefficient of variation
D^v	Dimensionless diffusion coefficient
D	Molecular diffusion coefficient, L^2/t
$E(\cdot)$	Expectation operator
erf	Error function
f_s	Shale fraction
F^s	Formation resistivity factor
k	Permeability, L^2
K	Dispersion coefficient, L^2/t
L	Medium length, L
N	Number of datum in a sample
N_{Pe}	Peclet number
N_{HP}	Number of Heller points
N_{GP}	Number of grid points or blocks
s	Standard error in Dykstra-Parsons coefficient
t^v	Time, volume injected as fraction of pore volume
u	Superficial velocity, L/t
v	Interstitial velocity, L/t
V_{DP}	Dykstra-Parsons coefficient
x	Distance parallel to flow, fraction of medium length
y	Distance transverse to flow

Greek:

α	Dispersivity, L
β	Exponent on velocity
Δ	Discrete change
λ	Correlation length
ϕ	Porosity, fraction

Subscripts:

D	Dimensionless
ℓ	Longitudinal
MA	Macroscopic
ME	Megascopic
o	Standard deviation

Superscript

^	Estimate

Literature Cited

1. Arya, Atul, Ph.D. Dissertation, The University of Texas at Austin, 1986.
2. Haldorson, Helge H., Ph.D. Dissertation, The University of Texas at Austin, 1983.
3. Jensen, Jerry Lee, Ph.D. Dissertation, The University of Texas at Austin, 1986.
4. Lake, Larry W., Alan J. Scott and Gary Kocurek, "Reservoir Characterization for Numerical Simulation," Final Report, sponsored by the U.S. Department of Energy/Bartlesville Project Office, June 1986.
5. Lake, Larry W., Mark A. Miller and Gary Kocurek, "A Systematic Procedure for Reservoir Characterization," Annual Report, sponsored by the U.S. Department of Energy/Bartlesville Project Office, 1987.
6. Lake, Larry W. In Numerical Simulation in Oil Recovery; Wheeler, Mary F., Ed.; Springer Verlag, in press.

RECEIVED April 7, 1988

PHASE BEHAVIOR, SURFACTANT DESIGN, AND ADSORPTION

Chapter 4

Multiphase Equilibria for Water–Carbon Dioxide–2-Propanol Mixtures at Elevated Pressures

J. R. DiAndreth[1] and M. E. Paulaitis[2]

[1]E. I. du Pont de Nemours and Company, Wilmington, DE 19898
[2]Department of Chemical Engineering, University of Delaware, Newark, DE 19716

Phase behavior involving three and four fluid phases is predicted for ternary mixtures of CO_2, water, and isopropanol based on phase equilibrium calculations with the Peng-Robinson equation of state and selected experimental measurements of phase compositions for L_1L_2G. $L_1L_2L_3$, and $L_1L_2L_3G$ equilibria. Although quantitative agreement with measured phase compositions was not obtained, the Peng-Robinson equation did predict all the regions of multiple equilibrium phases that were observed in our experiments including those regions not used to generate the equation-of-state parameters. The water-isopropanol-CO_2 system is representative of model surfactant systems containing CO_2 and water, and the complexity of the phase diagrams generated in our analysis suggests that the phase behavior for related surfactant systems can be quite complex.

Multicomponent phase equilibria involving three or more coexisting fluid phases is frequently encountered in liquefied natural gas processes (1), tertiary oil recovery by miscible gas displacement (2), and the use of surfactants in enhanced oil recovery (3).

In this paper, we describe multiphase behavior involving three and four equilibrium phases for ternary mixtures of water, carbon dioxide, and isopropanol at temperatures and pressures near the critical point of CO_2 . The complexity of the phase equilibria observed for such mixtures cautions against using limited experimental data to characterize phase behavior. We provide a more comprehensive analysis which involves coupling a thermodynamic model with limited experimental information to describe the global multiphase behavior exhibited by this ternary mixture. These ternary mixtures are representative of model surfactant systems containing CO_2 and water. Our approach has general utility for predicting complex phase behavior exhibited by many multicomponent mixtures of interest in enhanced oil recovery which includes the use of surfactants.

A review of the literature found only a few studies of multiphase behavior for related model surfactant systems. Fleck

0097–6156/88/0373–0082$06.00/0
© 1988 American Chemical Society

(4) investigated three- and four-phase equilibria for ternary
mixtures of water, carbon dioxide, and isopropanol. Phase
compositions were reported for only certain equilibrium phases
and the data showed appreciable scatter. Panagiotopoulos and
Reid (5) measured liquid-liquid-gas equilibrium for ternary
mixtures of water, CO_2, and n-butanol. Several aspects of the
multiphase behavior for this system were described qualitatively
with a cubic equation of state.

In a previous paper (6), we have reported quantitative
measurements of phase compositions and molar volumes for liquid-
liquid-gas, liquid-liquid-liquid, and liquid-liquid-liquid-gas
equilibria for ternary mixtures of water, carbon dioxide, and
isopropanol at 40°, 50°, and 60°C and elevated pressures. The
phase behavior observed at these conditions was found to be
quite complex.

At 60°C, three equilibrium phases were observed: a CO_2-
rich gas phase G, a water-rich liquid phase L_1, and a fluid
phase L_2, of intermediate density containing relatively high
concentrations of isopropanol. Three-phase, L_1L_2G equilibrium
exists over a range of pressures from 10.6 MPa to 12.9 MPa.
These pressures correspond to the L_1L_2 and GL_2 critical endpoints
that bound the three-phase region.

Three-phase L_1L_2G equilibrium was also observed at 50°C,
but at slightly lower pressures. The L_1L_2 critical endpoint
bounds the three-phase region at low pressure, as was observed
at 60°C. However, four coexisting fluid phases, rather than a
gas-liquid critical endpoint, were observed to bound the three-
phase region at higher pressures.

Four equilibrium phases were also observed at 40°C and 7.8
MPa. The fourth phase has a density intermediate to those of
the L_2 and G phases and is denoted L_3. No solid phases were
found at these conditions and gas hydrates would not be expected
at this relatively high temperature (7). A second three-phase,
$L_1L_2L_3$ region was also measured at 40°C and 8.4 MPa. This three-
phase equilibrium is distinct from the L_1L_2G three-phase region
observed at 50°C and 60°C. A third three-phase region, probably
corresponding to L_1L_3G equilibrium was observed at 40°C over a
pressure range from approximately 7.8 to 8.0 MPa.

In order to expand the scope of our experimental
observations we have predicted the phase behavior for this
ternary mixture over an extended range of temperatures and
pressures using the Peng-Robinson equation of state (8).
Although quantitative phase compositions, temperatures, and
pressures for such a highly non-ideal mixture cannot be expected
from this equation of state, a reasonably accurate description
of global phase behavior can be obtained.

Thermodynamic Model and Computer Algorithm

The Peng-Robinson equation of state (8) was chosen for our
thermodynamic model because of its fundamental basis,
reliability and wide use in vapor-liquid equilibrium
calculations. The pressure-explicit form of this equation for a
pure component is

$$P = \frac{RT}{v - b} - \frac{a(T)}{v(v + b) + b(v - b)} \tag{1}$$

where a and b are parameters related to physical properties of the pure component (e.g., critical properties and acentric factor). In this work, the temperature-dependent parameter a was obtained by fitting vapor pressures for each pure component. The parameter b was calculated from the critical temperature and pressure of each component. For mixtures, these parameters are defined by conventional composition-dependent mixing rules,

$$a_m = \sum_i \sum_j z_i z_j a_{ij} \qquad (2)$$

$$b_m = \sum_i z_i b_i \qquad (3)$$

where z_i is the mole fraction of component i and a_{ij} is defined by

$$a_{ij} = (1 - \delta_{ij}) \sqrt{a_{ii} a_{jj}} \qquad i \neq j \qquad (4)$$

The binary interaction parameter δ_{ij} is related to molecular interactions between components i and j, and is a characteristic parameter for each pair of mixture constituents.
Initial values of the binary interaction parameters were obtained from vapor-liquid equilibrium data for ternary mixtures (9). These interaction parameters were then adjusted to minimize the difference between calculated and measured phase compositions for the three-phase equilibria measured at 50°C. The values for the binary interaction parameters were $\delta_{CO_2-H_2O}$ = -0.053, δ_{CO_2-IPA} = 0.017, and δ_{IPA-H_2O} = $-.207$.
Figure 1 shows a comparison of calculated and experimental phase compositions at 40°, 50°, and 60°C. The fit of phase compositions for the L_1L_2G equilibrium at 60°C and 12.2 MPa is very good, while the fit of the $L_1L_2L_3$ phase equilibrium at 40°C and 8.4 MPa is semi-quantitative, at best. Since the three phases at 40°C correspond to a different three-phase equilibrium than the L_1L_2G equilibrium at 50°C, which was used to fit binary interaction parameters, the poorer fit at 40°C is not unexpected. Indeed, these results show that $L_1L_2L_3$ equilibrium can be predicted with parameters obtained from an entirely different three-phase region.
The computer algorithm for determining the compositions of three and four equilibrium phases at a fixed temperature and pressure is based on the successive substitution algorithm described by Heidemann (10). An initial estimate of the phase compositions was obtained from either the measured phase compositions at the temperature and pressure of interest, or from previous calculations at a nearby temperature and pressure. Trivial solutions giving two phases of identical composition would often result when the initial compositions were not close to the actual equilibrium compositions.
Three-phase equilibrium compositions were calculated over a range of pressures at a constant temperature. The equilibrium

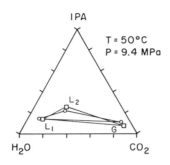

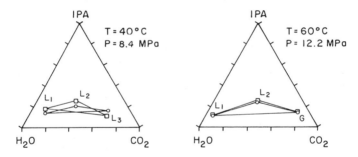

□ Experimental
○ Model Generated

Figure 1   Comparison of measured and model-generated
three- phase equilibrium compositions.

compositions at one pressure were used as starting values for
calculations at the next pressure. Near critical endpoints, the
phase compositions became sensitive to small changes in either
temperature or pressure. Critical endpoint temperatures and
pressures were estimated to be those temperatures and pressures
where an additional step of 0.0007 MPa or 0.01°C caused
convergence to a two-phase solution instead of a three-phase
solution. Four-phase equilibrium was calculated in a similar
manner with initial guesses determined from three-phase
equilibrium composition calculated at nearby temperatures and
pressures.

Calculated Multiphase Behavior

Phase compositions for liquid-liquid-gas, liquid-liquid-liquid,
and liquid-liquid-liquid-gas equilibria were calculated over a
range of temperatures from 20°C to 134°C and pressures from 4.7
MPa to 26.3 MPa. The predicted phase behavior includes all the
multiphase equilibrium observed experimentally (6).
    Calculated phase compositions for L_1L_2G equilibrium at 80°C
and pressures ranging from 14.0 MPa to 17.8 MPa are shown in
Figure 2. This isothermal pressure-composition prism is
constructed by stacking triangular phase diagrams each generated
at a different pressure. At low pressures, gas-liquid
equilibrium exists (this two-phase region is not depicted on the
phase diagram). As the pressure is increased, the liquid phase
splits into two liquid phases at the L_1L_2 critical endpoint
pressure. This L_1L_2 critical endpoint is the lower bound in
pressure for three-phase L_1L_2G equilibrium. As the pressure is
increased further, the composition of the L_2 phase changes
continuously from a water-rich phase similar in composition to
the L_1 phase, to a CO_2-rich phase similar in composition to the
gas phase. This L_2 phase eventually merges with the gas phase
at a GL_2 critical endpoint, representing the upper bound in
pressure for L_1L_2G equilibrium. At higher pressures, only gas-
liquid equilibrium exists. This transition from two-phase
behavior to three-phase behavior to two-phase behavior with
critical endpoints bounding the three-phase region corresponds
to experimental observations for L_1L_2G equilibrium at 60°C.
    The isothermal pressure-composition prism in Figure 3
represents calculated equilibrium phase compositions at 55°C.
The outline of the prism has been omitted to give an
unobstructed view of the (shaded) three- and four-phase regions.
At 9.8 MPa, four equilibrium phases are obtained: the L_1, L_2,
L_3, and G phases. At higher pressures, three-phase regions
corresponding to L_1L_3G and $L_1L_2L_3$ equilibria are obtained as the
four-phase quadrilateral splits diagonally from the L_1 phase to
the L_3 phase. At lower pressures, the four-phase quadrilateral
splits diagonally from the L_2 phase to the G phase, producing
three-phase regions corresponding to L_1L_2G and L_2L_3G equilibria.
    Of the five different multiphase regions depicted in Figure
3, four were observed in our experiments. Phase compositions
corresponding to the $L_1L_2L_3G$ equilibrium quadrilateral were
measured at 40°C and 7.8 MPa. Phase compositions for $L_1L_2L_3$
equilibrium were measured at 40°C and 8.4 MPa. Observations of
L_1L_3G equilibrium were made at 40°C and 7.9 MPa, but phase

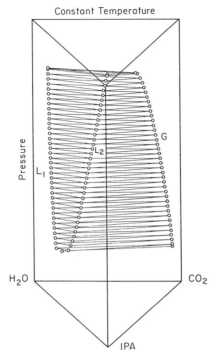

Figure 2   Model-generated pressure-composition phase diagram representing three-phase equilibrium at constant temperature.

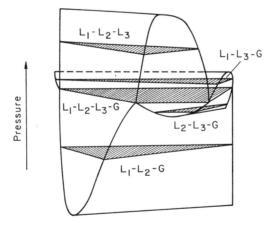

Figure 3   Model-generated pressure-composition phase diagram representing three- and four-phase equilibria at constant temperature.

compositions were not measured. Phase compositions for L_1L_2G
equilibrium were measured at 50° and 60°C. The range of
pressures and temperatures over which L_2L_3G equilibrium exists is
relatively small, as shown in Figure 3, and may explain why this
three-phase region was not observed in our experiments.

   A model-generated pressure-temperature (PT) projection for
this ternary system over the entire range of temperatures and
pressures studied is presented in Figure 4. This PT projection
includes the CO_2 vapor pressure curve, the CO_2-water liquid-
liquid-gas equilibrium curve, and the $L_1L_2L_3G$ equilibrium curve
and critical endpoint curves for the ternary mixture. Selected
temperatures and pressures corresponding to the critical
endpoint curves and $L_1L_2L_3G$ equilibrium curve are also presented
in Table I. Liquid-liquid and gas-liquid critical endpoint
curves bound the various three-phase regions which extend from
temperatures and pressures near the critical point of CO_2 (and
the CO_2-water liquid-liquid-gas equilibrium curve) up to the
tricritical point at 134°C and 26.3 MPa. This tricritical point
represents the intersection of the GL_2 and L_1L_2 critical endpoint
curves and corresponds to the temperature and pressure at which
all three equilibrium phases merge to form a single phase. A
lower temperature limit of 20°C was chosen for our calculations
because of the possible formation of gas hydrates, which is
beyond the scope of this study.

Table I. Boundaries of multiphase regions for the water-carbon
dioxide-isopropanol system

Temp., °C	20	40	58	60	62	80	100	120	134
GL_2	6.70	9.15		13.55		17.77	24.31	25.21	
GL_3		8.17	10.37 (UCEP)	10.66					
Tricritical points					10.89				26.27
$L_1L_2L_3G$	4.72	7.29							
L_2L_3	4.71	7.26		10.79					
L_1L_2	4.69	7.02		10.10		13.99	22.85	24.43	

   A second tricritical point is located at the intersection
of the GL_3 and L_2L_3 critical endpoint curves (62°C and 10.9 MPa).
This second tricritical point lies within the envelope formed by
the GL_2 and L_1L_2 critical endpoint curves. At temperatures
above 62°C, only a single three-phase region is obtained, L_1L_2G
as shown in Figure 2. Below 60°C additional three- and four-
phase equilibria are obtained. A secondary critical endpoint
(UCEP) representing the terminus of the $L_1L_2L_3G$ equilibrium curve

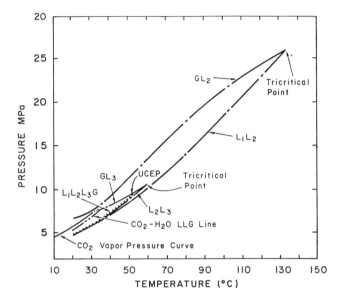

Figure 4   Model-generated P-T projection for ternary
mixtures of water, carbon dioxide, and
isopropanol.

is predicted at 58°C and 10.4 MPa. This secondary critical
endpoint is a GL_3 critical endpoint in equilibrium with both the
L_2 and L_1 phases. Below 58°C, multiphase behavior similar to
that shown in Figure 3 is obtained. The formation of the four
equilibrium phases below 58°C can be viewed as the intersection
in P-T-x space of two distinct three-phase regions corresponding
to L_1L_2G and L_2L_3G equilibria.

Summary

The full extent and variety of the phase behavior for water-
isopropanol-CO_2 mixtures observed experimentally and calculated
with the Peng-Robinson equation of state was not anticipated
based on known phase behavior for the constituent binary
mixtures or similar ternary mixtures. These results suggest
that multiphase behavior for related model surfactant systems
could also be complex. Measurements of all the critical
endpoint curves, the tricritical points, and secondary critical
endpoint for such systems would be tedious and are extremely
difficult. However, by coupling limited experimental data with
a thermodynamic model based on this cubic equation of state,
complex multiphase behavior can be comprehensively described.

Acknowledgments

        Financial support from the Exxon Research and Engineering
Company is gratefully appreciated.

Literature Cited

1. Luks, K.D., Merrill, R.C., and Kohn, J.P., Fluid Phase
   Equilibria, 14, 193, (1983).
2. Orr, Jr., F.M. and Taber, J.J., Science, 224, 563, (1984).
3. Smith, D.H., J. Colloid Sci., 102(2), 435, (1984).
4. Fleck, R. E., Ph.D. Thesis, University of California,
   Berkeley, (1967).
5. Panagiotopoulos, A.Z. and Reid, R.C., Fluid Phase
   Equilibria, 29, 525, (1986).
6. DiAndreth, J.R. and Paulaitis, M.E., Fluid Phase Equilibria,
   32, 261, (1987a).
7. Ng., H.-J. and Robinson, D.B., Fluid Phase Equilibria, 21,
   145, (1985).
8. Peng, D.-Y., and Robinson, D.B., Ind. Eng. Chem. Fundam.,
   15(1), 59, (1976).
9. Kander, R.G., Ph.D. Thesis, University of Delaware, (1987).
10. Heidemann, R.A., Fluid Phase Equilibria, 14, 55, (1983).

RECEIVED January 5, 1988

Chapter 5

Organized Surfactant Assemblies in Supercritical Fluids

John L. Fulton and Richard D. Smith[1]

Pacific Northwest Laboratory, Battelle Boulevard, Richland, WA 99352

Reverse micelle and microemulsion solutions are mixtures of a surfactant, a nonpolar fluid and a polar solvent (typically water) which contain organized surfactant assemblies. The properties of a micelle phase in supercritical propane and ethane have been characterized by conductivity, density, and solubility measurements. The phase behavior of surfactant-supercritical fluid solutions is shown to be dependent on pressure, in contrast to liquid systems where pressure has little or no effect. Potential applications of this new class of solvents are discussed.

Micelles and microemulsions are thermodynamically stable aggregates which are homogeneous on a macroscopic scale and form transparent solutions. Reverse (or inverse) micelles are small, dynamic aggregates of surfactant (amphiphilic) molecules forming shells around core regions containing a polar phase. Reverse micelle phases, as well as the water "swollen" microemulsion phase, can be considered subsets of a wide range of possible organized molecular structures which include a multitude of liquid crystalline and bicontinuous phases. In addition to potential applications in enhanced oil recovery (1,2), there is increasing interest in utilizing reverse micelles and microemulsions for separation of proteins from aqueous solutions (3,4), as reaction media for catalytic (5) or enzymatic (6) reactions, and as mobile phases in chromatographic separations (7,8).

Recently, the first observation of reverse micelles in supercritical fluid (dense gas) solvents has been reported (9) for the surfactant sodium bis(2-ethyhexyl) sulfosuccinate (AOT) in fluids such as ethane and propane. The properties of these systems have several attributes which are relevant to secondary oil recovery. In the supercritical fluid region, where the fluid temperature and pressure are above those of the critical point, the properties of the fluid are uniquely different from either the gas or liquid states.

[1]Correspondence should be addressed to this author.

0097–6156/88/0373–0091$06.00/0
© 1988 American Chemical Society

Carbon dioxide, ethane and propane are examples of pure components which can exist as supercritical fluids at typical oil well temperatures and pressures. In secondary oil recovery, an inexpensive and abundant fluid, such as CO_2, may be injected into the well to sweep out remaining oil (10). The addition of surfactant to this fluid may improve sweep efficiencies by further reducing capillary effects and by increasing viscosity to reduce sweep instabilities. As the sweep front moves through the porous bed, the surfactant-fluid phase behavior may be altered when intermixing with the oil and changes in pressure and temperature occur. Additionally, the original oil can contain significant amounts of dissolved, low molecular weight alkanes such as ethane, propane or butane. The presence of these gases may affect the phase behavior of the surfactant solution in the sweep fluid because intermixing of these two solutions occurs as the sweep front progresses. For these reasons, an understanding of the effect of pressure, temperature and concentration on the phase behavior of supercritical fluid-surfactant solutions is important in developing more efficient recovery methods.

The combination of the unique properties of supercritical fluids (viscosity, diffusion rates and solvent properties) with those of a dispersed micelle phase creates a whole new class of solvents. The physical properties of a supercritical fluid are variable between the limits of normal gas and near-liquid properties by control of pressure or temperature, and the solvent power of a supercritical fluid is different at each density. Typically, supercritical fluids have densities between 0.1 and 0.8 of their liquid density. Under these conditions, their diffusion coefficients are substantially greater than liquids. For example, the diffusivity of CO_2 above the critical temperature (31 °C) typically varies between 10^{-4} and 10^{-3} cm^2/s, whereas liquids typically have diffusivities of $<10^{-5}$ cm^2/s. Similarly, the viscosity of supercritical fluids mirrors the diffusivity, and is typically 10^1 to 10^2 times less than for liquids (11). The solvating power of a supercritical fluid at a given temperature is directly related to the fluid density (12). An interesting phenomenon associated with supercritical fluids is the apparent occurrence of a "threshold pressure" for solubility of a higher molecular weight solid (11). The solubility of the solute can increase by many orders of magnitude with only a small pressure increase near the critical temperature. These adjustable physical properties define some of the advantages of supercritical fluids in extractions, separations, and related applications. The reverse micelle phase adds to the unique properties of the supercritical fluid continuous phase what amounts to a second solvent environment, which is highly polar and may also be manipulated using pressure.

Figure 1 gives the pressure-density relationship for pure carbon dioxide in terms of reduced parameters (e.g., pressure, temperature, or density, divided by the appropriate critical parameter), which includes the two-phase vapor-liquid region. Similar diagrams with only minor differences occur for other single-component fluid systems. Isotherms for various reduced temperatures in Figure 1 show the variations in density which can be expected with changes in pressure. Thus, the density of the supercritical fluid will typically be 10^2 to 10^3 times greater than that of normal gases at atmospheric pressure. Consequently,

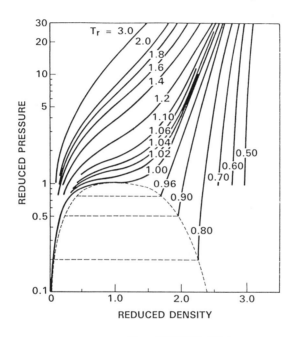

Figure 1. Pressure-density behavior for a pure supercritical fluid (CO_2) in terms of reduced parameters. The area below the dotted line represents the two-phase gas-liquid equilibrium region.

molecular interactions are increased due to the shorter
intermolecular distances. Compared with low density gases the
nearly "liquid-like" density of a supercritical fluid greatly
enhances solubilizing capabilities. Table I gives the critical
parameters for a number of common supercritical fluid solvents of
interest for microemulsion formation. In conjunction with the
general pressure-temperature-density relationships illustrated in
Figure 1, these data provide a vast amount of practical guidance
for fluids and their gas-liquid equilibria.

Table I.   Common SFC Solvents

Compound	Boiling Point (C)	Critical Temperature (C)	Critical Pressure (bar)	Critical Density (g/cm^3)
CO_2	-78.5[a]	31.3	72.9	0.448
NH_3	-33.4	132.4	112.5	0.235
H_2O	100	374.2	218.3	0.315
N_2O	-88.6	36.5	71.7	0.45
Ethane	-88.6	32.3	48.1	0.203
Ethylene	-102.7	9.2	49.7	0.218
Propane	-42.1	96.7	41.9	0.217
n-Butane	-0.5	152.0	37.5	0.228
n-Pentane	36.1	196.6	33.3	0.232

[a] Sublimes

For a pure supercritical fluid, the relationships between
pressure, temperature and density are easily estimated (except very
near the critical point) with reasonable precision from equations
of state and conform quite closely to that given in Figure 1. The
phase behavior of binary fluid systems is highly varied and much
more complex than in single-component systems and has been well-
described for selected binary systems (see, for example, reference
13 and references therein). A detailed discussion of the different
types of binary fluid mixtures and the phase behavior of these
systems can be found elsewhere (13). Cubic equations of state have
been used successfully to describe the properties and phase
behavior of multicomponent systems, particularly for hydrocarbon
mixtures (14). The use of conventional equations of state to
describe properties of surfactant-supercritical fluid mixtures is
not appropriate since they do not account for the formation of
aggregates (the micellar pseudophase) or their solubilization in a
supercritical fluid phase. A complete thermodynamic description of
micelle and microemulsion formation in liquids remains a
challenging problem, and no attempts have been made to extend these
models to supercritical fluid phases.

The surfactant AOT forms reverse micelles in non-polar fluids
without addition of a cosurfactant, and thus it is possible to
study simple, water/AOT/oil, three component systems. To determine
micelle structure and behavior in water/AOT/oil systems,
investigators have studied a wide range of properties including
conductivity (15), light (16), and neutron (17) scattering, as well
as solution phase behavior (16). From information of this type one
can begin to build both microscopic models and thermodynamic

descriptions of these macroscopically homogeneous but microscopically heterogeneous, micellar solutions.

In a previous paper we reported our initial observations of reverse micelles and microemulsions in supercritical fluids (9). We reported that reverse micelles in a supercritical alkane systems can solubilize a highly polar dye (Malachite Green) and that a high molecular weight protein (Cytochrome C, MW = 12,384) can be solubilized in supercritical propane containing a reverse micelle phase.

In the studies described here, we examine in more detail the properties of these surfactant aggregates solubilized in supercritical ethane and propane. We present the results of solubility measurements of AOT in pure ethane and propane and of conductance and density measurements of supercritical fluid reverse micelle solutions. The effect of temperature and pressure on phase behavior of ternary mixtures consisting of AOT/water/supercritical ethane or propane are also examined. We report that the phase behavior of these systems is dependent on fluid pressure in contrast to liquid systems where similar changes in pressure have little or no effect. We have focused our attention on the reverse micelle region where mixtures containing 80 to 100% by weight alkane were examined. The new evidence supports and extends our initial findings related to reverse micelle structures in supercritical fluids. We report properties of these systems which may be important in the field of enhanced oil recovery.

Experimental Section

Sodium bis(2-ethyl-hexyl) sulfosuccinate (AOT) was obtained from Fluka (>98%, "purum") and was further purified according to the method of Kotlarchyk (18). In the final step, the purified AOT was dried in vacuo for eight hours. The molar water-to-AOT ratio, W = [H$_2$O]/[AOT], was taken to be 1 in the purified, dried solid (18). Solutions of 50 mM AOT in iso-octane had an absorbance of less than 0.02 A.U. at 280 nm, which compares favorably with AOT purified by HPLC (6). Potentiometric titration indicated that acid impurities were less than 0.2% mole percent (6). The purified AOT was analyzed by mass spectrometry using 70 eV electron ionization of the sample from a direct probe introduction. Two trace impurities were identified: 2-ethyl-1-hexanol and maleic acid. The ethane and propane were both "CP" grade from Linde. The iso-octane (GC-MS grade) was used as received from Burdick and Jackson. Distilled, deionized water was used throughout.

The phase behavior of the AOT/water/supercritical fluid systems was studied in a high-pressure stainless steel view cell having a 3/4-in. diameter by 3-in. cylindrical volume, capped on both ends with 1-in. diameter by 1/2-in. thick sapphire windows. Silver plated metal "C" ring seals (Helicoflex) formed the sapphire to metal seal. The fluid mixtures were agitated with a 1/2-in. long Teflon-coated stir bar driven by a magnetic stirrer (VWR, Model 200). The insulated cell was heated electrically. Temperature was controlled to ±0.1 °C with a platinum resistance probe and a three-mode controller (Omega, No. N2001) monitored with a platinum resistive thermometer (Fluka, No. 2180A, ±0.3 °C accuracy). The fluid pressure was measured with a bourdon-tube type pressure gauge (Heise, ± 0.3 bar accuracy). While stirring, the fluid was allowed to equilibrate thermally for 10 min. before each new reading. Much longer observation periods (~ one day) were

used to access the phase stability of selected systems, although
equilibria were established rapidly in the systems reported.
The procedure for finding a point on the two-phase boundary of
the n-alkane/AOT/water systems was as follows. A weighed amount of
solid AOT was placed in the view cell, and then after flushing air
from the cell with low pressure alkane, the cell was filled to
within 10 bar of the desired pressure with a high pressure syringe
pump (Varian 8500). This AOT/alkane solution was diluted with pure
water by injecting successive 27 μL increments of water until the
two-phase boundary was reached. A hand operated syringe pump (High
Pressure Equipment, No. 87-6-5) was used to slowly inject the water
through a metering valve into the supercritical-reverse micelle
solution. By keeping the water in the syringe pump at a constant
pressure slightly above the view cell pressure, the amount of
injected water could be determined from the vernier scale on the
screw of the pump. The same procedure was used to study phase
behavior in the liquid iso-octane system. At each temperature,
four different AOT concentrations (0.020, 0.050, 0.075, and 0.150
M) were prepared to study phase behavior in the range of pressures
from 100 to 350 bar.

The accuracy of the location of the phase boundary, determined
by the above method, was verified using a slightly different
technique. Before pressurizing with the alkane fluid, the weighed
AOT sample was placed in the cell, along with a predetermined
amount of water. The view cell was pressurized to within 20 bar of
the expected single phase pressure and then stirred for 10 min.
The fluid pressure was then increased by 10 bar by adding the
alkane followed by stirring for 10 min. until a single phase system
was obtained. The phase boundaries for five systems were found to
agree within ± 5% of the values from the previous measurement
technique.

The solution conductivity was measured using a Yellow Springs
Instrument conductivity meter (YSI Model 34) with a high pressure
conductivity cell (cell constant of 0.0044 cm^{-1}). The high
pressure cell consisted of ten stacked, stainless steel disc
electrodes (10-mm diameter discs), insulated with Teflon washers.
The meter is particularly well suited for use with this type of
cell because capacitance errors are minimized by the active circuit
and electrode over-potential is eliminated by measurement
potentials of less than 1 volt.

The solubility of AOT in supercritical ethane and propane was
determined by sampling an equilibrium cell using chromatographic
techniques. An excess of solid AOT was loaded into a 17 mL high
pressure vessel. The fluid was saturated with AOT by recirculation
through the solid bed of AOT using a magnetically coupled gear pump
(Micropump, No. 182-346). Using a HPLC injection valve, 100-μL
samples of this solution were introduced into a UV absorbance
detector (ISCO V[4]). The mobile phase, maintained at constant flow
rate, was pure liquid ethane or propane at 300 bar and 25 °C. The
amount of AOT in the 100 μL sample was determined by integrating
the absorbance peaks (monitored at 230 nm) following calibration
using solutions of known concentration.

A high pressure vibrating tube densimeter (Mettler-Paar DMA
512) was used to measure the density of the AOT/water/supercritical
ethane solutions. By recirculating water from a thermostated water
bath through the water-jacketed measuring cell, the temperature of
the cell could be controlled to ± 0.01 °C. The micelle solutions
were prepared by loading measured amounts of AOT and water into a

50 mL high pressure vessel placed in the water bath. After the vessel was filled with supercritical ethane, the solution was mixed and recirculated through the vibrating tube sensor by means of the magnetically coupled gear pump. The temperature and pressure were measured using the previously described instruments. The partial molal volume of AOT, $\bar{v}_2$, in supercritical ethane was calculated from the expression,

$$\bar{v}_2 = v - y_1 \ (\partial v/\partial y_1)_{T,P}$$

where v is the specific volume of the solution and y_1 is the ethane mole fraction. The measured AOT concentration was converted to AOT mole fraction using an iterative procedure. Initially the value of $\bar{v}_2$ for pure AOT solid was used to estimate y_2, allowing a new value of $\bar{v}_2$ to be calculated from which a better estimate of y_2 could be determined.

Results and Discussion

A simple visual experiment in which polar dyes or proteins which are insoluble in the pure fluid are solubilized by supercritical fluid-surfactant solutions is convincing evidence for the existence of a reverse micelle phase. A colored azo dye, malachite green [p,p'-(p-phenylmethylidene)bis(N,N-dimethylaniline)] is very soluble in a 0.075 M AOT/supercritical ethane solution at 37 °C and 250 bar when the water-to-AOT ratio, W, is above 3. Supercritical propane reverse micelles at 103 °C and 250 bar can solubilize substantial amounts of high molecular weight proteins such as Cytochrome C. These polar substances were determined to have negligible solubility in the pure fluid and in the water saturated fluid. In the binary solvent of AOT and fluid (where we assume W ≃ 1 due to the difficulty of completely drying the AOT) these polar substances are only sparingly soluble. However, by increasing W to 3 or above, the solubility of polar compounds is greatly increased.

Solubilization of Cytochrome C in propane/AOT/water solutions is particularly convincing evidence for micelle formation in supercritical fluids because it excludes the possibility of a simple ion-pair mechanism of solubilization. It seems likely that this large, water soluble enzyme is solvated by the highly hydrophobic fluids only if the polar functional groups on the surface of the protein are shielded from the fluid by surfactant molecules.

AOT readily forms reverse micelles in nonpolar solvents without the addition of cosurfactants in part because the twin alkane tail groups of the surfactant molecule provide a very favorable geometric packing for inverted structures. The surfactant's relatively high molecular weight (444.5) and anionic head group result in very low solubility of the monomeric form in low molecular weight, nonpolar fluids. The dissolution of the micellar form is much more favorable because the polar head groups are shielded from the nonpolar fluid phase. A measure of solubility in the supercritical fluid phase can be obtained from the solubility of the surfactant in the near critical liquids at 25 °C where, although the temperatures are much lower, the fluid densities are near the upper limit of those normally obtained for the supercritical fluid phase. The solubility of AOT in various fluids at 25 °C is shown in Table II. AOT is insoluble in pure liquid and supercritical CO_2 at temperatures from 25 to 100 °C

and at pressures up to 400 bar. AOT is very soluble in liquid
ethane above 200 bar and is likewise very soluble in supercritical

Table II. AOT Solubility in Fluids at 25 °C

Fluid	AOT Solubility @ 25 °C	MW	T_C (°C)	P_C (Bar)	Polarizability 10^{-24} cm^3
CO_2	Insoluble, surfactant remains solid up to 200 bar	44	31.3	72.9	2.9
Xenon[a]	Melts at 100 bar at a fluid density = 1.76 g/cm^3	131	16.6	57.6	4.2
Ethane	Very soluble above 200 bar. Surfactant melts at 100 bar.	30	32.3	48.1	4.5
Propane	Very soluble at pressures above 10 bar	44	96.7	41.9	6.3
CO_2-10% IPA	Very soluble	45	60.8[b]	90.0[b]	3.2
SF_6	Insoluble, surfactant remains solid up to 200 bar	146	45.5	37.1	6.5

[a] Supercritical at 25 °C.
[b] Estimated by method given in Reference 19.

ethane (at T = 37 °C and at a slightly higher pressure of 250 bar).
In liquid propane at 25 °C, AOT solubility is very high at
pressures above the propane vapor pressure of 9 bar and remains
very soluble in supercritical propane, T = 103 °C, above 100 bar.
At a pressure of 100 bar, AOT begins to melt in supercritical
xenon, a phenomena which is also observed in ethane at about the
same pressure. Although solubility at higher xenon pressures has
not been investigated, the behavior mimics AOT/ethane behavior up
to 100 bar. The order of increasing solubility appears to be CO_2,
xenon, ethane, and propane, which follows the order of increasing
polarizability. The more polarizable fluids might be expected to
better solvate the surfactant monomer (which has a large permanent
dipole moment) as well as the large micellar structures which have
locally high polarizabilities. However, AOT is insoluble in sulfur
hexafluoride which has a relatively large polarizability, and
similar behavior is observed for freon 13, indicating that such an
explanation is overly simplistic. By adding 10% iso-propyl alcohol
(IPA) to the liquid CO_2 (by volume), AOT becomes very soluble and
is also soluble in this binary fluid at 80 °C and 200 bar. We do

not yet know whether the solubilized AOT is in the micellar form in the IPA-CO_2 solution but the IPA may act as a cosurfactant being locally concentrated at the surfactant interface and thus further shielding the ionic head groups from the nonpolar, predominately CO_2, continuous phase.

The pressure dependence of the phase behavior of supercritical fluid solutions containing a reverse micelle phase is striking and can be illustrated by a description of the solvation process from view cell studies. The dissolution of 1 g of AOT solid (W $\approx$ 1) into 25 mL of supercritical ethane or propane proceeds in four distinct stages. At low pressures the AOT solid is in equilibrium with a low density fluid containing a small or negligible amount of dissolved solid. At somewhat higher pressures (80 to 100 bar) the AOT begins to "melt", forming a system with three phases: solid AOT, a viscous AOT liquid with a small amount of dissolved fluid, and a fluid phase containing dissolved surfactant. At moderate pressures a two-phase system exists consisting of a viscous, predominantly AOT liquid in equilibrium with a fluid containing appreciable amounts of surfactant. Finally, at higher pressures (typically >120 bar) a single, reverse micelle-containing phase is created with the AOT completely solvated by the fluid.

The ternary phase diagrams for supercritical ethane, propane and liquid iso-octane surfactant solutions are shown in Figures 2, 3, and 4, respectively. The region of interest in this study is the alkane rich corner of the phase diagram represented from 80 to 100% alkane and less than 10% water by weight. Each phase diagram shows the location of the phase boundaries separating the single- and two-phase regions at several different pressures in the range of 100 to 350 bar. The areas to the right of these boundaries are regions where a single, reverse micelle phase exists; to the left of these lines, a two-phase system exists containing a liquid and a gas phase. This liquid phase is predominantly water containing dissolved surfactant most likely in the form of monomer or normal micelle aggregates. The phase boundary lines also define the maximum water-to-surfactant ratio, W_O. At a given pressure W_O appears to be nearly constant over the range of AOT concentrations studied. The supercritical ethane data shown in Figure 2 (at 37 °C) are 5 °C above the ethane critical temperature; the supercritical propane data are at the same reduced temperature (T/T_c) as the ethane (6 °C above the critical temperature of propane). To compare the phase behavior of an alkane liquid with that of supercritical propane, the phase diagram for liquid iso-octane at 103 °C and various pressures is shown in Figure 4.

The phase boundary lines for supercritical ethane at 250 and 350 bar are shown in Figure 2. The surfactant was found to be only slightly soluble in ethane below 200 bar at 37 °C, so that the ternary phase behavior was studied at higher pressures where the AOT/ethane binary system is a single phase. As pressure is increased, more water is solubilized in the micelle core and larger micelles can exist in the supercritical fluid continuous phase. The maximum amount of water solubilized in the supercritical ethane-reverse micelle phase is relatively low, reaching a W value of 4 at 350 bar.

In contrast to ethane, the maximum amount of solubilized water in the supercritical propane-reverse micelle system is much higher, having a W_O value of 12 at 300 bar and 103 °C. Again, the W_O values increase as pressure increases from a W_O value of 4 at 100 bar to 12 at 300 bar, as shown in Figure 3. The phase behavior in

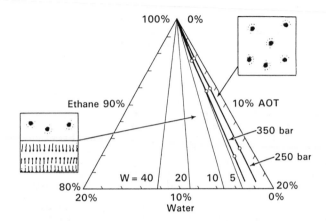

Figure 2. Ethane-rich corner of the ethane/AOT/water ternary phase diagram (weight %) at 37 °C and at two pressures, 250 and 350 bar.

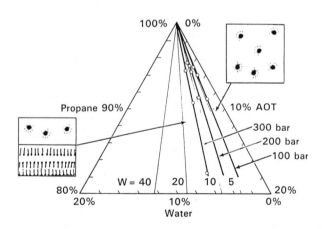

Figure 3. Propane-rich corner of the propane/AOT/water ternary phase diagram (weight %) at 103 °C and at three pressures, 100, 200 and 300 bar.

the supercritical fluid surfactant systems is markedly different than that of the liquid iso-octane reverse micelle system. In the liquid iso-octane system there is no substantial effect of pressure on the phase behavior at the temperature studied.

The critical micelle concentration (CMC) defines the minimum amount of surfactant monomer required to form the reverse micelle phase and may be considered to represent the solubility of the surfactant monomer (although the CMC is much less clearly defined than in normal micelle systems). In a reverse micelle solution this small amount of monomeric surfactant exists in equilibrium with the bulk of the surfactant in the form of micellar aggregates. For example, the CMC of AOT in liquid iso-octane is ~6 x 10^{-4}M. The solubility of surfactant monomer in a particular solvent is dependent on specific solvent-solute forces. The dominant intermolecular interactions between a polar surfactant molecule and alkane solvent molecules are the dipole-induced dipole and the induced dipole-induced dipole forces. In supercritical fluids, the magnitudes of these forces are strongly dependent on the pressure and temperature conditions of the fluid which determine the intermolecular distances (19). At similar molecular densities, hexane and iso-octane are expected to be better solvents for polar surfactant molecules because their polarizabilities (12×10^{-24} and 17×10^{-24} cm^3, respectively) and, hence, their induced dipoles are greater than those for ethane and propane (4.4×10^{-24} and 6.3×10^{-24} cm^3, respectively). Even so, AOT exhibits very high solubility in supercritical ethane and propane at moderate densities, as shown in Figures 5 and 6. For ethane, the solubility is much higher than one would expect for a high molecular weight, polar molecule in a low molecular weight fluid. This high solubility is readily explained in terms of formation of AOT aggregates; i.e., a reverse micelle phase dispersed in the fluid. It seems apparent from the solubility data that at moderate pressures the surfactant monomer is soluble above the CMC in ethane and propane, although these data show no evidence of changes in solubility due to the CMC.

As indicated in Figures 5 and 6, there is a nearly linear relationship between the log[AOT] solubility and the fluid density over several order of magnitude of AOT concentration. This type of behavior would be expected for the solubility of a non-aggregate forming, solid substance in a supercritical fluid (11). The solubility and phase behavior of solid-supercritical fluid systems has been described by Schneider (20) and others, and such behavior can be predicted from a simple Van der Waal's equation of state. Clearly, this approach is not appropriate for predicting surfactant solubilities in fluids, because it does not account for the formation of aggregates or their solubilization in a supercritical fluid phase.

In Figures 5 and 6, one might expect to see two different solubility regions. At low fluid densities where intermolecular forces are reduced and the surfactant concentration is below the CMC, the solubility should increase gradually as the density increases. At higher densities, above the CMC, the solubility should increase rapidly because the total surfactant solubility is dominated by the saturation concentration of micelles in the fluid. This type of behavior is not apparent in Figures 5 and 6, perhaps because the CMC is below 10^{-4}M.

An alternative explanation is that the CMC for AOT in supercritical fluids is density dependent. This might be expected

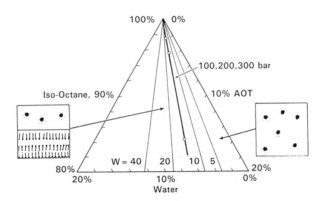

Figure 4. Iso-octane rich corner of the iso-octane/AOT/water ternary phase diagram (weight %) at 103 °C and at three pressures, 100, 200 and 300 bar.

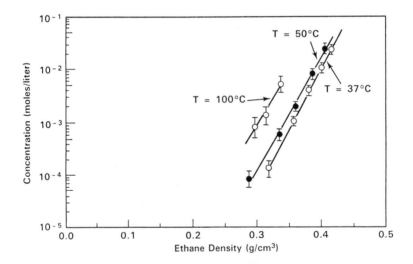

Figure 5. Solubility of AOT in supercritical ethane at 37, 50 and 100 °C, W = 1.

because the CMC can be sensitive to temperature and the nature of the continuous phase. As we have noted, at high dilution there is typically a nearly linear relationship between log [solubility] and fluid density for solid solutes. If the AOT monomer conforms to this behavior, the CMC might be expected to have a similar relationship with fluid density (ρ); i.e., log (CMC) $\propto$ ρ. Clearly, further studies are required to resolve these points.

The effect of temperature on AOT solubility in ethane is also shown in Figure 5. The range of fluid densities studied was limited at higher temperatures by the pressure constraints of our apparatus. In our initial correspondence it was shown that the minimum ethane density necessary to support reverse micelles (at W ~1) had a nearly linear inverse relationship with temperature extending from the near-critical liquid (at 23 °C) to well into the supercritical region (>100 °C). (The previous experiments utilized an AOT concentration of ~ 2 x 10^{-2} moles/liter, and corresponds to a solubility measurement in which the fluid density necessary for solvation was estimated by using the density of the pure fluid. The results are in good agreement with the present more extensive measurements.) The solubility of AOT is greater in propane than in ethane even, at similar temperatures, although the greater slope of the log [AOT] vs. ρ data suggests that the differences are small at higher densities.

In Figure 7 the conductivities of solutions containing reverse micelles dispersed in supercritical propane at 103° C are compared with conductivities of solutions containing reverse micelles in liquid iso-octane at 25 and 103 °C, and pressures ranging from 75 to 350 bar. The AOT concentrations were approximately 37 and 80 mM at W ~ 1. In all cases the conductivities of these solutions are very low, below 10^{-6} mhos/cm. This evidence is entirely consistent with a reverse micelle structure forming in a nonpolar supercritical fluid. Reverse micelle solutions formed in supercritical propane are more conducting than those formed in liquid iso-octane at the same temperature, pressure, and AOT and water concentrations. Part of this difference can be explained by the higher mobility of ions in the lower viscosity propane. The viscosity of propane at 103 °C varies from 0.07 cp to 0.09 cp between 175 to 350 bar, whereas the viscosity of iso-octane is 0.5 cp at these conditions. The difference in measured conductivity between propane and iso-octane solutions at 103 °C is not as large as would be expected based solely on the factor of six difference in viscosity of the two fluids. This indicates that other factors, such as differences in the concentration of surfactant monomer, may be important. For the supercritical propane solutions, conductivity decreases at higher pressures. The pressure dependence of conductivity in propane can be entirely explained in terms of reduced ionic mobility as the viscosity of the fluid increases at higher pressures.

As shown in Figure 7, adding surfactant to propane increases the conductivity by several orders of magnitude over the binary system of propane saturated with pure water. The predominant contribution to conductance in these solutions is anticipated to be from dissociated surfactant monomer in the continuous phase or from micelles containing one or more ionized molecules. The degree of dissociation is quite low, but should be slightly higher in the liquid alkane solutions due to the somewhat larger dielectric constant.

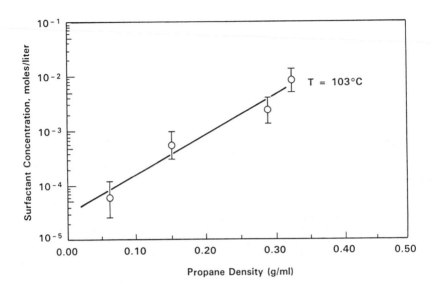

Figure 6. Solubility of AOT in supercritical propane at 103 °C, W = 1.

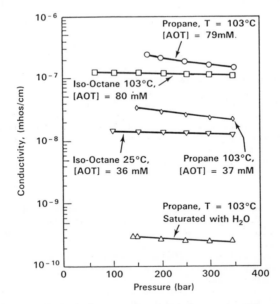

Figure 7. Conductivity of reverse micelle phases in supercritical propane and liquid iso-octane at various pressures, W = 1.

Measurements of supercritical ethane density versus the AOT concentration shown in Figure 4 (T = 37 °C, P = 250 bar) indicate that the properties of the supercritical continuous phase resemble those of the pure fluid. The dispersed micelle phase does not appear to increase the critical temperature or critical pressure of the binary solution to the point of inducing a phase change in the system. There is a small increase in density as surfactant is added to the system which confirms the visual observation that a second liquid phase of much higher density is not formed.

From the data in Figure 8, the partial molal volume of an AOT molecule in a micellar aggregate dispersed in supercritical ethane at 37 °C and 250 bar is estimated to be -43.0 ± 30 cm^3/mole. A negative partial molal volume for a solute in a supercritical fluid is not surprising since lower molecular weight solutes such as naphthalene in ethylene near the critical point can have a partial molal volume of -3000 cm^3/mole (21). This behavior is due to the locally higher solvent density around the higher molecular weight, polarizable solute molecule (22,21). From the estimate of the partial molal volume of AOT in supercritical ethane, the system may be seen to be composed of a micellar structure surrounded by an ethane shell of density greater than the bulk fluid, and this entire structure is dispersed in the continuous, supercritical ethane phase.

Conclusions

The existence of a reverse micelle phase in supercritical fluids has been confirmed from solubility, conductivity and density measurements. The picture of the aggregate structure in fluids is one of a typical reverse micelle structure surrounded by a shell of liquid-like ethane, with this larger aggregate structure dispersed in a supercritical fluid continuous phase.

The reverse micelle phase behavior in supercritical fluids is markedly different than in liquids. By increasing fluid pressure, the maximum amount of solubilized water increases, indicating that these higher molecular weight structures are better solvated by the denser fluid phase. The phase behavior of these systems is in part due to packing constraints of the surfactant molecules and the solubility of large micellar aggregates in the supercritical fluid phase.

An understanding of the phase behavior of surfactant-supercritical fluid solutions may be relevant to developing efficient secondary oil recovery methods because oil displacing fluids, such as a CO_2/surfactant mixture, may be supercritical at typical well conditions. In addition, the original oil in the well may contain dissolved gases such as ethane, propane, or butane, which may effect the phase behavior of the surfactant solution used to sweep out remaining oil.

A number of other important potential applications of a micellar phase in supercritical fluids may utilize the unique properties of the supercritical fluid phase. For instance, polar catalyst or enzymes could be molecularly dispersed in a nonpolar gas phase via micelles, opening a new class of gas phase reactions. Because diffusivities of reactants or products are high in the supercritical fluid continuous phase, high transport rates to and from active sites in the catalyst-containing micelle may increase reaction rates for those reactions which are diffusion limited.

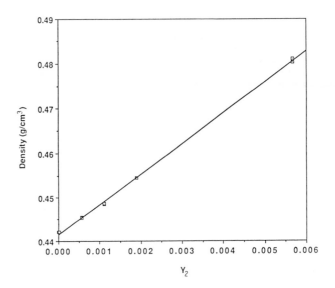

Figure 8. Density of AOT-supercritical ethane solutions at
37 °C and 240 bar.

The recovery of product or catalyst from the micelle core may be
simplified because the micelle size, and even its existence, are
dependent on fluid pressure, in contrast to liquid systems where
pressure has little or no effect.

Acknowledgment

        We thank the Department of Energy, Office of Basic Energy
Sciences, for support of this work through Contract DE-AC06-76RLO
1830.

Literature Cited

1. Neogi, P. In Microemulsions: Structure and Dynamics; Friberg, S. E.; Bothorel, P. Eds.; CRC Press: Boca Raton, 1987, pp. 197-210.
2. Langevin, D. In Reverse Micelles; Luisi, P. L., Straub, B. E., Eds.; Plenum Press: New York, 1984; pp 287-303.
3. Luisi, P. L. Angew. Chem. Ind. Engl. 1985, 24, 439-450.
4. Goklen, K. E.; Hatton, T. A. In Separation Science and Technology; Bell, J. T.; Watso, J. S., Eds.; Marcel Dekker: New York, 1987, pp. 831-841.
5. Leong, Y. S.; Candau, F. J. of Phys. Chem. 1982, 86, 2269-2271.
6. Luisi, P. L., Meier, P., Imre, V. E., Pande, A. In Reverse Micelles; Luis,i, P. L.; Straab, B. E., Eds.; Plenum Press: New York, 1984; pp. 323-337.
7. Hernandez-Torres, M. A.; Landy, J. S.; Dorsey, J. G. Anal. Chem. 1986, 58, 744-747.
8. Gale, R. W.; Smith, R. D.; Fulton, J. L. Anal. Chem. 1987, 59, 1977-1979.
9. Gale, R. W.; Fulton, J. L.; Smith, R. D. J. Am. Chem. Soc. 1987, 109, 920-921.
10. Orr, F. M.; Taber, J. J. Science 1984, 224, 563-569.
11. McHugh, M. A.; Krukonis, V. J. Supercritical Fluid Extraction; Butterworths: Boston, 1986.
12. Smith, R. D.; Frye, S. L.; Yonker, C. R.; Gale, R. W. J. Phys. Chem. 1987, 91, 3059-3062.
13. Streett, W. B. In: Chemical Engineering at Supercritical Fluid Conditions; Paulaitis, M. E.; Penninger, J.M.L.; Gray, R. D.; and Davidson, P., Eds.; Ann Arbor Science: Ann Arbor, 1983.
14. Peng, D.; Robinson, D. B. Ind. Eng. Chem., Fundam. 1976, 15, 59-64.
15. Eicke, H. F.; Kubik, R.; Hasse, R.; Zschokke, I. In Surfactants in Solution; Mittal, K. L.; Lindman, B.; Eds.; Plenum Press: New York, 1984, pp. 1533-1549.
16. Zulauf, M.; Eicke, H. F. J. Phys. Chem. 1979, 83, 480-486.
17. Kotlarchyk, M.; Huang, J. S.; Chen, S. H. J. Phys. Chem. 1985, 89, 4382-4386.
18. Kotlarchyk, M.; Chen, S.; Huang, J. S.; Kim, M. W. Physical Review A. 1984, 29, 2054-2069.
19. Prausnitz, J. M. Molecular Thermodynamics of Fluid-Phase Equilibria; Prentice-Hall: New Jersey, 1969.
20. Schneider, G. M. Angew. Chem. Int. Ed. Engl. 1978, 17, 716-727.
21. Eckert, C. A.; Ziger, D. H.; Johnston, K. P.; Kim, S. J. Phys. Chem. 1986, 90, 2738-2746.
22. Yonker, C. R.; Frye, S. L.; Kalkwarf, D. R.; Smith, R. D. J. Phys. Chem. 1986, 90, 3022-3026.

RECEIVED January 5, 1988

Chapter 6

Temporary Liquid Crystals in Microemulsion Systems

Stig E. Friberg[1], Zhuning Ma[1], and Parthasakha Neogi[2]

[1]Chemistry Department, Clarkson University, Potsdam, NY 13676
[2]Chemical Engineering Department, University of Missouri—Rolla, Rolla, MO 65401

The transport processes between and in W/O microemulsions in contact with water were determined by direct analysis at different times of the concentrations of the components at different distances from the interface between the microemulsion and the water.
The results demonstrated pronounced variation of the transport rate in different parts of the microemulsions when in contact with water. This variation caused an interface to appear within the oil phase and also gave rise to temporary liquid crystals at high cosurfactant/surfactant content.
The presence of high concentrations of salt changed the microemulsion region considerably, but had only limited effect on the transport properties. The main effect was that the duration of the liquid crystal presence was shortened and that an isotropic liquid middle phase was formed.

Mass transfer in surfactant systems is important in many areas. One such field that is particularly important is the surfactant flooding of oil fields (1-4), which is an attractive candidate for tertiary oil recovery. Consequently diffusional processes of aggregates such as micelles and microemulsions (5-9) have been studied extensively. In particular, contacting studies where two phases of varied constituents (surfactant, cosurfactant, oil, water, electrolyte) are in contact and results interpreted on the basis of mass transfer and phase diagrams have become the standard method for studying transport in such systems (10-18).

Previously (11-12), we had contacted water and water in oil (W/O) microemulsion and found that transient lamellar liquid crystals (11) were formed even though these were not located in the pseudoternary phase diagram. The interesting feature was that these phases could be precipitated either in the oil phase or in the water phase, although the controlling factors could not be established. Because of its considerable significance in oil recovery (4) the mechanism of the precipitation of liquid crystals is under detailed investigation. The two possible reasons which govern the formation and location of liquid crystals are those of chemical potentials and diffusion rates. It is evident that the phenomenon is similar to the salting out process, however it is the competative diffusional rates that govern the local concentrations and take the compositions outside the one phase areas. A lamellar liquid crystalline phase may then form. We found a clarification of the relation between diffusion rates and the occurrence of temporary liquid crystals to be of pronounced importance. With this study we present results showing the concentration changes with time, which give rise to temporary liquid crystals.

Experimental

Materials. The surfactant sodium dodecyl sulfate (SDS) from BDH Chemical Ltd., Poole, England, was twice recrystallized from absolute ethanol. The cosurfactant, pentanol, the methanol, the salt, sodium chloride and the Karl Fisher reagent, which was used for water determination, were all from Fisher Scientific Company and certified. The hydrocarbon, t-butylbenzene, with purity of 99% was from Aldrich Chemical Company, Inc. All of those were used without further purification. Twice distilled water was used.

W/O Microemulsion Regions. These were determined in the following systems:
1. Sodium dodecyl sulfate, pentanol and water; (Fig. 1A).
2. Sodium dodecyl sulfate, pentanol and aqueous solution of sodium chloride with concentration of 0.5 M (Fig. 1B) and 1 M (Fig. 1C).
3. Sodium dodecyl sulfate, water and pentanol/hydrocarbon (t-butylbenzene) (1:1 by weight) (Fig. 1D). The W/O microemulsion regions were found by titration with water.

Diffusion of Components. The diffusion process was investigated in the following manner: W/O microemulsions with a minimum water content were carefully layered on top of water in amounts to give a final composition of a W/O microemulsion with a maxiumum water content. The layered samples were thermostated at 30°C and the variations in layer height and interface appearance were observed regularly. Even spaces were marked on the outside of the test samples and the resulting layers were analyzed at different times.
    Analysis was as follows: Water was titrated with Karl Fischer reagent using a Metrohm Herisau titrimeter consisting of potentiograph E536, Dosimat E535 and E549. The amount of hydrocarbon was determined from its UV spectrum with a Perkin-Elmer 522 spectrophotometer. Water, hydrocarbon, and pentanol were vaporised in a vacuum oven, leaving SDS as the remainder. The pentanol content was obtained as the balance, and checked using gas chromatograph-2500.

Results

The microemulsion regions for different conditions are given in Fig-

ures 1A-D. The typical triangular region found with no salt present,
Fig. 1A, is shrunk and the water solubilization maximum moved towards
higher surfactant/cosurfactant ratios with increased salt content
(Figs. 1B-C). Addition of hydrocarbon caused a reduction in the maximum water solubilization (Fig. 1D) all in accordance with earlier
results (18).

The diffusion experiments for the nonsalt compositions (Fig. 1E)
showed a fast equilibration of the surfactant concentration with
equal concentration of surfactant in the entire system after 30 days
at the lowest surfactant/(cosurfactant + surfactant) weight ratio,
0.14, Fig. 2A. Thereafter the concentration in the lower part was
higher than in the upper part, a fact that is to be viewed against
the former low cosurfactant concentration, Fig. 2B, and its high
water content, Fig. 2C. The increase in water content in the upper
layers ceased after 20 days, Fig. 2C, at the time when the liquid
crystal began to form in layer 6, Fig. 3. During the first 7 days
the aqueous solution was turbid and an interface appeared within the
oil phase, Fig. 3. This interface moved upwards in the oil phase and
disappeared after 36 days.

With increased surfactant/(cosurfactant + surfactant) ratio
(0.22), the liquid crystal was formed immediately and the interface
in the oil layer now lasted only 12 days, (Fig. 4). The formation of
a liquid crystal impeded the transport of surfactant to the lower
part, Fig. 5A. In this case, the surfactant concentration remained
lower in the bottom layers during the entire duration of the experiment; more than 2 months. The transport of cosurfactant to lower
parts, Fig. 5B, and water from the layers below the liquid crystal,
Fig. 5C, were not influenced to a great degree by the enhanced amount
of liquid crystal.

For the highest surfactant/(cosurfactant + surfactant) ratio
(0.35), the liquid crystal also formed early, Fig. 6, in layers 5 and
6. This resulted in the concentration changes being focused towards
the bottom layers; the higher reservoir of surfactant and cosurfactant in the top layers resulted in small changes of their concentrations, Figs. 7A-C. The slowest process was the disappearance of the
birefringence; a time of 9 months was needed for that to happen, Fig.
6.

In the system with 0.5 M NaCl, Fig. 8, the series with the lowest surfactant/cosurfactant ratio showed an initial uptake of between
50 and 60% of the water during the first 15 days followed by an
extremely slow period (100 days) during which the remaining water was
absorbed into the microemulsions. No liquid crystals were observed.
The samples with the surfactant/(cosurfactant + surfactant) weight
ratio equal to 0.35 showed an initial formation of liquid crystal at
the interface lasting approximately 10 days, Figs. 9A,B. The liquid
crystal was present irrespective of whether all the sodium chloride
was in the water or distributed in equal concentration in the two
original layers. The water volume was reduced faster for the microemulsion containing salt than that without salt when contacted with
water containing salt. (Cfr Fig. 9A and Fig. 9B; Fig. 4A and Fig.
4B). The absence of salt gave a huge region of liquid crystals lasting more than 80 days, Fig. 10. It is interesting to notice that
the presence of the huge liquid crystalline layer had but little
influence on the concentration changes with time, (Cfr Figs. 11A-C

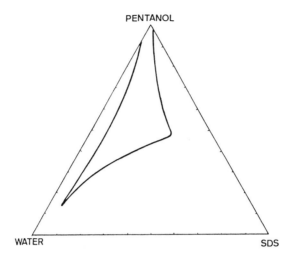

Figure 1A. The solubility region for the pentanol solution in the system: Water, sodium dodecyl sulfate (SDS) and pentanol.

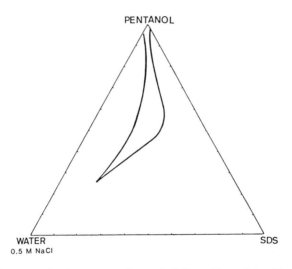

Figure 1B. Replacing water by a 0.5 M sodium chloride solution gave a smaller solubility region.

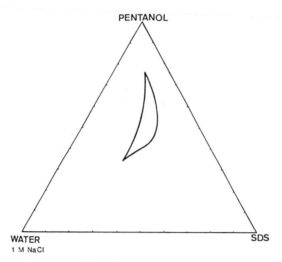

Figure 1C. 1 M sodium chloride solution gave a strong reduction
of the solubility region.

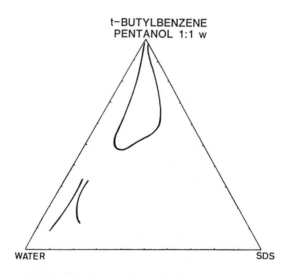

Figure 1D. Part of the isotropic liquid solutions in the system:
Water, sodium dodecyl sulfate and pentanol/t-butylebenzene
solution at a 1/1 weight ratio.

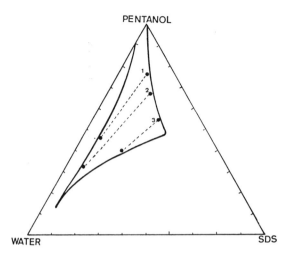

Figure 1E. Compositions at points 1 to 3 were layered on water in amounts to give final combined compositions at the left end-points of the dashed lines.

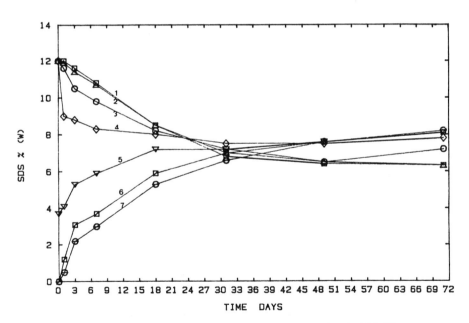

SDS% OF EVERY LAYER AT POINT 1 in FIG.1E

Figure 2A. The concentration of sodium dodecyl sulfate in layers versus time (Composition 1, Fig. 1E). Layer 1 is the top part of the sample, layer 7 the bottom part. The interface was initially in the upper part of layer 5.

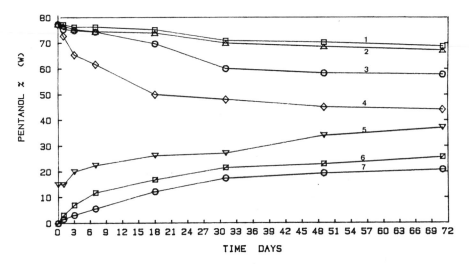

PENTANOL% OF EVERY LAYER AT POINT 1 in FIG.1E

Figure 2B.   The concentration of pentanol in layers versus time
(Composition 1, Fig. 1E).   Layer 1 is the top part of the sample,
layer 7 the bottom part.   The interface was initially in the upper
part of layer 5.

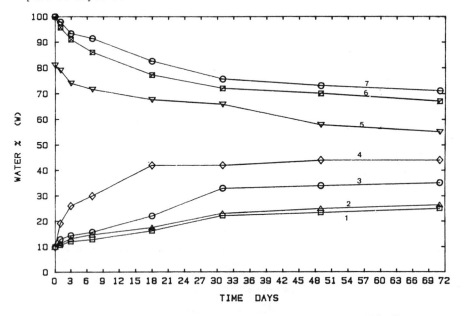

WATER % OF EVERY LAYER AT POINT 1 in FIG.1E

Figure 2C.   The concentration of water in layers versus time
(Composition 1, Fig. 1E).   Layer 1 is the top part of the sample,
layer 7 the bottom part.   The interface was initially in the upper
part of layer 5.

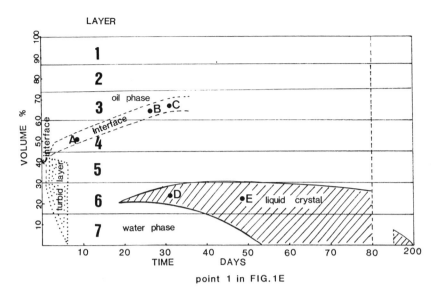

point 1 in FIG.1E

Figure 3. After layering the composition at point 1, Fig. 1E on water the interface was found rising upward in the W/O microemulsion (A,B,C). After 20 days a birefringent layer was formed in the aqueous part slowly disappearing in 200 days.

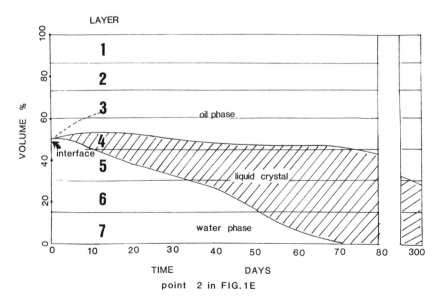

point 2 in FIG.1E

Figure 4. After layering the composition at point 2, Fig. 1E on water the interface was found rising upward in the W/O microemulsion (A,B,C). The birefringent layer in the aqueous phase was formed immediately and lasted more than 300 days.

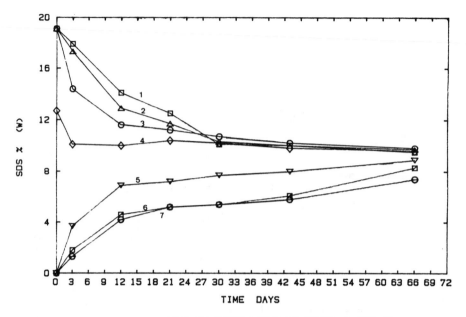

SDS% OF EVERY LAYER AT POINT 2 in FIG.1E

Figure 5A. The concentration of sodium dodecyl sulfate in layers versus time (Composition 2, Fig. 1E). Layer 1 is the top part of the sample, layer 7 the bottom part. The interface was initially in the middle of layer 4.

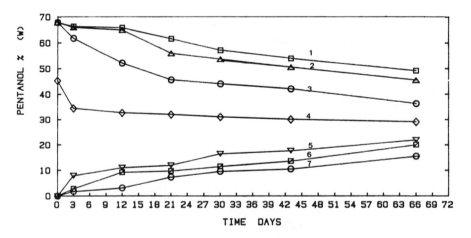

PENTANOL% OF EVERY LAYER AT POINT 2 in FIG.1E

Figure 5B. The concentration of pentanol in layers versus time (Composition 2, Fig. 1E). Layer 1 is the top part of the sample, layer 7 the bottom part. The interface was initially in the middle of layer 4.

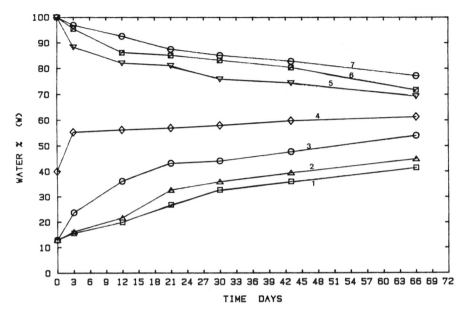

WATER% OF EVERY LAYER AT POINT 2 in FIG.1E

Figure 5C.   The concentration of water in layers versus time (Composition 2, Fig. 1E).   Layer 1 is the top part of the sample, layer 7 the bottom part.   The interface was initially in the middle of layer 4.

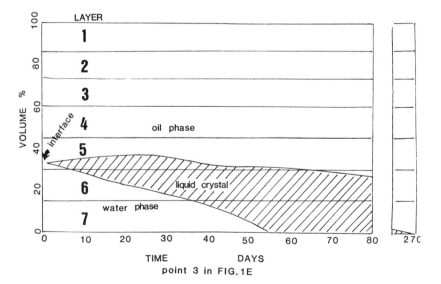

point 3 in FIG.1E

Figure 6.   A birefringent layer was formed in the aqueous part immediately, slowly disappearing in 270 days.   The interface was originally in the lower part of layer 5.

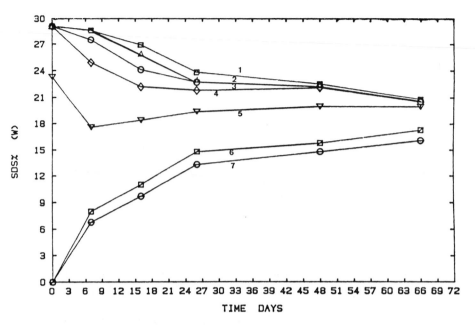

SDS% OF EVERY LAYER AT POINT 3 in FIG.1E

Figure 7A. The concentration of sodium dodecyl sulfate in layers versus time (Composition 3, Fig. 1E). Layer 1 is the top part of the sample, layer 7 the bottom part. The interface was initially in the middle of layer 5.

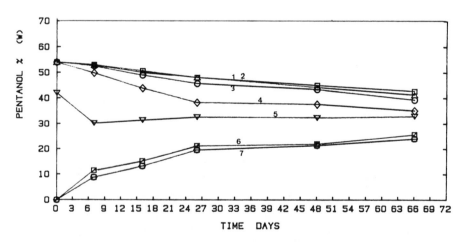

PENTANOL% OF EVERY LAYER AT POINT 3 in FIG.1E

Figure 7B. The concentration of pentanol in layers versus time (Composition 3, Fig. 1E). Layer 1 is the top part of the sample, layer 7 the bottom part. The interface was initially in the middle of layer 5.

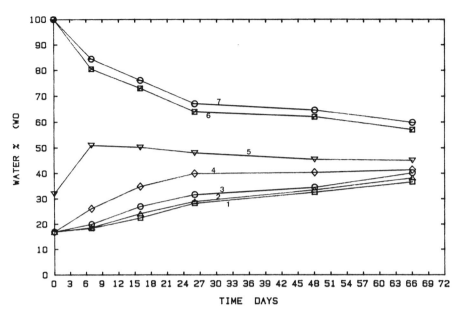

WATER% OF EVERY LAYER AT POINT 3 in FIG.1E

Figure 7C.   The concentration of water in layers versus time
(Composition 1, Fig. 1E).   Layer 1 is the top part of the sample,
layer 7 the bottom part.   The interface was initially in the lower
part of layer 5.

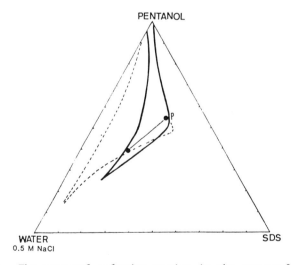

Figure 8.   The pentanol solution region in the system 0.5 M NaCl
aqueous solution, sodium dodecyl sulfate (SDS) and pentanol.   The
dashed line shows the corresponding area in the system with the
water (Fig. 1A).

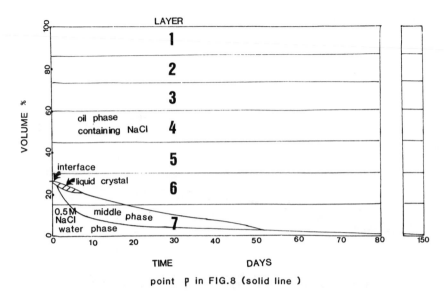

point P in FIG.8 (solid line)

Figure 9A.   After layering the composition at point P, Fig. 8 on
water a birefringent layer was formed in the aqueous part after 20
days, disappearing in 8 days point P. It was replaced by a fairly
extensive isotropic liquid middle phase, which was gradually
reduced to zero in 52 days.   The aqueous phase was slowly depleted
lasting more than 150 days.

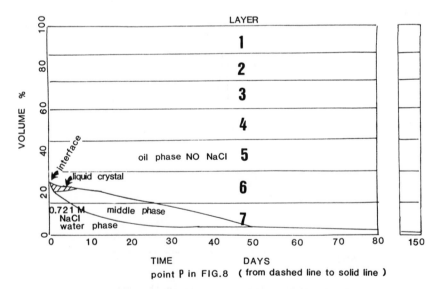

point P in FIG.8 (from dashed line to solid line)

Figure 9B.   With no electrolyte in the pentanol solution the
behavior was similar to the system in Fig. 9A.

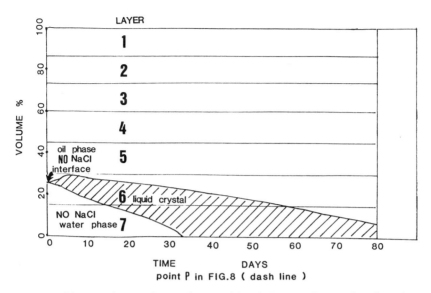

Figure 10. Without electrolyte a birefringent layer developed in the aqueous phase.

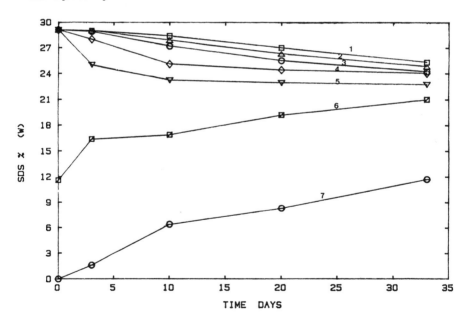

Figure 11A. The concentration of sodium dodecyl sulfate in layers versus time (Composition P, Fig. 8). Layer 1 is the top part of the sample, layer 7 the bottom part. The interface was initially in the upper part of layer 6.

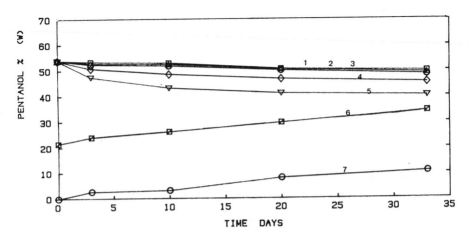

PENTANOL% OF EVERY LAYER AT POINT P in FIG.8 (solid line)

Figure 11B.   The concentration of pentanol in layers versus time
(Composition P, Fig. 8).   Layer 1 is the top part of the sample,
layer 7 the bottom part.   The interface was initially in the upper
part of layer 6.

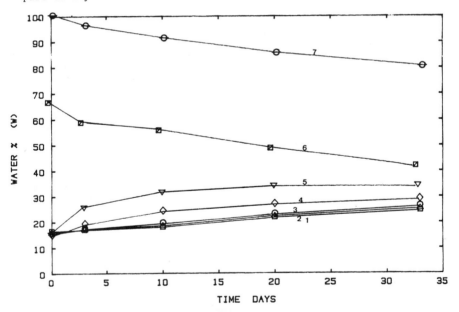

WATER % OF EVERY LAYER AT POINT P   in FIG.8 (solid line)

Figure 11C.   The concentration of water in layers versus time
(Composition P, Fig. 8).   Layer 1 is the top part of the sample,
layer 7 the bottom part.   The interface was initially in the upper
part of layer 6.

and Figs. 12A-C) except for the fact that the pentanol transport into
the NaCl aqueous solution (Fig. 11B) was slower than in to pure water
(Fig. 12B).

For the 1 M NaCl system the solubility region was further
reduced, Fig. 13, and the water solubilization maximum found at even
higher surfactant/cosurfactant ratio. The series with the lower
ratios of surfactant to cosurfactant showed an uptake of the aqueous
solution somewhat similar to the series in the system with 0.5 M
NaCl. The series with the surfactant/(cosurfactant + surfactant)
ratio equal to 0.4 gave an initial liquid crystal formation lasting
for 2-3 days folllowed by a middle phase lasting a longer time. The
liquid crystalline and the middle phase layer were both more pronounced for the sample with initial salt concentration equal in the
water and in the microemulsion, Fig. 14A, than for the sample with
all the salt in the water, Fig. 14B.

The hydrocarbon system was combined with water to form the W/O
microemulsions marked in Fig. 15. The surfactant/(cosurfactant +
surfactant) weight ratio 0.25 gave a liquid crystal with initial fast
extension for three days followed by a new fast growth between 10 and
17 days and a subsequent decline to zero in 40 days, Fig. 16. These
changes were reflected in the concentration changes in the layers
around the liquid crystal (Figs. 17A-D). The water concentration
showed a rapid growth in layer 3, the first three days caused by a
reduction of the surfactant, cosurfactant and the hydrocarbon content.

The conditions in the series with surfactant/(cosurfactant +
surfactant) ratio (0.30), Fig. 18, were similar but the duration of
the liquid crystalline phase was shorter. The concentration changes
were also very similar, Figs. 19A-C.

Discussion

The results gave direct information about the reason for the appearance of the liquid crystal when a water-poor W/O microemulsion is
contacted with water. They also explain the appearance of an interface within the microemulsion layer, a not so normal phenomenon
considering the fact that the phases above and below this interface
is a W/O microemulsion with similar composition.

The formation of a liquid crystal was influnced by dilution of
the W/O microemulsion by hydrocarbon. The results showed that layers
of liquid crystals are formed also in the presence of hydrocarbon and
lasting a considerable time, 40 days, Fig. 16. In comparison, the
non-hydrocarbon systems gave more extended duration, the composition
in Fig. 4 would give a liquid crystal lasting at least two years.

Another important factor is the salinity of the aqueous phase.
The presence of high concentrations of electrolyte usually destabi-
lizes a liquid crystalline phase (18) of a charged surfactant and a
long chain alcohol; the present results show the temporary liquid
crystals to exist only for a few days, when the water was 1 M or 1.7
M NaCl solution, Figs. 14A,B. After that time the liquid crystal was
replaced by an isotropic liquid middle phase (2,3).

The fundamental phenomenon of interest is the explanation for
the appearance of the liquid crystal. It is provided by the diagrams
showing concentration changes in the different layers versus time

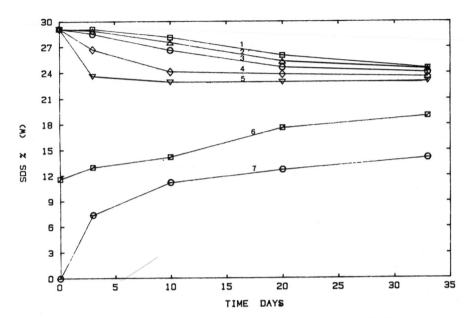

SDS% OF EVERY LAYER AT POINT P in FIG.8 (dashed line)

Figure 12A. The concentration of sodium dodecyl sulfate in layers versus time (Composition P, Fig. 8; No electrolyte). Layer 1 is the top part of the sample, layer 7 the bottom part. The interface was initially in the upper part of layer 6.

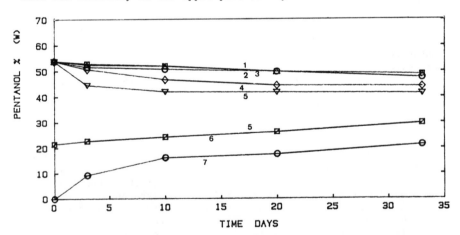

PENTANOL% OF EVERY LAYER AT POINT P in FIG.8 (from dashed line to solid line)

Figure 12B. The concentration of pentanol in layers versus time (Composition P, Fig. 8; No electrolyte). Layer 1 is the top part of the sample, layer 7 the bottom part. The interface was initially in the upper part of layer 6.

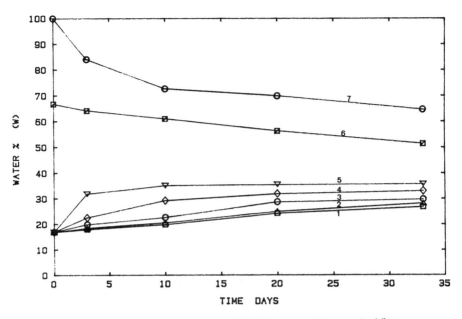

WATER% OF EVERY LAYER AT POINT P in FIG.8(from dashed line to solid line)

Figure 12C.   The concentration of water in layers versus time (Composition P, Fig. 8; No electrolyte).   Layer 1 is the top part of the sample, layer 7 the bottom part.   The interface was initially in the upper part of layer 6.

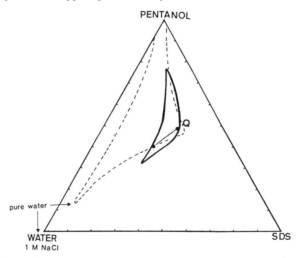

Figure 13.   The pentanol solution region in the system 0.5 M NaCl aqueous solution, sodium dodecyl sulfate (SDS) and pentanol.   The dashed line shows the corresponding area in the system with water (Fig. 1A).

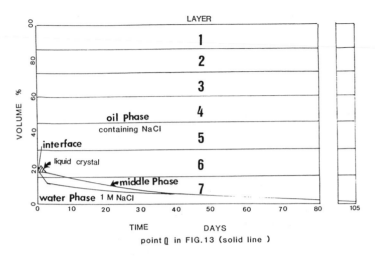

Figure 14A.   After layering the composition at point Q, Fig. 13,
on 1 M NaCl a birefringent layer lasting 3 days was found at the
top of the aqueous phase.   It was replaced by a fairly extensive
isotropic liquid middle phase, which was gradually reduced to zero
in 40 days.   The aqueous phase was slowly depleted lasting more
than 105 days.

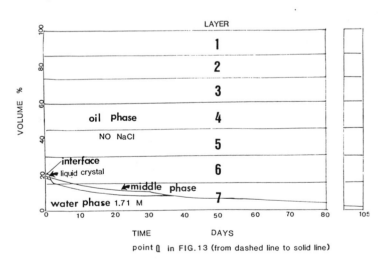

Figure 14B.   After layering the composition at point Q, Fig. 13,
on 1.71 M NaCl a birefringent layer lasting 8 days was found at
the top of the aqueous phase.   It was replaced by a fairly extensive isotropic liquid middle phase, which was gradually reduced to
zero in 37 days.   The aqueous phase was slowly depleted lasting
more than 105 days.

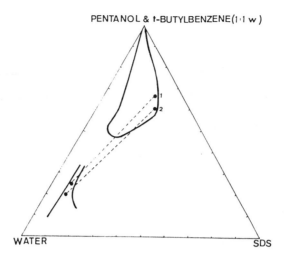

Figure 15. The solubility region for the W/O microemulsion (top) and part of the O/W microemulsion region in the system water, sodium dodecyl sulfate, pentanol and t-butylbenzene.

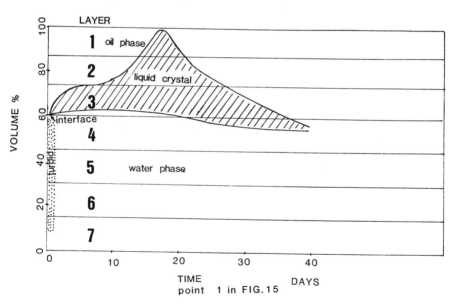

Figure 16. After layering the W/O microemulsion composition in point 1, Fig. 15, on water to give a total composition at the end-point of the dashed line from point 1 a birefringent layer developed in the oil phase reaching a maximum in 18 days and depleted in 40 days.

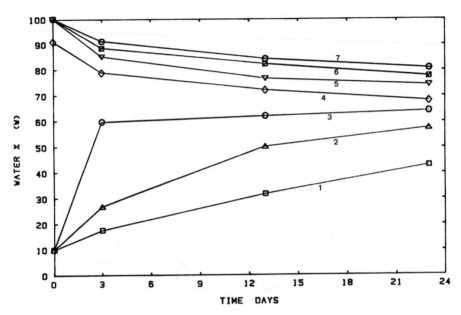

WATER% OF EVERY LAYER AT POINT 1 in FIG.15

Figure 17A. The water concentration at different heights for the
conditions in Fig. 16. The interface was initially between layers
3 and 4.

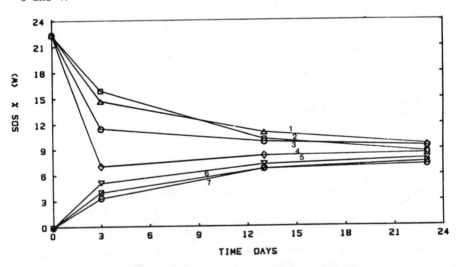

SDS% OF EVERY LAYER AT POINT 1 in FIG.15

Figure 17B. The sodium dodecyl sulfate concentration at different
heights for the conditions in Fig. 16. The interface was
initially between layers 3 and 4.

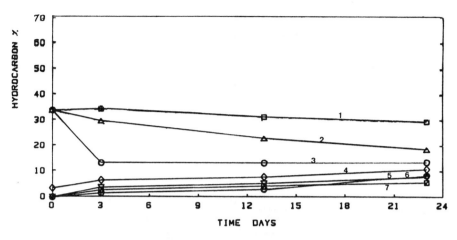

HYDROCARBON% OF EVERY LAYER point 1 in FIG.15

Figure 17C.   The hydrocarbon concentration at different heights
for the conditions in Fig. 16.   The interface was initially
between layers 3 and 4.

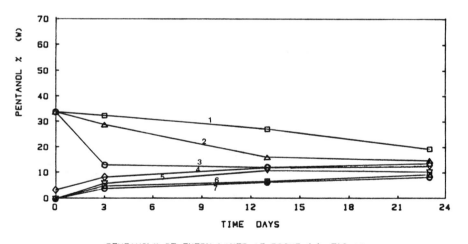

PENTANOL% OF EVERY LAYER AT POINT 1 in FIG.15

Figure 17D.   The pentanol concentration at different heights for
the conditions in Fig. 16.   The interface was initially between
layers 3 and 4.

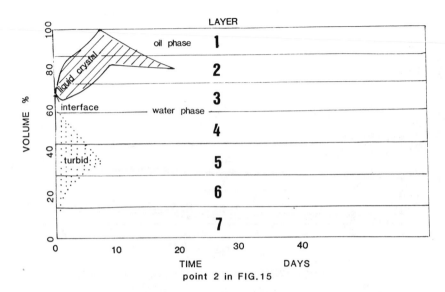

Figure 18.   After layering the W/O microemulsion composition in point 1, Fig. 15, on water to give a total composition at the mid-point of the dashed line from point 1 a birefringent layer developed in the oil phase reaching a maximum in 8 days and depleted in 40 days.

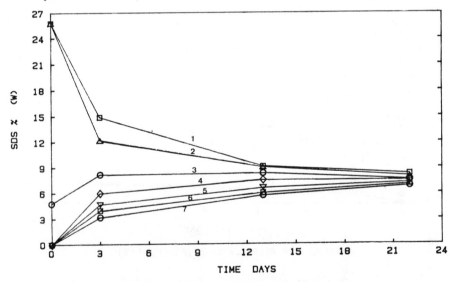

Figure 19A.   The sodium dodecyl sulfate concentration at different heights for the conditions in Fig. 16.   The interface was initially in layer 3.

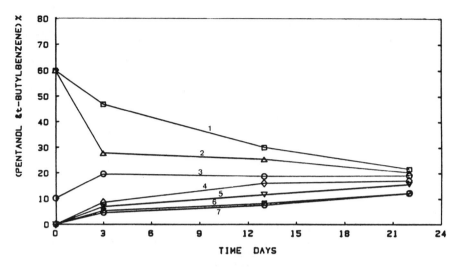

COS & HYDROCARBON OF EVERY LAYER at point 2 in FIG.15

Figure 19B. The pentanol and hydrocarbon concentration at different heights for the conditions in Fig 16. The interface was initially in layer 3.

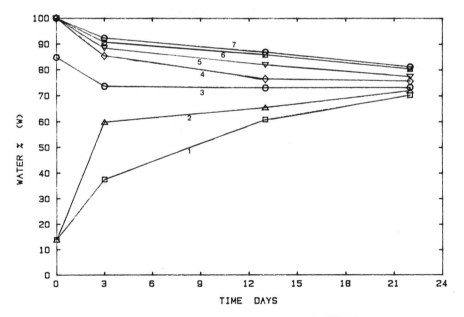

WATER% OF EVERY LAYER AT POINT 2 in FIG.15

Figure 19C. The water concentration at different heights for the conditions in Fig. 16. The interface was initially between layers 3 and 4.

exemplified by Figs. 2A,B. This diagram shows an initial rapid
increase of surfactant concentration in the aqueous layers 5-7, Fig.
2A, while the changes in cosurfactant concentration were more modest,
Fig. 2B.

A more rapid transport of the surfactant than the cosurfactant
to the lower layers gives a higher surfactant/cosurfactant ratio in
the aqueous part and vice versa in the upper layers. The result is
compositions to the right of the straight line between water and the
original microemulsion composition for the lower layers and a corresponding deviation to the left of the line for the upper layers.
Fig. 20A illustrates this behavior. In the beginning of the
experiment the upper four layers and part of the fifth layer
consisted of the W/O microemulsion while the layers 6 and 7, as well
as 1/3 of the layer 5 were water, Figs. 20A,3. After 71 days the
composition in layers 6 and 7 was moved from the original pure water
to positions to the right of microemulsion solubility region, Fig.
20B.

A position to the right of the microemulsion area means the
presence of a lamellar liquid crystal as has been repeatedly demonstrated by Ekwall (19). The temporary appearance of liquid crystals
when W/O microemulsions are brought into contact with water have,
with this result, been given a satisfactory explanation. The faster
transport of the surfactant into the aqueous layers gives rise to
temporarily higher surfactant concentrations and the stability limits
for the water rich W/O microemulsions phase are exceeded towards the
liquid crystalline phase in water.

The upper layers, 1-4 on the other hand, were moved to positions
to the left of the line between the original W/O microemulsion composition and water (Fig. 1E) and it is interesting to notice the fact
that after 71 days, they remained exactly inside the phase limit of
the microemulsion region.

This was not the case earlier, when the interface appeared in
the W/O microemulsion, Fig. 3. Compositions within this interfacial
layer were analyzed after 9, 26 and 30 days, (A, B, and C, Fig. 3) as
well as within the birefringent layer (D, E, Fig. 3). The compositions A, B, and C are all outside the microemulsion region to the
left; an obvious consequence of rapid depletion of surfactant due to
its faster transport out of the microemulsion. The composition outside the solubility limit explains the interface found in the oil
phase, Fig. 3.

A third factor of interest is the influence of the lamellar
phase on the diffusion process per se. Our present results gave no
consistent indication that the birefringent layer influenced the
transport rates. When stating that fact it is essential to realize
that the birefringent layer by no means should be interpreted as a
sign of liquid crystalline phase of the thickness shown in figures
such as 3, 4, 6 and others. The birefringent layer is in all probability a dispersion of the lamellar liquid crystal in an aqueous
solution. Such a dispersion should show a lower diffusion coefficient than an aqueous solution, but the transport rate into the W/O
microemulsion phase may be determined by the reorganization of the
association structures at the interface.

The analysis of concentrates of the compounds could in principle
be used to determine diffusion coefficients for the individual compo-

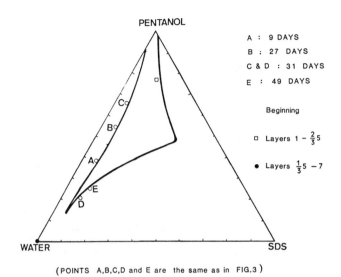

(POINTS A,B,C,D and E are the same as in FIG.3)

Figure 20A. The compositions at the internal interface (A,B,C, Fig. 3) were located outside the W/O microemulsion solubility region (A, 9 days; B, 27 days and C, 31 Days). The composition of the birefringent layer (D,E,F, Fig. 3) was also outside the solubility region but towards high surfactant concentration (D, 31 days and E, 49 days).

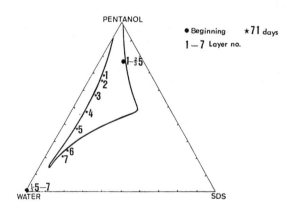

Figure 20B. At 71 days the internal interface in Fig. 3 had disappeared and all the W/O microemulsion compositions were now within the solubility region (1-5, 71 days), but the birefringent layer still persisted (Fig. 3) and layers 6 and 7 were still outside the solubility limit.

nents taking the interdependence of diffusion coefficients into consideration.
The fluxes of component α = 1-3 are described by three equations

$$S_\alpha = -C\sum_{\beta=1}^{3} D_{\alpha\beta}\frac{\partial\psi_\beta}{\partial \ell}$$

with α values equal to 1, 2 and 3, respectively.

S = flux mol/area sec
C = moles of component 1-3
$D_{\alpha\beta}$ = diffusion coefficient of α with dependence
    on β
ψ_β = mol fraction of component β
ℓ = distance

Such an analysis is appropriate and useful for non-associated solutions in which the diffusion coefficients are constant and hence, meaningful. In the present system the association structures vary in a pronounced manner in the W/O microemulsion region (21-23) and the transfer from one phase to another of a component may strongly influence the flue. With this in mind no attempts were made to calculate the $D_{\alpha\beta}$.

Acknowledgments

This research was supported by Department of Energy, Office of Basic Sciences, Grant #DOE DE AC02 83ER13081.

Literature Cited

1. Healy, R.L.; Reed, R.N. In Improving Oil Recovery by Surfactant and Polymer Flooding; Shah, D.O.; Schechter, R.S., Eds.; Academic Press, Inc.: New York, 1977; p 383.
2. Sharma, M.K.; Shah, D.O. In Macro- and Microemulsions; Shah, D.O., Ed.; ACS: Washington, DC, 1985; p 149.
3. Miller, C.A.; Qutubuddin, S. In Interfacial Phenomenon in Apolar Media; Parfitt, G.D.; Eicke, H.F., Eds.; Marcel Dekker, Inc.: New York, 1986.
4. Neogi, P. In Microemulsions; Friberg, S.E.; Bothorel, P.; Eds.; CRC Press: Baton Rouge, Louisiana, 1987; p 197.
5. Evans, D.F.; Mukherjee, S.; Mitchell, D.J.; Ninham, B.W. J. Colloid Interface Sci. 1985, 93, 184.
6. Lam, A.C. Diffusion in Micellar and Microemulsion Systems; Ph.D. Thesis, Department of Chemical Engineering, University of Texas at Austin, Texas, 1986.
7. Lindman, B.; Kamenka, N.; Kathopoulis, T.M.; Brun, B.; Nilson, P.G. J. Phys. Chem. 1980, 84, 2485.
8. Lindman, B.; Stilbs, P.; Moseley, M.E. J. Magnetic Resonance 1980, 40, 401.
9. Mackay, R.A.; Dixit, N.S.; Agarawal, R.; Seiders, R.P. J. Disp. Sci. & Techn. 1983, 4, 397.

10. Friberg, S.E.; Mortensen, M.; Neogi, P. Sep. Sci. & Techn. 1985, 20, 613.
11. Friberg, S.E. In Proc. Eng. Foundation Conf.; Brown, F., Ed.; Santa Barbara, California, 1978; p 99.
12. Friberg, S.E.; Podzimek, M.; Neogi, P. J. Disp. Sci. & Techn. 1986, 7, 57.
13. Cash, R.L., Jr.; Cayias, J.S.; Hayes, M.; MaCallister, D.J.; Schares, T.; Schechter, R.S.; Wade, W.H. Spontaneous Emulsification - A Possible Mechanism for Enhanced Oil Recovery; SPE Paper, 1975; 5562.
14. Lam, A.C.; Schechter, R.S.; Wade, W.H. Soc. Pet. Eng. J. 1983, 23, 781.
15. Benton, W.J.; Miller, C.A.; Fort, T., Jr. J. Disp. Sci. & Techn. 1982, 3, 1.
16. Raney, K.A.; Benton, W.J.; Miller, C.A. In Macro- and Microemulsions; Shah, D.O., Ed.; ACS: Washington, DC, 1985; p 193.
17. Miller, C.A.; Mukherjee, S.; Benton, W.J.; Natoli, J.; Qutuddin, S.; Fort, T., Jr. AIChE Symp. Ser. 212; 1982; Vol. 78, p 29.
18. Friberg, S.E.; Buraczewska, I. Progr. Colloid Polym. Sci. 1978, 63, 1.
19. Ekwal, P. In Advances in Liquid Crystals; Brown, G.W., Ed.; Academic Press: New York, 1975; Vol. 1, p 1.
20. Crank, J. The Mathematics of Diffusion; Oxford Univ. Press: 1985.
21. Sjoblom, E.; Friberg, S.E. Light Scattering and Electron Microscopy Determinations of Association Structures in W/O Microemulsions; Journal of Colloid and Interface Sci: 1978, Vol. 67, No. 1.
22. Clausse, M.; Rayer, R. Colloid and Interface Science; Kerker, M., Ed.; Academic Press: New York, 1976; Vol. 1, p 217.
23. Lapczinska, I.; Friberg, S.E. Microemulsions Containing Nonionic Surfactants; 48th Natl. Colloid Symp.: Houston, Texas, 1974.

RECEIVED January 29, 1988

Chapter 7

Foam Stability:
Effects of Oil and Film Stratification

D. T. Wasan, A. D. Nikolov[1], D. D. Huang[2], and D. A. Edwards

Department of Chemical Engineering, Illinois Institute of Technology,
Chicago, IL 60616

Foam stability in the presence of Salem crude oil and
pure hydrocarbons is investigated as a function of chain
length of α-olefin sulfonates and electrolyte concenra-
tion. Interactions between aqueous foam films and
emulsified oil droplets are observed using transmitted
light, incident light interferometric and differential
interferometric microscopic techniques. Foam destabilization factors are identified including the pseudoemulsion film tension and the surface and interfacial
tension gradients. Results from foam-enhanced oil recovery experiments in Berea Sandstone cores are presented using the combined gamma ray/microwave absorption technique to measure dynamic fluid satruation
profiles.

Three phase foam stability, as has been discussed by numerous authors
is of great practical significanct (1-14). However, despite the
recognized significance, the mechanisms by which oil affects foam
stability are still under investigation.

The effect of oil upon foam stability has been explained in
rather general terms through the mechanism of oil spreading
phenomena, but the reason why the oil droplets spread and the film
between the oil droplets and air bubbles breaks has not been discussed.

Our objective in this study is to elucidate the complex phenomena occurring during the process of three phase foam thinning, to
identify the interaction mechanisms between the oil droplets, the
thinning foam film and the Plateau-Gibbs borders and the role of
surface and interfacial tension gradients in foam stability, and to
examine the implications upon crude oil displacement by foam in
pourous media.

[1]Current address: University of Sofia, 1126 Sofia, Bulgaria
[2]Current address: Polaroid Corporation, Bedford, MA 01730

During the process of three phase foam thinning, three distinct films may occur: foam films (water film between air bubbles), emulsion films (water between oil droplets) and pseudoemulsion films (water film between air and oil droplets) (Figure 1). To study the behavior of these films and particularly the oil droplet-droplet, oil droplet-air bubble and oil droplet-foam frame interactions it is necessary to utilize numerous microscopic techniques, including transmitted light, microinterferometric, differential interferometric and cinemicrographic microscopy.

EXPERIMENTAL

Surfactant-Oil-Electrolyte Systems. In this study we used as surfactants alpha-olefin sulfonates C_{12}, C_{14}, and C_{16} (anionic surfactants, product of Ethyl Corp.) and Enordet AE 1215-30 (nonionic surfactant, product of Shell Development Co.). For all measurements, the surfactant concentration was chosen at 3.16×10^{-2} mol/1, several times above the critical micelle concentration (cmc). These particular surfactants (and concentrations) were chosen on the basis of industrial applications (6,7,15).

As oil phases we used n-octane (Fisher Sci. Co. reagent grade Lot 746833 class IB) and n-dodecane (Fisher Sci. Co. purified grade Lot 852154), both chosen for their well-defined structure and Salem crude oil, chosen for its practical value.

Electrolyte was chosen as NaCl, at two concentrations 0.17 mol/1 (1 wt%) and 0.51 mol/1(3 wt%).

In each study the oil-water system was preequilibrated for one week.

Macroscopic Observations of Three Phase Foam Structure. In Figures 2 and 3 are shown foam drainage results in the presence of Salem crude oil with the surfactants C_{12}AOS and C_{16}AOS at 1 wt% NaCl. obtained from the transmitted light microscope. Surfactant chain length is clearly seen to be a significant factor (compare Figure 2 with Figure 3).

In Figure 2 the process of coalescence between a samll air bubble (white circular object entering the movie frame from the upper middle portion of the frame) and an oil lens on large bubble surface (the thick dark edge of the large white object in the upper portion of the movie frame is the oil lens on the large bubble surface) is illustrated for the C_{12} AOS system. After a certain thickness the "pseudoemulsion" film formed between the oil lens and the air bubble surface ruptures and the oil phase spreads on the surface, forming an oil "bridge" between small and large bubble surfaces.

The "pseudoemulsion" film is therefore unstable for the C_{12} AOS system.

In Figure 3, movie frames capturing the foam structure and dynamics of the C_{16} AOS system reveal that the pseudoemulsion films formed between the oil droplets and air bubble surface are stable enough to allow the eventual migration of the droplets to the Gibbs-Plateau border where at a certain capillary radius they coalesce (Figure 3). Further increasing the curvature radii of the Plateau borders results in a rupturing of the pseudoemulsion film in the borders. Rupture of the pseudoemulsion film allows the oil to

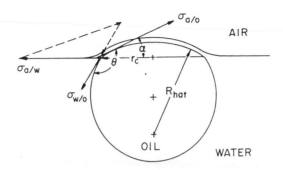

Figure 1. Pseudoemulsion film between oil droplet and air/water surface.

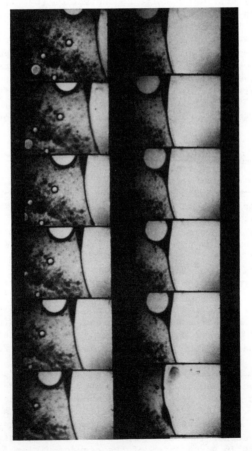

Figure 2. Spreading of Salem crude oil between two bubble surfaces for C_{12}AOS with 1 wt% NaCl.

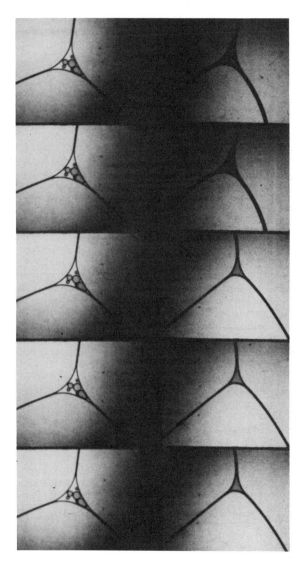

Figure 3. Two dimensional foam drainage for $C_{16}AOS$ with 1 wt% NaCl.

spread, disturbing the mechanical equilibrium between the foam lamellae and border such that the entire foam frame breaks (Figure 4). Therefore, the interaction between the oil droplets in the Plateau borders and liquid film plays an important role in determining foam stability.

Macroscopic three phase foam observations therefore reveal that C_{16} AOS produces a more stable foam in the presence of oil than C_{12} AOS.

<u>Microscopic Observations of Three Phase Foam Structure</u>. We will now discuss observations of the microscopic oil droplet-foam interactions, and observations regarding the attachment of oil droplets to the air-water surface.

<u>Oil Droplet-Foam Film Interaction</u>. The commonly known mechanism in the literature by which oil droplets affect foam stability is, as previously mentioned, the mechanism of oil droplet spreading (<u>8</u>). It is suggested that during the process of foam lamella thinning, the oil droplets are squeezed between the film surface and spread on one of the film surfaces (in the form of lenses) eventually then spreading also on the second film surface. Finally, it is assumed that an "island" of oil is formed which breaks the lamella (thick film).

To test this hypothesis, we designed two microscopic interferometric experiments (<u>16-17</u>). The first experiment was to form a film with a 0.02 cm radius from aqueous surfactant solutions of C_{12} AOS and C_{16} AOS which were preequilibrated with Salem crude oil.

During the time of preequilibration part of the crude oil formed a stable oil in water emulsion (more pronounced for C_{16} AOS). Prior to extracting the liquid from the double concave meniscus, it was observed that floating lenses of crude oil were spread upon the foam surfaces.

After withdrawing liquid from the film, the oil droplets and the floating lenses on the film surfaces migrated from the area of film thinning to the meniscus. Following the droplet migration (after a film thickness of 100nm) the process of thinning displayed two different phenomena, depending upon the electrolyte concentration. At an electrolyte concentration less than 1 wt %, the film had two thickness transitions. The first from 50nm to 35nm and the second from 35nm to 19nm, which is the final equilibrium thickness. The steps of these two transitions were approximately equal and independent of the film radius (Figure 5). At one-to-one electrolyte concentration higher than 1 wt % the film thickness changed by a single step transition and was dependent upon the film radius (Figure 6).

For C_{12} AOS there was found only one transition of the foam film both for the systems with and without electrolyte.

The second type of microscopic interferometric experiment was devised to study the thinning of an emulsion film. For an emulsion film the dispersed oil phase may be in contact with the film surfaces for a longer time due to the diminished interfacial tension, therefore the thinning phenomena of emulsion films differs from that of foam films.

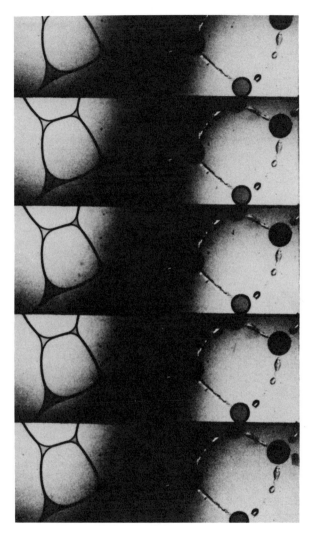

Figure 4. Breakup of foam frames in the presence of Salem crude oil.

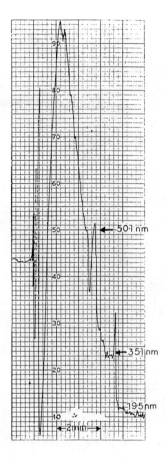

Figure 5. Photocurrent vs. time interferogram of the foam film
formed with $C_{16}AOS$.

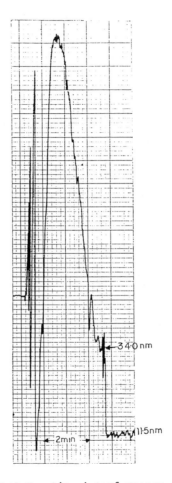

Figure 6. Photocurrent vs. time interferogram of the foam film formed with C_{16}AOS and 1 wt% NaCl.

Using both n-octane and n-dodecane as the dispersed phase in place of the air, we observed that for both C_{12} AOS and C_{16} AOS, with and without electrolyte, the time-scale of thinning increased at least two-fold over the time-scale of the foam thinning.
For C_{16} AOS without electrolyte, again there were two thickness transitions although the stepwise transitions seemed somewhat greater than those for the foam film. The final thickness of the emulsion film was also greater than that of the foam film (about 20nm). After about 10 minutes following the final film thickness, the film ruptured.
At 1 wt % electrolyte for C_{12} AOS there was one step transition which was dependent upon film radius as in the case of the foam film. Unlike the case without electrolyte the emulsion film ruptured during the film transition.
Identical film transition results were obtained for C_{12}AOS.

Study of Oil Attachment to the Surface. The process of oil droplet penetration to the air/water surface and the spreading of the oil depends upon the rupture of the pseudoemulsion film which separates the oil (Figure 1).
The study of the behavior of this pseudoemulsion film with curvature is essential for understanding the role of spreading phenomena in three phase foam stability.
Three different configurations of oil droplets were observed by using differential (DI) interferometry (18,32) (Figure 7). Firstly there are oil droplets separated with a thick film from the air-water surface. This configuration of oil was very common for C_{16}AOS: approximately 60% of the oil for this surfactant was present in this configuration, separated by the pseudoemulsion film. For C_{14}AOS this oil amount was approximately 40% and less than 20% for C_{12}AOS. A second and third configuration of the oil phase at the surface is a droplet separated by a thin pseudoemulsion film (Figure 7b,c). In the case of C_{12}AOS the oil spreading configuration (Figure 7c) was predominant.
The advantage of DI for analyzing not only qualitatively, but also quantitatively the droplet geometry at the air/water surface is that the height of the hat at various local positions may be determined and used to calculate the curvature radius at these local positions. Agreement, within 10% ($\pm$ 2μm) of the curvature radii then indicates spherical shape. The curvature radius may then be used with the radius of the contact to determine the angle α (Figure 7).
Having determined the angle α by DI, we may then determine the tension of the air/oil "film" (see Figure 1) $\sigma_{a/o}$ by a force balance in the vertical direction.

$$\sigma_{a/o} = \frac{\sigma_{w/o} \, Sin(\theta-\alpha)}{Sin\alpha} \quad (1)$$

where θ is the angle between the air/oil and water/oil surfaces, given by

$$\theta = Sin^{-1} (\frac{r_c}{R_d}) + \alpha \quad (2)$$

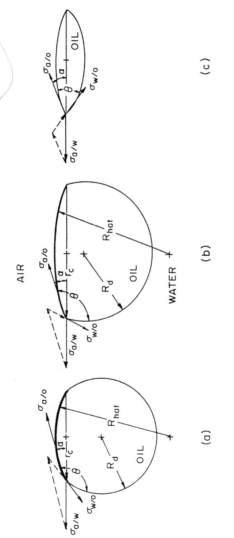

Figure 7. Three oil droplet configurations at an air/water surface; a) thick pseudoemulsion film; b) thin pseudoemulsion film;
c) oil droplet spread on surface.

where R_d is the droplet radius. (The water/oil interfacial tension $\sigma_{w/o}$ was determined by the spinning drop method, and found to be 1.21 dyne/cm at $25°C$ for Salem crude oil/water/C_{16}AOS).

Finally, having determined $\sigma_{a/o}$ by the geometrical parameters obtained by DI we may calculate the air/water surface tension by

$$\sigma_{a/w}^2 - 2\,\sigma_{a/o}\,\sigma_{a/w}\,\cos\alpha + \sigma_{a/o}^2 - \sigma_{w/o}^2 = 0 \qquad (3)$$

Table 1. Film Parameters for
C_{16}AOS/Salem Crude Oil System at
1% NaCl

	Thick Film	Thin Film	Lens
$\theta°$	173.0	155.0	14.8
$\alpha°$	0.3	1.15	0.7
$\sigma_{w/o}$	1.21	1.21	1.21
$\sigma_{a/w}$	25.62	25.62	25.62
$\sigma_{a/o}$	26.96	26.57	24.20
$S^{a/o}$	0	<0	>0

In Table 1, measured and calculated parameters are presented.

In this context it should be noted that the surface tension $\sigma_{a/w}$ will not, in general be the tension existent in the absence of oil, due to the solubilization of oil on the air/water surface. This point is particularly critical for determining the classical spreading coefficient

$$S = \sigma_{a/w} - \sigma_{w/o} - \sigma_{a/o} \qquad (4)$$

Here it is clear that surface tension data taken for $\sigma_{a/w}$ in the absence of oil, or by a technique which disturbs the oil present at the surface, thus providing a value for $\sigma_{a/w}$ larger than the actual value, may result in a spreading coefficient not only of incorrect magnitude but also of incorrent sign.

This new DI technique therefore presents a method for determining $\sigma_{a/o}$ and relating the pseudoemulsion film tension to spreading phenomena and to the rather complex interactions which exists between the oil droplets and the air/water surface.

DISCUSSION

As has been shown in the previous section, the stability of foam, emulsion and pseudoemulsion films manifest stratification phenomena, curvature phenomena and Marangoni phenomena. We will first discuss the microstructure of the thinning films due to micellar interactions, which we have observed through stratification phenomena. We will then discuss the observed behavior of the pseudoemulsion films with curvature and finally the role of Marangoni effects in the stabilization of the foam structure in the presence of oil.

Microstructure of Film Formed from Micellar Solution.   The dependence
of foam stability upon surfactant concentration is well known.
Specifically, above a certain surfactant concentration (after cmc),
the stability of a foam increases sharply with surfactant concentration. This fact is used in industry, where stable foams are created
with surfactant concentrations far above the cmc.

At these high surfactant concentrations spheroidal micelles are
formed in the bulk phase (22), which, above a certain surfactant concentration may achieve a regular periodic structure. This periodic
structuring occurs due to a spatial ordering of the micelles in
periodic potential minima arising from the balance between the
attractive and repulsive long range micellar interactions (23).

We have seen evidence of a similar structuring of micelles in
thin foam and emulsion films containing $C_{16}AOS$, in the form of stepwise transitions or stratification phenomena (see Figure 5).
Numerous authors have observed a similar phenomenon, Bruil and
Lyklema (24), Kruglyako et al (25), Manev et al. (16,26) and Wasan
et al. (23,28). Some of these authors have established that the
thickness of the transitions and the number of transitions are a
function of surfactant concentration. The stratification phenomena
have been observed for foam (16,24-29) and emulsion (23-24) films
with both anionic (10,19,22) and nonionic (28) surfactants.

Wasan et al. (27-28) explained the process of stratification on
the basis of a micelle-latticing structure model. In Figure 8 a
schematic of the latticing model for film thinning is provided. By
fluctuations in the structure of the micellar lamellae (i.e. the
individual "rows" of micelles in Figure 8), the film can change its
thickness by stepwise transitions, each of which are equal to the
micellar-lamellae thickness. According to this model the number of
transitions will depend upon the micelle concentration.

With this model, which is in fact consistent with the structuring of micelles in the bulk phase, an explanation may be offered for
the existence of only one transition for $C_{12}AOS$ and two transitions
for $C_{16}AOS$, **viz.**, the cmc for $C_{12}AOS$ is higher than that of $C_{16}AOS$
such that only one micellar layer may exist in the $C_{12}AOS$ film.

Using this model, two important conclusions may be made con-
cerning the stability of a foam. Firstly, after a certain surfactant concentration (above the cmc) the stability of the foam (or
emulsion) film will increase due to the additional stabilizing force
of the micellar periodic structure in the film. Secondly, increasing the electrolyte concentration will decrease the repulsive forces
between the micelles in the film thus inhibiting the formation of a
micellar periodic structure in the film core. The effect of electrolyte will of course differ for anionic and nonionic surfactants.
For similar micellas concentrations, the effect of electrolyte will
be more severe for the anionic surfactant as it will suppress the
electrostatic double-layer repulsive forces acting between the
micelles. For nonionic surfactant, the repulsive force between
micelles is a steric force rather than an electrostatic force, such
that electrolyte has less of an effect.

It is whorthwhile discussing another relevant property of the
micelles in the aqueous phase; that is the process of solubilization

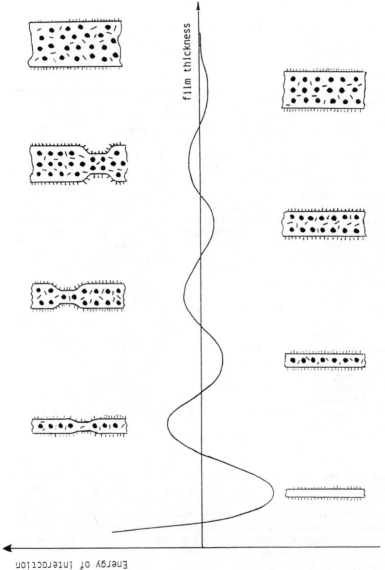

Figure 8.   Film thinning in the presence of micelles.

of oil from the oil phase into the micelles. This solubilization
may play an important role in oil loss (as well as a role in micellar
interaction).

The effect of solubilization upon film thinning is relevant because of the relation between solubilization and micellar structure.
The micelles swell with the solubilization of oil thus increasing
the thickness of the transitions and final equilibrium thickness of
the thinning film (recall that the C_{16}AOS foam film had a final thickness of 19nm whereas the emulsion film had a final thickness of 20nm).
Transitions are also less regular and occur with greater ease (29)
as seen in Figures 9 and 10. These data show that there is an increase both in the number of transitions (from 5 to 6), and in the
magnitude of the step-wise transition in the presence of oil.

"Pseudoemulsion" Film with Curvature. We observed previously that
two types of pseudoemulsion films (thick and thin) may occur when oil
droplets attach to an air/water surface in the presence of 1 wt %
electrolyte (see Figure 7 and Table I). An explanation of the two
film types may be given on the basis of the DLVO theory.

DLVO theory predicts one maximum separating two minima for the
disjoining pressure-distance isotherm (30), which suggests the
possible existence of two types of films: 1) Common thick film
achieved by the balance of electrostatic and dispersion forces and
2) Newton film (thin film) with approximately a bilayer structure.
Increasing the electrolyte concentration depresses the electrostatic
repulsive forces and causes transition from the thick to the thin
film. The electrostatic effect will of course vary with difference
in the structure of the double layer.

To confirm the marked effect of the electrolyte concentration
upon the pseudoemulsion film stability we increased the NaCl concentration to 3 wt % and used DI to investigate the configuration of
Salem crude oil droplets in the presence of C_{16}AOS surfactant (since
this surfactant yields the most stable foam).

In the absence of electrolyte it was observed that the oil droplets were large (radium about 150 μm) with a strong deformation due
to gravitational effects. Very little oil was spread upon the surface in this case. But at the highest electrolyte concentration
there could only be observed spread-oil, observable by the irregular
three phase contact line at the air/water surface.

It is surprising to note that most of the small oil droplets
were separated by thick, rather than thin films. On the basis of
DLVO theory one would expect the large capillary force between the
highly curved small droplets and the air/water surface to induce
thinning, encouraging thinner rather than thicker films. A possible
explanation is that the small droplets deform very little as they
spproach the air/water surface (large capillary pressure) thus hydro-
dynamically inhibiting the thinning process and furthermore surface
tension gradients, magnified on the surface due to the large surface
curvature also resist film thinning. Therefore whereas the large
droplets have a propensity to continue thinning beyond the metastable
thick film, the smaller droplets tend to stabilize, separated from
the air bubble with a thick film.

To support this hypothesis we performed a model investigation
with C_{16}AOS using n-dodecane and 1 wt% NaCl. The oil was

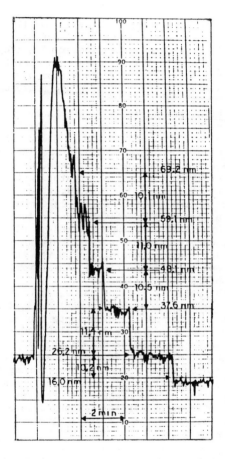

Figure 9.   Photocurrent vs. time interferogram of Enordet AE
1215-30 at 5.2 10^{-2} mole/l.

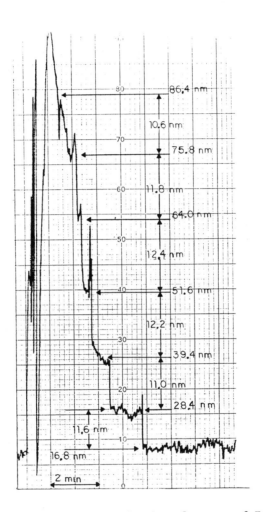

Figure 10. Photocurrent vs. time interferogram of Enordet AE 1215-30 at 5.2 10^{-2} mole/l pre-equilibrated with n-decane.

preequilibrated with C_{16}AOS for one week. The oil droplets were
attached to the air/water surface by injecting the droplets into the
aqueous phase.

Ten minutes following injection of the droplets a photograph
was taken. The small droplets (diameter less than 100μm) were
separated with a thick film while on the hat of certain of the
larger droplets there could be seen a dimple illustrative of a thinning film.

Marangoni Effects in Foam Stability. To estimate the effect of
interfacial tension gradients upon foam stability we used the maximum droplet pressure technique (19). The oil phases chosen were
n-octane and n-dodecane and the surfactants used were C_{12}AOS, C_{16}AOS
and Enordet AE 1215-30. The electrolyte concentration was 1 wt%.
All measurements were performed at 25°C.

In Figure 11 is shown the dynamic interfacial tension as s
function of droplet frequency for two different oil phases (n-octane
and n-dodecane). The frequency range of droplet formation was chosen
based upon a visual observation of the rate of droplet migration
into the Plateau-Gibbs borders. The surfactants, based upon this
Figure may be classified as possessing either a small or a large rate
of change of surface tension with frequency. For both C_{12}AOS and
C_{16}AOS the dynamic interfacial tension change with frequency is
larger in n-dodecane than n-octane. The oil phases have the
opposite effect upon Enordet AE1215-30.

To establish a correlation between the interfacial tension
gradient measurements given above with three phase foam stability
we used the qualitative Barch method for determining foam stability,
due to the physicochemical difficulties in providing an accurate
measure of three phase foam stability.

In a 25 ml cylinder, 10 ml aqueous surfactant solution was pre-
equilibrated with 2 ml oil phase for one week. Strong agitation
produced a three phase foam, which was then observed from the moment
of formation to collapse.

In Figure 12 the process of collapse versus time for the three
surfactant solutions and two oil phases is presented. From these
results we conclude that C_{12}AOS in the presence of n-octane and n-
dodecane and Enordet AE 1215-30 in the presence of n-dodecane are
low stability foams. Higher stability foams are formed by C_{16}AOS
in the presence of n-dodecane and Enordet AE 1215-30 in the presence
of n-octane.

These high stability foams correlate directly with the results
based upon interfacial tension gradient measurements, confirming
the significance of Marangoni phenomena (31) in three phase foam
stability.

Similar three phase foam stability tests have been performed
for C_{12}AOS and C_{16}AOS in the presence of Salem (light oil) and
Wilmington (heavy) crude oils, and in both cases higher stability
has been observed for C_{16}AOS.

Of course the Marangoni effect is not the only stabilizing
factor in the three phase foam. Another critical factor is droplet
size. Smaller droplet size is accomplished by lower interfacial
tension, wherefore it is found that C_{16}AOS yields more stable foam

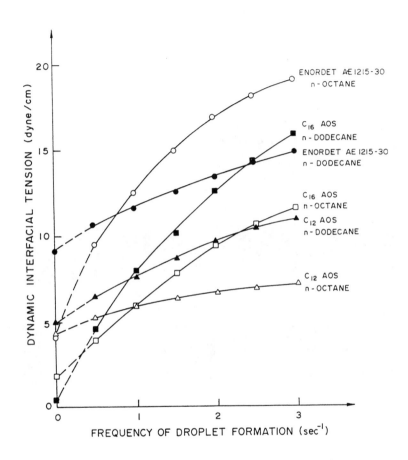

Figure 11. Dynamic interfacial tension vs. droplet frequency.

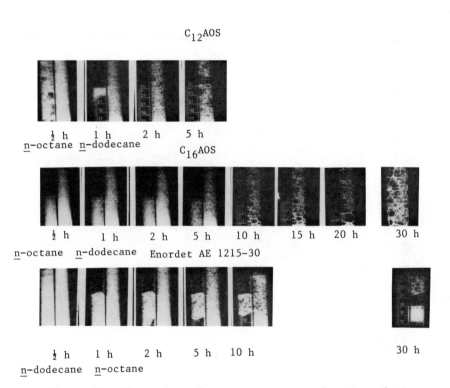

Figure 12.   Three phase foam structure as a function of time.

than Enordet AE 1215-30 as seen in Figure 12 since $C_{16}AOS$ gives a lower interfacial tension than Enordet AE 1215-30 (see Figure 11).

There appear then three primary mechanisms for stabilizing (or destabilizing) a three phase foam. The first derives from the micelle structuring in the film and depends directly upon surfactant concentration and electrolyte concentration. The second is a surface tension gradient (Marangoni) mechanism which relates to the short range intermolecular interactions and the rate of surface expansion. And the third is an oil droplet size effect which depends upon the magnitude of the dynamic interfacial tension.

In summary, the results of our thin film drainage study as well as our investigation of oil spreading mechanisms and frequency dependence of dynamic interfacial tension all suggest that the $C_{12}AOS$ system, which displays the most unstable foam behavior in the presence of oil, should not perform as effectively as the $C_{16}AOS$ system in oil displacement experiments in porous media.

In the following section we present results of our foam-enhanced oil recovery experiments in Berea Sandstone cores to assess the performance of the three α-olefin sulfonates discussed above.

FOAM-ENHANCED OIL RECOVERY

The Berea cores (4 in. x 0.75 in. x 12 in.) used as porous media were vacummed to displace interstitial air and then saturated with 1% NaCl aqueous solution. About six pore volumes of the same solution was injected to stabilize the clay as well as to determine the absolute permeability and porosity (average permeability = 380 millidarcy, average porosity = 0.19). The porous medium was then flooded with Salem crude oil (viscosity at room temperature = 6 cp) to irreducible water content. Waterflood (1% NaCl) was carried out to waterbreak-through. The surfactant solution (3.16×10^{-2} mol/ℓ, 1% NaCl) was then injected into the porous medium until the rock adsorption was completed.

Two types of displacement experiments were conducted to analyze foam behavior in porous media. In one set of experiments nitrogen gas was injected following the surfactant solution to form the foam in situ with an imposed pressure differential of 45 psi. In the second set of experiments the nitrogen gas was injected at a constant flow rate of 10 ft/day. During the constant flow rate experiments the dynamic fluid saturations along the core length were measured by the combined gamma ray/microwave technique. A computer controlled table moves the core to the desired scanning area. The produced fluids and the gas breakthrough times were recorded.

Tables 2 and 3 show the results of the foam-enhanced oil recovery tests for the constant pressure and the constant flow rate experiments respectively. The foam-enhanced oil recovery results from both sets of experiments are similar in that the recovery efficiency and gas breakthrough time increased in the same order: $C_{12}AOS$, $C_{14}AOS$ and $C_{16}AOS$.

The oil saturation profiles and pressure data for the constant flow experiments are shown in Figures 13-16.

In the case of $C_{12}AOS$, Figure 13 reveals that at 1 pore volume of surfactant solution injected, the oil saturation at the first

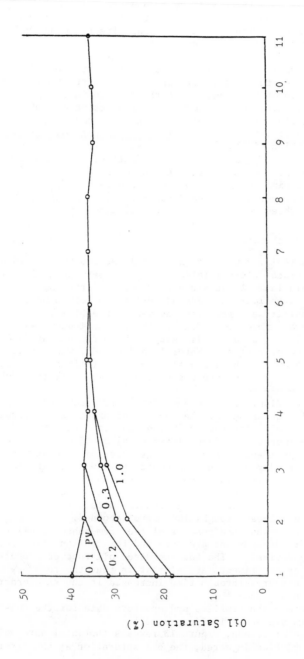

Figure 13. Oil saturation during foam-enhanced oil recovery with $C_{12}AOS$.

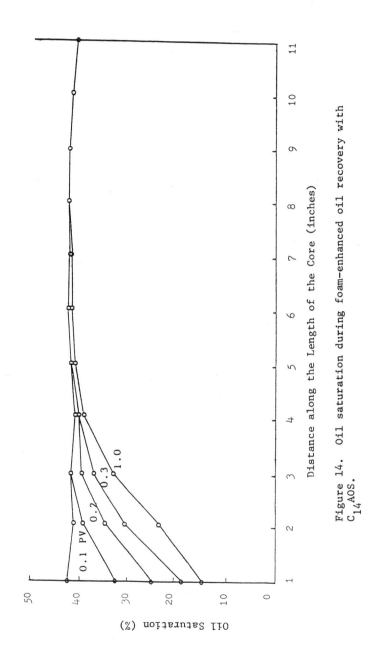

Figure 14. Oil saturation during foam-enhanced oil recovery with C_{14}AOS.

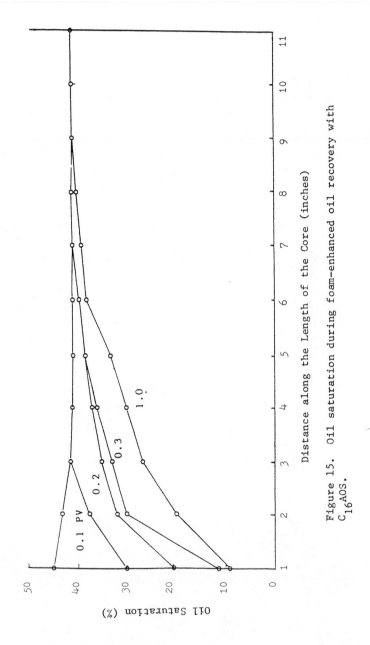

Figure 15. Oil saturation during foam-enhanced oil recovery with C$_{16}$AOS.

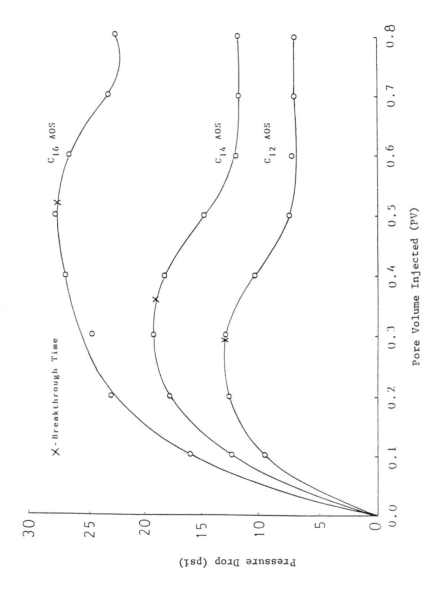

Figure 16. Pressure data for foam-enhanced oil recovery with $C_{12}AOS$, $C_{14}AOS$ and $C_{16}AOS$.

point dropped below 20%, while saturation for positions 2-4 dropped to 0.28 - 0.36. The oil saturation for the rest of the core remained virtually unchanged during the foam flooding process. By comparison, Figure 14 demonstrates that the oil saturation profile for C_{14}AOS is different than that of C_{12}AOS. The curves are steeper and hence the frontal speed of oil in this case is higher. At 1 pore volume of surfactant solution injected, the oil saturation at the first point dropped to about 15% and to about 38% by position 4. In the case of C_{16}AOS the oil saturation at the first point dropped below 10% and to about 30% at position 4. The C_{16}AOS system displays the fastest propagation rate of oil (Figure 15).

Table 2. Oil Recovery Results from Produced Fluid Analyses for C_{12}, C_{14} and C_{16} AOS under Constant Pressure

Experiment	1	2	3
	C_{12}	C_{14}	C_{16}
Initial Oil Saturation	73.0	71.5	74.0
Waterflood Oil Saturation	42.1	39.8	41.5
Surfactant Flood Oil Saturation	39.0	37.0	39.5
% Recovery*	4.3	3.9	2.7
Foam Flood Oil Saturation	28.5	22.5	23.5
% Recovery*	14.4	20.2	21.6
Break Through Time (min)	23.0	35.0	39.0

* Percent oil recovery is based on the initial oil saturation

Table 3. Oil Recovery Results from Produced Fluid Analyses for C_{12}, C_{14} and C_{16} AOS under Constant Flow Rate

Experiment	4	5	6
	C_{12}	C_{14}	C_{16}
Initial Oil Saturation	74.0	73.8	75.2
Waterflood Oil Saturation	42.1	43.2	43.5
Surfactant Flood Oil Saturation	39.8	40.6	41.7
% Recovery*	3.1	3.5	2.4
Foam Flood Oil Saturation	33.9	33.1	30.6
% Recovery*	8.0	10.2	14.8
Break Through Time (min)	36.0	48.0	62.0

* Percent oil recovery is based on the initial oil saturation

Figure 16 shows the pressure drop across the core as a function of pore volume of nitrogen gas injected. The highest pressure drop is always observed before the gas breakthrough (it is worth noting, for the C_{16}AOS system, the faster propagation rate of oil is accompanied by a more rapid increase in the pressure drop).

Our foam-enhanced oil recovery experiments in Berea Sandstones showed for the first time a striking correlation between basic foam properties and oil displacement efficiencies of three different foaming agents: $C_{12}AOS$, $C_{14}AOS$ and $C_{16}AOS$.

CONCLUDING REMARKS

The interactions between an oil phase and foam lamellae are extremely complex. Foam destabilization in the presence of oil may not be a simple matter of oil droplets spreading upon foam film surfaces but may often involve the migration of emulsified oil droplets from the foam film lamellae into the Plateau borders where critical factors, such as the magnitude of the Marangoni effect in the pseudoemulsion film, the pseudoemulsion film tension, the droplet size and number of droplets may all contribute to destabilizing or stabilizing the three phase foam structure.

The stability of emulsion and foam films have also been found dependent upon the micellar microstructure within the film. Electrolyte concentration, and surfactant type and concentration have been shown to directly influence this microstructure stabilizing mechanism. The effect of oil solubilization has also been discussed. The preceding stabilizing/destabilizing mechanisms for three phase foam systems have been shown to predict the effectiveness of aqueous foam systems for displacing oil in enhanced oil recovery experiments in Berea Sandstone cores.

Finally we should comment that it is necessary to employ in the calculation of the spreading coefficient (which is often used as a stability criterion) accurately measured values of the various tensions operative in the "pseudoemulsion" film to determine whether oil is spreading or nonspreading in the three phase foam structure. We have found that the differential interferometric technique to be particularly useful in this regard (32).

ACKNOWLEDGMENTS

This study was supported by the National Science Foundation and partly by the U. S. Department of Energy.

LITERATURE CITED

1. Lau, H.C. and O'Brien, S.M.: New paper submitted SPE 15668, April, 1986.
2. Minssaiux, L.: JPT (Jan. 1974), 100–108.
3. Farouq, Ali, S.M., and Selby, R.L.: Oil & Gas J. (Feb. 3, 1986), 57–63.
4. Bernard, G.G., and Holm, L.W.: Soc. Petr. Eng. J. (Dec. 1915), 295–300.
5. Raza, S.H. Soc. Petr. Eng. J. (Dec. 1970), 328–336.
6. Lau, H.C., and O'Brien, S.M.: paper SPE 14391, presented at the 1985 SPE Annual Technical Conference and Exhibition, Las Vegas, Sept. 22–25.
7. Hu, P.C., Tuvell, M.E., and Bonner, G.A.: Paper SPE/DOE 12660 presented at the Fourth Symposium on Enhanced Oil Recovery, April 15–18, 1984.
8. Prince, A.: Seminar presented at the Illinois Institute of Technology, (June, 1985).

9. Dunning, N., Eakin, J.L. and Walker, C.L.: Monogr. 11, U.S. Bureau of Mines (1961) 38-47.

10. Pletnev, M. Yu., Trapeznikov, A.A.: In "Foams Their Generation and Applications," presented at the second All-union conference [in USSR (in Russian)] Shebekino (ed.) (1979) 33-52.

11. Roberds, K., Axberg, C., Osterlund, R.: J. Coll. Int. Sci., (1977) 62, 264-272.

12. Ross, S.: Chem. Eng. Prog. (1967) 41-47.

13. Perri, J.M.: Foams Theory and Industrial Applications, J.J. Bikerman (ed.), Reinhold, New York (1953) 195-211.

14. Fried, A.N.: U.S. Bur. Mines, Rep. Inv. 5866. (1961).

15. Shell Chemical Company, "Enordet EOR Surfactants," Houston. (1984).

16. Manev, E.D., Sazdanova S.V., Rao, A.A. and Wasan, D.T.: J. Disp. Sci. Tech. (1982) 3, 435-463.

17. Rao, A.A., Wasan, D.T. and Manev, E.D.: Chem. Eng. Commun. (1982 15, 63-81.

18. Nikolov, A., Kralchevsky, P., Ivanov, I.: J. Coll. Inter. Sci. (1986) 112, 122-132.

19. Kao, K., Ph.D. Thesis in progress, Illinois Institute of Technology, Chicago (1987).

20. Huang, D.D., Nikolov, A.D., and Wasan, D.T.: Langmuir (1986) 2, 672-683.

21. Wasan, D.T., Perl, J.P., and Milos, F.S.: paper SPE8327 presented at the 1979 Fall Technical Conference of the Soc. Pet. Eng., Las Vegas, N.M.

22. Reiss-Husson, F. and Luzzati, V.: J. Phys. Chem. (1964) 68, 3504-3511.

23. Efremov, I.F.: "Periodic Colloid Structure" in "Surface and Colloid Science," E. Matijevic (ed.), Wiley Inter. Sci., New York (1976) 85.

24. Bruil, H.G., and Lyklema, J.: J. Nature Phys. Sci.(1971) 223, 19-20.

25. Kruglyakov, P.M. and Rovin, I.G.: "Physical Chemistry of Black Hydrocarbon Fioms - Biomolecular Lipid Membrane," Nauka, Moscow, (1978) (in Russian).

26. Manev, E.D., Sazdanova, S.V. and Wasan, D.T.: J. Disp. Sci. Tech. (1984)5, 111-117.

27. Nikolov, A.D. and Wasan, D.T.: "Layered Structures in Thin Liquid Films: Micellar Interactions and Microstructure Effects on Film Stability," (in preparation).

28. Wasan, D.T. and Nikolov, A.D.: "Micelles Interaction in Foam Film Formed from Nonionic Surfactants," (in preparation).

29. Wasan, D.T. and Nikolov, A.D., paper presented at the Sixth International Symposium on Surfactants in Solution, New Delhi, August, 1986.

30. Verwey, E.J. and Overbeek, J.G.: "Theory of Stability of Lypophobic Colloids", Elsevier, Amsterdam (1948).

31. Marangoni, C.: Nuovo Cinento, (1978) Ser 3, 50, 97, 192.

32. Nikolov, A.D., Wasan, D.T., Huang, D.D., and Edwards, D.A., SPE Preprint 15443, paper presented at the SPE Meeting, New Orleans, La., October 5-8, 1986.

RECEIVED March 3, 1988

Chapter 8

Surfactants for Carbon Dioxide Foam Flooding

Effects of Surfactant Chemical Structure on One-Atmosphere Foaming Properties

John K. Borchardt[1], D. B. Bright[1], M. K. Dickson[1], and S. L. Wellington[2]

[1]Westhollow Research Center, Shell Development Company, P.O. Box 1380, Houston, TX 77251–1380
[2]Bellaire Research Center, Shell Development Company, P.O. Box 481, Houston, TX 77001

A one atmosphere foam test has been designed to permit the study of large numbers of surfactants and the identification of promising candidates for evaluation under reservoir conditions. The inclusion of an oil phase in these experiments is a key feature which allows the effect of oil composition on surfactant foaming to be studied. Classes of surfactants studied included alcohol ethoxylates, alcohol ethoxysulfates, alcohol ethoxyethylsulfonates and alcohol ethoxyglycerylsulfonates. The ability to test large numbers of surfactants permits the relationship of surfactant chemical structure and physical properties to foaming properties to be studied. Surfactants which performed well in the 1 atmosphere foaming experiment were also good foaming agents in sight cell and core flood experiments (1,2) performed in the presence of CO_2 and reservoir fluids under realistic reservoir temperature and pressure conditions. Therefore, it appears that the one atmosphere foaming experiment is a useful screening test.

Core floods and high pressure sight cell experiments are unsuitable for screening large numbers of surfactants as mobility control agents because of the long duration of the experiments. The high cost of the required experiment prevents the performance of a large number of tests. Care must be taken to achieve reproducible foam generation in both core floods and sight cell foaming experiments. For example, in sight cell studies the rate of generation, "foam" cell geometry, and stability of supercritical CO_2 "foams" can vary from one sight cell to another probably due to variations in the nominally identical sintered glass tubes used to generate the foam. Therefore, comparison of supercritical CO_2 "foaming" properties of various surfactants may best be performed in the same sight cell

0097–6156/88/0373–0163$06.00/0
© 1988 American Chemical Society

apparatus. This prevents the simultaneous performance of experiments.

The unsuitability of core floods and sight cell experiments for screening large numbers of surfactants led to the consideration of a one atmosphere foaming test. The advantages of a one atmosphere foaming experiment are that a reproducible experiment can be designed, the test design allows the evaluation of a large number of surfactants in a relatively short time, and the evaluation can be performed in the presence of reservoir brine and oil at formation temperature. The capability to determine comparative foaming properties in the presence of an oil phase and the effect of oil phase composition on foaming are particularly important features of the one atmosphere experiment. Previous one atmosphere test designs did not allow for the study of the effect of oil phase composition.

However, there are strong potential objections to the use of a one atmosphere foaming experiment to evaluate surfactants. These objections must be considered to determine the relevance of one atmosphere foaming experiments to surfactant performance in a reservoir. Bikerman[3] has noted that while shaking is the simplest method of producing foam, "the height and volume of the foam obtained depend on the details of the shaking procedure...and thus cannot be used to characterize the foaminess of a liquid in a (reasonably) absolute manner...the foam heights reproduced are specific for the test procedure selected and have no general validity."(3)

The purpose of the one atmosphere experiment described herein is to determine the relative not the absolute foaming properties of surfactants and to determine the best candidates for evaluation under realistic reservoir conditions. Therefore, the dependence of foam volumes on the experiment design are not of concern as long as the one atmosphere experiments are performed in a reproducible manner and the test design is such as to distinguish between poor, mediocre, and good candidates for testing under reservoir conditions.

Cell size and uniformity are also important variables when studying foams. However, space limitations preclude discussions of these variables herein. Other surfactant properties critical to the success of an EOR process are surfactant adsorption and thermal stability. These questions, under study in our laboratory, are not considered in the short-term one atmosphere foam test experiment and therefore will not be discussed herein.

Experimental Section

Surfactants Studied
Ethoxylated surfactants were chosen for study based on predicted foaming properties, thermal and chemical stability, and adsorption characteristics. Only foaming properties are discussed herein.

Our naming system for the alcohol based surfactants is:

1.   class designation using a 2-4 letter acronym
     AE = alcohol ethoxylate
     AES = alcohol ethoxysulfate

AESo = alcohol ethoxyethyl sulfonate
AEGS = alcohol ethoxyglyceryl sulfonate
2.   carbon number range in the hydrophobe (R)
3.   average number of ethylene oxide units

AE and AES samples were commercial or developmental ENORDET surfactants from Shell Chemical Company except for AES 810-2.6 supplied by GAF Corporation. AESo and AEGS surfactants were experimental research samples synthesized in our laboratories or were supplied by Koninlijke/Shell Laboratorium Amsterdam with the exception of AESo 911-2.5, 911-3.25, 911-4, and 1215-12 obtained from Diamond Shamrock Corporation.

Experimental Section

1 Atmosphere Foaming Test(4)
This static test involves the generation of foams in the presence or absence of hydrocarbon phases at temperatures from 24°C (77°F) to 90°C (194°F). Sometimes warming was required to prepare the 0.5% surfactant solutions in brine. The surfactant solution (10cc) was placed in a clean tared 25cc graduated cylinder. The hydrocarbon phase: decane, decane/toluene (1:1 by volume), stock tank oil, or supercritical CO₂-extracted stock tank oil (3.0cc) was then added. In tests using typical west Texas stock tank oil, carbon dioxide was then passed over the surface of the liquid to remove most of the air from the headspace. Samples were shaken after temperature equilibration, allowed to stand for 24 hours, and shaken again. Foam volume was then determined at set times. This method was chosen over continuous monitoring in order to perform more experiments simultaneously. Detailed kinetics of foam decay were not required to determine relative foaming properties of surfactants (see above).
    The one atmosphere foam test was found to be quite reproducible when performed by a single operator. Average deviations in foam volumes were 0.3-0.7cc. This was more than sufficient to distinguish excellent foaming agents from good ones and good foaming agents from poor ones. However, rank ordering of surfactant foam volumes within these categories (particularly the poor foaming agents) was occasionally complicated by experimental errors in foam volumes.
    The following synthetic brines were used in the first series of tests:

Brine Designation	%w		[Ca+²]
	NaCl	CaCl₂	(ppm)
0.5X	5.19	0.38	1372
1.0X	10.38	0.76	2745
1.5X	15.57	1.14	4118
2.0X	20.76	1.52	5490

Preparation of the synthetic west Texas brine is described in reference 4. This brine had the following composition prior to filtration:

NaCl	40.38g/1000cc
NaHCO$_3$	2.00g/1000cc
Na$_2$SO$_4$	3.06g/1000cc
CaCl$_2$	7.02g/1000cc
MgCl$_2$	2.25g/1000cc

The west Texas stock tank oil and stock tank oil extracted with CO_2 for eight hours at 40°C and a pressure of approximately 2200 psig were used in other experiments. Compositions of these oils are given below:

Analysis	Stock Tank Oil	CO$_2$-Extracted Stock Tank Oil
Median Carbon Number[a]	16	26
% asphaltenes	0.78	1.52
% sulfur	1.95	2.67
25°C Surface Tension (dynes/cm)	26.7	29.6
25°C Density (g/cc)	0.8588	0.9140
Total Acid and Base Number	0.33	0.85

a) Determined using a high pressure liquid chromatographic technique.

<u>High Pressure Sight Cell Studies</u>
A high pressure windowed test cell was charged with a 0.5% solution of surfactant in 1.0X brine. The cell was heated to 75°C and pressurized with CO_2 to a pressure of 2500 psig (1.7237 X 10^7 Pa). A 1:1 volume ratio of liquid and CO_2 was used. The charged cell was then agitated until its contents became thoroughly mixed. As soon as the fluids became static (ca 1 min), the foam height was measured. A second measurement was made 30 minutes later.

<u>Results</u>

<u>Effect of Surfactant Structure on Foam Volume</u>
At 75°C (167°F), a representative U.S. Gulf Coast formation temperature, AES, AEGS, and AESo surfactant classes produced fairly stable foams in short-term tests (see <u>Table I</u>). Examination of these data indicated that the AEGS surfactants generally gave the

greatest relative foam stability at higher salinity. Representative data plotted in Figure 1 indicated that at 75°C in the presence of decane, AEGS surfactants produced more stable foams than AESo, AES, and AE surfactants having very similar hydrophobes and comparable EO contents.

Effect of Number of EO Groups
Examination of the data summarized in Table I indicated that, at a constant number of carbon atoms in the hydrophobe, foam stability generally increased as the number of ethylene oxide groups was increased. The effect of a change in EO level on foam volume in the presence of a hydrocarbon phase was generally greater at lower EO levels (Figure 2). For some oils such as west Texas stock tank oil, the foam volume reached a maximum at ca 20-30 moles EO per mole AE and then decreased.

The effect of EO level on the 75°C 10 minute foam volume (in 1.0X brine in the presence of decane) produced by AEGS and AESo surfactants having the same hydrophobe is shown in Figure 1. The AEGS foam volume increased more rapidly with increased EO level than did foam volumes of AESo surfactants in the absence of an oil phase.

The value of the 1.0X:1.5X brine foam volume ratio at 75°C may be taken as a measure of the sensitivity of surfactant foaming properties to aqueous phase salinity. Values of this ratio determined at 75°C in the presence of decane are summarized below:

Surfactant	10 min Foam Volume Ratio 1.0X:1.5X Brine[a]
AES 911-2.5	1.17
AES 911-5	0.80
AES 911-8	4.0
AES 1215-3	0
AES 1215-6	0.40
AESo 1215-3	--[b]
AESo 1215-6	0.8
AESo 1215-12	2.27
AESo 1215-16	5.71
AEGS 1215-3	1.50
AEGS 1215-7	1.75
AEGS 1215-12	2.25
AEGS 1215-18	2.50

a)   AESo and AEGS surfactants were experimental research samples.
b)   Ratio = 2.3/0

Table I. 75° (167°F) One Atmosphere Foaming Studies

	0.5X Brine			1.0X Brine			1.5X Brine			2.0X Brine		
Surfactant	No Oil	D	D/T	No Oil	D	D/T	No Oil	D	D/T	No Oil	D	D/T
Alcohol Ethoxylates												
AE 1215-7	0.7	0	0	0	0	0	0	0	0	0	0	0
AE 1215-12	2.8	0.2	0	0	0	0	0	0	0	0	0	0
AE 1215-18	0.5	3.3	0	0.8	0.8	0	0.7	2.4	0	0	0	0
Alcohol Ethoxysulfates												
AES 911-2.5S	19.9	6.0	3.0	21.3	7.0	1.0	21.7	6.0	1.0	21.8	4.0	0
AES 911-5S	3.1	3.6	8.0	5.0	4.0	1.2	7.3	5.0	0.3	15.9	3.0	0
AES 911-8S	--	--	--	0.4	7.2	--	1.6	1.8	--	--	--	--
AES 1213-6.5A	11.4	5.0	1.0	10.2	4.0	0	20.8	6.0	0	20.4	1.0	0
AES 1213-12A	6.6	4.0	4.0	4.8	6.0	0.1	9.8	6.0	0	3.6	2.0	0
AES 1215-3S	2.0	3.0	0	13.0	0	0	12.2	0	0	5.7	0	0
AES 1215-6S	5.7	6.4	1.9	6.0	2.0	0	18.4	5.0	0	19.9	1.0	0
AES 810-2.6A	3.1	4.0	1.0	8.8	3.0	2.0	6.2	3.0	0	18.8	1.0	0

10 Minute Foam Volume (cc)

Alcohol Ethoxyethylsulfonates

AESo 1215-1	5.0	2.0	0	3.5	0	0	3.1	0	0	1.2	0	0
AESo 1215-3	18.2	8.7	2.7	11.3	2.3	0	6.5	0	0	4.2	0	0
AESo 1215-6	9.6	8.8	2.0	5.2	3.2	0.4	6.5	0.4	0	2.2	0	0
AESo 1215-12	9.2	8.6	1.8	7.6	3.4	0	10.0	1.5	0	6.3	0.8	0
AESo 1215-16	2.0	16.0	2.0	1.9	4.0	0.4	1.6	0.7	0.2	9.0	0.1	0
AEGS 1215-3	5.0	10.0	4.0	10.2	3.0	0.5	6.2	2.0	0	5.2	0	0
AEGS 1215-7	15.0	5.0	1.0	1.0	7.0	0	3.0	4.0	0	7.2	3.0	0
AEGS 1215-12	4.8	12.0	6.0	3.6	9.0	1.2	3.7	4.0	0.6	6.0	3.9	0
AEGS 1215-18	2.4	13.0	4.0	4.2	10.0	2.0	4.6	4.0	0	6.0	3.0	0

NOTE: D = decane; D/T = 1.1 by volume blend of decane and toluene. AESo and AEGS surfactants were experimental research samples synthesized for this study.

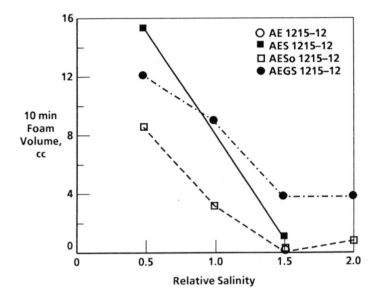

Figure 1.   Effect of relative salinity on foam volume.

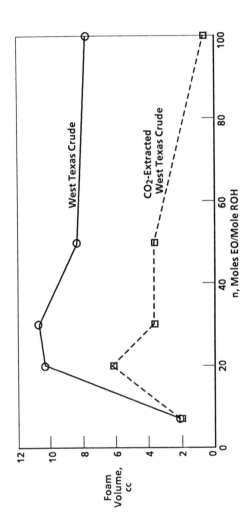

Figure 2. Effect of number of ethoxy groups on foam volume produced by AE 1215-n surfactants.

    This foam volume ratio generally increased with increasing
surfactant EO content, for example AEGS 1215-3 < AEGS 1215-7 < AEGS
1215-12 < AEGS 1215-18. At 25°C, this ratio for AES surfactants
was less sensitive to salinity and EO content than at 75°C. The
limited hydrolytic stability of AES surfactants(4-8) renders the
relevance of the short-term 75°C data to long-term low pH field
applications questionable. Cloud point data indicated that the
75°C test temperature was above the cloud point of these AE surfactants (which ranged from 49-66°C in 10% NaCl brine and decreased
with increasing aqueous phase salinity).

Temperature Effects on Foaming
Results summarized in Table I indicated that foam stability generally decreased with increasing aqueous phase salinity. The effect
of surfactant functional group on salt sensitivity may be considered by comparing the behavior of AE, AESo, and AEGS surfactants. Surfactants being compared have similar hydrophobes and
approximately the same number of EO groups in the hydrophile.
(Differences in the method of synthesis of some of these surfactants can result in different distribution of the number of hydrophobe carbon atoms and of the EO content about the average values.
Also the identity and concentration of impurities may vary. These
changes could effect the observed results.) The ratio of foam
volume in 2.0X brine to that in 0.5X brine at 75°C (Table II)
served as a measure of surfactant salt sensitivity. This ratio
declined in the order: AEGS 1215-12 > AESo 1215-12 ~ AE 1215-12.
This decline indicates that AEGS 1215-12 foam stability was less
sensitive to the presence of salts in the aqueous phase than the
other classes of surfactants studied.

Effect of Oil on Foaming - Refined Hydrocarbons
Unlike previous one atmosphere foam test designs, the present test
permits the effect of the oil phase on surfactant foaming properties to be determined. Refined hydrocarbons were used as model oil
phases. Results summarized in Tables I and Figure 3 indicated that
the presence of hydrocarbons decreased the foam stability. Examination of Table I indicated that the presence of a hydrocarbon
substantially reduced the 75°C foam volumes produced by AES and
AESo surfactants.
    At higher EO levels, the foam volume produced by AEGS and AESo
surfactants were less adversely affected by the presence of an oil
phase than were other surfactants studied (Table I, Figure 1).
This behavior was likely due to the formation of an oil/water
emulsion which stabilized the fluid films between gas bubbles.
Although foam volumes were smaller, at 75°C in three different
brines, the sensitivity of AE and AES surfactants to the presence
of decane decreased with increasing surfactant ethylene oxide
content.
    In contrast, the presence of decane generally had little
detrimental effect on the AEGS 10-minute foam volume under these
test conditions. However, the AEGS surfactants did exhibit significantly reduced foam volumes in the presence of a 1:1 blend of
decane and toluene.

Table II. Effect of an Oil Phase and Aqueous Phase Salinity
on Surfactant Foaming Properties

	Foam Volume Ratio Decane/No Oil			
	0.5X	1.0X	1.5X	2.0X
AES 911-2.5	0.30	0.33	0.28	0.18
AES 911-5	1.16	0.80	0.68	0.19
AES 911-8	ND	18.0	1.1	ND
AES 1213-6.5A	0.44	0.39	0.29	0.05
AES 1213-12A	0.61	1.25	0.61	0.56
AES 1215-3	1.50	0	0	0
AES 1215-6	1.12	0.33	0.27	0.05
AESo 911-2.5	0.84	0.32	0.10	0.04
AESo 911-3.25	0.27	0.29	0.47	0.01
AESo 911-4	0.44	1.21	1.30	0.01
AESo 1215-1	0.40	0	0	0
AESo 1215-3	0.48	0.20	0	0
AESo 1215-6	0.92	0.62	0.06	0
AESo 1215-12	0.93	0.45	0.15	0.13
AESo 1215-16	8.0	2.11	0.44	0.01
AEGS 911-8	18.3	11.5	4.75	2.60
AEGS 1215-3	2.0	0.29	0.32	0
AEGS 1215-7	0.33	7.0	1.33	0.42
AEGS 1215-12	2.50	2.50	1.08	0.65
AEGS 1215-18	5.42	2.38	0.87	0.50

NOTE: ND = not determined. AESo and AEGS surfactants were experimental research samples synthesized for this study.

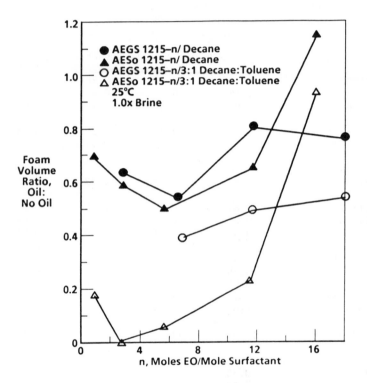

Figure 3.   Effect of oil phase and number of ethoxy groups
on foam volume produced by sulfonate surfactants.

Increasing the aqueous phase salinity appeared to increase foam sensitivity to the presence of a hydrocarbon phase. This behavior may be due to increased surfactant partitioning into the oil phase. This can be quantified by determining the ratio of foam volume in the presence of decane to that in the absence of an added hydrocarbon (Table II, Figure 3). With few exceptions, this ratio decreased with increasing aqueous phase salinity. The values of this ratio for AEGS surfactants declined less with increasing aqueous phase salinity than for other surfactants.

Effect of Oil Composition
The ratio of foam volume in the presence of a 1:1 (by volume) blend of decane and toluene vs that in the presence of decane alone is a measure of the sensitivity of surfactant foaming properties to the aromatic content of the oil phase. Since the values of this ratio were less than unity, aromatic species as represented by toluene had a more negative effect on foam stability than aliphatic hydrocarbons such as decane. Toluene sensitivity appeared to increase with increasing aqueous phase salinity and to decrease with increasing surfactant EO content. Again, this may reflect surfactant partitioning behavior. AEGS, AESo, AES, and AE surfactants did not differ greatly in toluene sensitivity in 0.5X brine. When the aqueous phase salinity was increased, AEGS foam volumes were the least affected by the presence of toluene in the hydrocarbon phase.
     For most of the AES surfactants studied, the ratio of foam volume in the presence of added decane to that in its absence was relatively constant over the 0.5X-1.5X salinity range but decreased significantly when the solvent was 2.0X brine.
     Limited data on two compounds (the sodium and ammonium AES 911-2.5 salts) in 1.0X and 1.5X brines at 75°C indicated no obvious dependence of decane foaming sensitivity to the identity of the AES counterion.

Effect of Stock Tank Oil
Procedures of these 40°C (104°F) experiments are described in the Experimental Section. Tests were performed at a representative west Texas formation temperature using a typical west Texas stock tank oil and a synthetic brine having a composition typical of west Texas injection waters. Results are summarized in Table III. The ratio of foam volume after 30 minutes at 40°C to that after 1 minute was used as an indication of foam stability. The surfactants which produced the greatest initial (1.0 minute) foam volumes also exhibited the greatest foam stability over the thirty minute test period. Because test temperature and salinity were different than used in earlier experiments, results in the presence of west Texas stock tank oil cannot be compared to results described above. However, trends in foam stability were consistent with those described above. Average stability of the foams produced by the AEGS and AES surfactant classes was greater than that of the AE foams.
     The 30-minute foam volume in the presence of stock tank oil vs EO content of a series of C_{12-15} alcohol ethoxylates was plotted in Figure 2. Foam volume was a maximum at approximately 30 moles

Table III. Surfactant Foaming in the Presence of West Texas Field Stock Tank Oil at 40°C

Surfactant	30 min Foam Volume (cc)		Foam Volume Ratio 30 min:1 min	
	STO	CO_2 Extracted STO	STO	CO_2 Extracted STO
AE 911-8	1.1	6.4	0.1	0.4
AE 911-12	5.3	9.7	0.4	0.5
AE 911-20	8.3	6.3	0.5	0.3
AE 1213-6.5	1.0	3.8	0.2	0.5
AE 1215-9	2.2	5.6	0.3	0.6
AE 1215-12	6.8	7.4	0.6	0.6
AE 1215-18	9.9	9.0	0.7	0.5
AE 1215-30	10.4	1.6	0.7	0.1
AE 1215-50	8.9	--	0.7	--
AE 1215-84	7.2	--	0.8	--
AE 1415-7	2.1	2.0	0.4	0.4
AE 1415-20	10.3	6.1	0.7	0.6
AE 1415-30	10.7	3.6	0.8	0.4
AE 1415-50	8.3	3.6	0.7	0.3
AE 1415-100	7.7	0.7	0.7	0.1
AES 911-2.5S	19.8	20.7	0.9	0.9
AES 1215-3S	3.8	13.4	0.3	0.7
AESo 911-2.5S	20.0	20.7	0.9	0.9
AEGS 911-8	20.2	20.8	0.9	0.9
AEGS 1215-12	10.1	12.5	0.7	0.7
AEGS 1215-18	11.9	--	0.8	--

NOTE: STO = stock tank oil. See experimental section for test procedures. S = sodium couterion. AESo and AEGS surfactants were experimental research samples synthesized for this study.

EO/mole of parent alcohol and then gradually declined. Somewhat different behavior was observed using supercritical CO$_2$-extracted stock tank oil which had a higher carbon number 26 vs 16), a higher asphaltene content (1.52% vs 0.78%), and a higher combined acid and base number (0.85% vs 0.33) than the unextracted oil (see above). The maximum foam volume was observed at a lower ethylene oxide content (for C$_{12-15}$, 20 moles of EO vs 30 moles of EO for unextracted oil). The decline in foam volume with further increases in EO content was much more rapid in the presence of CO$_2$-extracted oil. This behavior was also observed for C$_{14-15}$ alcohol ethoxylates and C$_{9-11}$ alcohol ethoxylates.

AE surfactants appeared to exhibit two modes of behavior. Generally foam volume did not decrease for AE surfactants containing less than ca 20 moles EO/mole surfactant when CO$_2$-extracted stock tank oil was used instead of stock tank oil (Figure 2). However, at EO levels above 30 moles/mole surfactant, foam volume in the presence of CO$_2$-extracted stock tank oil was less than in the presence of stock tank oil.

These results imply that since residual crude oil composition changes as it undergoes extraction by injected CO$_2$, the optimum CO$_2$ mobility control agent may change during the course of the CO$_2$ flood.

High Pressure Sight Cell Studies

Sight cell studies have been performed at 75°C and 2500 psig CO$_2$ pressure using AESo and AEGS surfactants. The foam stability, as indicated by the 30 minute:1 minute foam volume ratio, as a function of the surfactant EO content is shown in Figure 4. In one atmosphere foaming tests in the presence of decane, the 10 minute foam volume was taken as a measure of foam stability (see above) and plotted against surfactant EO content in Figure 4. The similar geometry of the AESo and AEGS curves in Figure 4 indicated that increasing the test pressure from 1 atmosphere to 2500 psig CO$_2$ did not alter the effect of surfactant chemical structure on foam stability. At both test pressures, increasing the EO content from 6 to 12 moles per mole AESo surfactant did not have a substantial effect on foam stability. However, at both 1 atm and 2500 psig CO$_2$ pressure, an increase from 12 to 16 moles EO per mole AESo surfactant resulted in a significant increase in foam stability. The AEGS foam stability at both 1 atmosphere and 2500 psig CO$_2$ increased steadily with increasing EO content of the surfactant.

Core Floods

The surfactants (AEGS) which performed best in the one atmosphere foaming experiments in the presence of oil, both refined and crude, prevented gravity override and viscous instabilities enabling high pressure CO$_2$ to displace all the oil in tertiary first contact miscible core floods in a piston-like manner. (1) In the absence of surfactant, gravity override was clearly observed. This would lead to a lower volumetric sweep efficiency, higher produced gas:oil ratios, and lower oil recovery at equivalent CO$_2$ volume injection. (See "A CT Study of Surfactant-Induced Mobility Control for Carbon Dioxide" by S. L. Wellington and H. J. Vinegar, this book.)

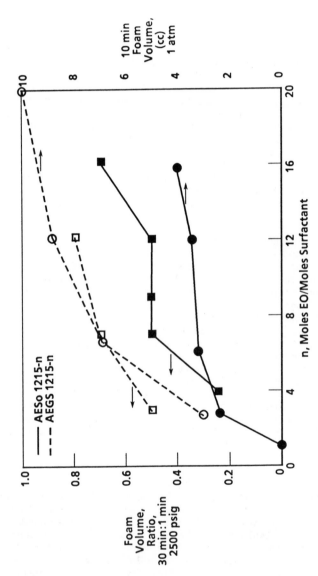

Figure 4.   Effect of number of ethoxy groups on foaming properties of sulfonate surfactants at 1 atmosphere and at 2500 psig.

At the surfactant concentrations employed in the core flood(1), there was no evidence of severe permeability reduction that would cause substantially reduced injectivity.

The surfactant did not cause viscous forces to dominate during immiscible tertiary carbon dioxide injection. Apparently, the unmobilized oil reduced the foam stability while the surfactant reduced the interfacial tension and therefore the CO_2-brine capillary pressure sufficiently to allow gravity effects to dominate the flood.(9)

CONCLUSIONS

Surfactant foaming properties are related to surfactant chemical structure parameters such as hydrophobe size, ethylene oxide chain length, and hydrophile functional group.

Increasing the test pressure from one atmosphere to 2500 psig CO_2 did not alter the effect of surfactant chemical structure on relative foaming performance.

Surfactant foaming properties are related to oil phase composition. The composition of the residual oil will change in the course of a CO_2 EOR project. The optimum CO_2 mobility control agent may thus change during the course of the project.

Of the surfactants tested, AEGS surfactants produced the most persistent foams at high salinity and elevated temperatures in the presence of synthetic and crude oils (in one atmosphere experiments).

Acknowledgments

The authors wish to acknowledge the important contributions of David Haseltine, Craig Yates and Tanya Balthazar who performed many of the one atmosphere foaming experiments and of Eugene F. Lutz, J. Dan Paiz and T. A. B. M. Bolsman who synthesized some of the test surfactants. The authors would like to thank Shell Development Company for permission to publish this work.

Literature Cited

1.   Wellington, S. L. and Vinegar, H.J. "CT Studies of Surfactant-Induced CO_2 Mobility Control," paper SPE 14393 presented at the 60th Annual Technical Conference and Exposition of the Society of Petroleum Engineers of AIME, Las Vegas, September 22-25, 1985.
2.   Wellington, S. L., Reisberg, J., Lutz, E. F., and Bright, D. B. U. S. Patent 4502538, (1985).
3.   Bikerman, J. J. "Foams," Springer-Verlag, New York (1973), p. 85.
4.   Borchardt, J. K., Bright, D. B., Dickson, M. K., Wellington, S. L. Paper SPE 14394 presented at the 60th Annual Technical Conference and Exhibition of the Society of Petroleum Engineers of AIME, Las Vegas, September 22-25, 1985.
5.   Patton, J. T. "Enhanced Oil Recovery by CO_2 Foam Flooding," U. S. DOE Report, Contract No. DE-AC21-78MC03259 (February 1980).

6.   Heller, J. P. "Reservoir Application of Mobility Control
      Foams in CO_2 Floods, paper SPE/DOE 12644 presented at the
      SPE/DOE Fourth Joint Symposium on Enhanced Oil Recovery,
      Tulsa, Oklahoma, April 15-18, 1984.
7.   Schmitt, K. D. "The Hydrolytic Stability of Ethoxylated Alkyl
      Sulfates," paper INDE-14 presented at the 184th National
      Meeting of the American Chemical Society, Kansas City,
      Missouri, September 12-17, 1982.
8.   Bernard, G. G., Holm, L. W., and Harvey, C. P. Soc. Pet.
      Eng. J., 1980, 281-292.
9.   Wellington, S. L. and Vinegar, H. J. J. Pet. Technol., 1987,
      36, 885-898.

RECEIVED January 29, 1988

Chapter 9

Structure–Property Relationships for Mobility-Control Surfactants

John K. Borchardt

Westhollow Research Center, Shell Development Company, P.O. Box 1380, Houston, TX 77251–1380

Correlation equations relating surfactant chemical structure to performance characteristics and physical properties have been established. One atmosphere foaming properties of alcohol ethoxylates and alcohol ethoxylate derivatives have been related to surfactant hydrophobe carbon chain length, ethylene oxide content, aqueous phase salinity, and temperature. Similar correlations have been established for critical micelle concentration, surfactant cloud point, and surfactant adsorption.

Foam exhibits higher apparent viscosity and lower mobility within permeable media than do its separate constituents.(1-3) This lower mobility can be attained by the presence of less than 0.1% surfactant in the aqueous fluid being injected.(4) The foaming properties of surfactants and other properties relevant to surfactant performance in enhanced oil recovery (EOR) processes are dependent upon surfactant chemical structure. Alcohol ethoxylates and alcohol ethoxylate derivatives were chosen to study techniques of relating surfactant performance parameters to chemical structure. These classes of surfactants have been evaluated as mobility control agents in laboratory studies (see references 5 and 6 and references therein). One member of this class of surfactants has been used in three field trials.(7-9) These particular surfactants have well defined structures and chemical structure variables can be assigned numerical values. Commercial products can be manufactured in relatively high purity.

Core floods and high pressure sight cell studies are unsuitable for evaluating large numbers of surfactants as mobility control agents because of the long duration of properly designed

0097–6156/88/0373–0181$07.00/0

experiments and the cost of the equipment required to run a large number of experiments simultaneously. The unsuitability of core floods and sight cell experiments for screening large numbers of surfactants has led various research groups to develop one atmosphere foam tests. A reproducible one atmosphere foam test has a major advantage; large numbers of surfactants can be evaluated rapidly in the presence of reservoir fluids at formation temperatures. The evaluation of large numbers of surfactants is necessary to determine structure - property relationships which can be used to design new surfactants having improved performance properties.

However, there are potential objections to the use of a one atmosphere foaming experiment. These must be considered to determine the relevance of these experiments to surfactant performance in a formation. Bikerman has stated that in one atmosphere experiments "the height and volume of the foam obtained depend on the details of the shaking procedure...and thus cannot be used to characterize the foaminess of a liquid in a (reasonably) absolute manner...the foam heights reproduced are specific for the test procedure selected and have no general validity."(10)

The purpose of the one atmosphere experiment is to determine the relative, not the absolute, foaming properties of a series of surfactant in order to define the best candidates for more lengthy evaluation under realistic reservoir conditions. Quantitative determination of one atmosphere foaming properties is no more useful than the determination of relative foam volumes because of the difference in test conditions between one atmosphere foaming experiments and core floods. In a core flood or a high pressure sight cell experiment, one is studying the behavior of a supercritical CO_2 emulsion in a porous media under dynamic conditions (core floods) or in bulk under static conditions (sight cell experiments) compared to a bulk foam under static conditions in a three phase (aqueous surfactant solution, oil, and CO_2 vapor) one atmosphere test. These differences mean that the one atmosphere experiment cannot be used to predict the quantitative performance of mobility control surfactants under reservoir conditions.

Studies of two series of alcohol ethoxylate derivatives indicated the rank order of one atmosphere foam volumes and supercritical CO_2 emulsion volumes at 2500 psig were the same.(5) The surfactant which gave the greatest dispersion volumes and stability in both tests was used in a supercritical CO_2 core flood.(6) Results indicated excellent performance as use of the surfactant decreased CO_2 gravity override resulting in increased volumetric sweep efficiency and improved oil recovery. Therefore, the dependence of the foam volume on experiment design is not of concern as long as the one atmosphere tests are reproducible and the test is able to distinguish between poor, moderate, and good candidates for evaluation under reservoir conditions.

Experimental Section

Surfactants

Alcohol ethoxylates and alcohol ethoxylate derivatives were chosen for study based on their predicted foaming properties, thermal and chemical stability, salinity tolerance, and adsorption characteristics. Table 1 illustrates the classes of surfactants used and the shorthand surfactant naming system employed. Except when noted, surfactants were developmental ENORDET® surfactants from Shell Chemical Company or were research samples synthesized in our laboratory or at Koninlijke/Shell Laboratorium, Amsterdam. AES 810-2.6A was obtained from GAF Corporation.

Warming of the sample was sometimes required to prepare 0.5% (by weight) surfactant solutions in brine. A typical west Texas field brine was prepared at 40°C from two stock solutions. Solution 1 contained 2.00 g $NaHCO_3$ and 3.06 g Na_2SO_4 in 500 cc deionized water. Solution 2 contained 40.38 g NaCl, 9.30 g $CaCl_2 \cdot 2H_2O$, and 4.80 g $MgCl_2 \cdot 6H_2O$ in 500 cc deionized water. Equal volumes of the two solutions were mixed immediately before use, filtered through a 0.45 micrometer filter twice, and saturated with carbon dioxide. Other brine compositions are given in the footnotes of the appropriate data tables.

The composition of the west Texas stock tank oil and of the stock tank oil extracted with CO_2 for eight hours at 40°C (2200 psig) are given in reference 5.

One Atmosphere Foaming Experiment(5)

The surfactant solution (10 cc) was placed in a clean 25 cc graduated cylinder. The hydrocarbon phase (3 cc) was added and CO_2 passed over the surface of the liquid to remove most of the air from the headspace. The tubes were sealed, shaken, allowed to equilibrate at test temperature, and shaken again. After 24 hours at test temperature, the tubes were shaken for 60 seconds and foam volume determined as a function of time.

The one atmosphere foam test was quite reproducible when performed by a single operator. Average deviations in foam volumes were 0.3-0.7 cc. This was adequate to distinguish excellent, good, and poor foaming agents. Rank ordering of the surfactants within these categories was occasionally complicated by experimental uncertainties. Foam volume errors were much greater, ±20%, when the sample surfactant solution was evaluated by six different operators. Therefore, when comparing surfactants, care was taken that all experiments were performed by the same operator.

Foam cell size and size distribution are important variables when studying foams. Other surfactant properties crucial to the success of an enhanced oil recovery process include critical

Table 1

Test Surfactant Chemical Structure

Structure	Designation	Carbon Number Values of Alkyl Group R	Average Value of N
$R-(OCH_2CH_2)_n-OH$	AE (Alcohol Ethoxylate)	9-11, 12-15, 12-13, 14-15	6.5, 7, 8, 12, 18, 20, 30, 50, 84, 100
$R-(OCH_2CH_2)_n-OSO_3 M^+$	AES (Alcohol Ether Sulfate)	8-10, 9-11, 12-13, 12-15	2.5, 2.6, 3, 5, 6, 6.5, 12
$R-(OCH_2CH_2)_n-SO_3 Na^+$	AESo (Alcohol Ethoxyethysulfonate)	9-11, 12-15	1, 2.5, 3, 3.25, 4, 6, 12, 16
$R-(OCH_2CH_2)_n OCH_2 \underset{OH}{C}HCH_2 SO_3 Na^+$	AEGS (Alcohol Ethoxyglycerylsulfonate)	9-11, 12-15	3, 7, 8, 12, 18

micelle concentration (which affects surfactant adsorption), adsorption, thermal stability, interfacial tension, and salt tolerance. The critical micelle concentration, salt tolerance, and interfacial tension will be discussed herein. Due to space limitations and the use of the one atmosphere experiment to only screen surfactants and predict relative performance under reservoir conditions, the other variables will not be discussed herein.

A randomized complete block experiment design could determine surfactant differences as well as or better than multiple correlation analysis. However, one of the objectives of this work was to try to obtain additional information from pre-existing sets of data. For this, multiple correlation analysis is particularly well suited.

Results

Effect of Ethoxylate Chain Length on Foam Volume

Examination of the data summarized in Table 2 indicated that, for a constant hydrophobe carbon number (HCN), foam stability generally increased as the number of ethylene oxide (EO) groups in the surfactant was increased. This was true for alcohol ethoxylates and three classes of alcohol ethoxylate derivatives. Data reported in references 5 and 11 is consistent with this observation.

Foaming properties can be quantitatively related to surfactant chemical structure, surfactant physical properties, and test conditions using the technique of multiple correlation analysis.(11) The current studies were restricted to linear correlation equations to permit the analyses to be performed on a small microcomputer. While non-linear equations having higher correlation coefficients than obtained herein can be developed, theoretical insights are often limited due to the complexity of the various terms of such equations. The quality of the correlations were assessed using the correlation coefficient (r^2) criteria of Jaffe:(12)

$r^2 > 0.99$	excellent
$r^2 > 0.95$	satisfactory
$r^2 > 0.90$	fair
$r^2 < 0.90$	poor

The foaming properties of a large number of alcohol ethoxylates have been reported in the literature(13) (Table 3). No oil phase was present in these tests. Surfactants were obtained from two different suppliers.

Table 2

Surfactant Foaming in the Presence of West Texas Stock Tank Oil
at 40°C

| | 30 min Foam Volume (cc) | |
Surfactant	STO	CO_2 Extracted STO
AE 911-8	1.1	6.4
AE 911-12	5.3	9.7
AE 911-20	8.3	6.3
AE 1213-6.5	1.0	3.8
AE 1215-9	2.2	5.6
AE 1215-12	6.8	7.4
AE 1215-18	9.9	9.0
AE 1215-30	10.4	1.6
AE 1215-50	8.9	--
AE 1215-84	7.2	--
AE 1415-7	2.1	2.0
AE 1415-20	10.3	6.1
AE 1415-30	10.7	3.6
AE 1415-50	8.3	3.6
AE 1415-100	7.7	0.7
AES 911-2.5	19.8	20.7
AES 1215-3	3.8	13.4
AESo 911-2.5	20.0	20.7
AEGS 911-8	20.2	20.8
AEGS 1215-12	10.1	12.5
AEGS 1215-18	11.9	--

NOTE: STO = stock tank oil. See experimental section for test procedures.
SOURCE: Reproduced with permission for ref. 5. Copyright 1985 Society of Petroleum
Engineers.

Table 3

Foam Volume of Alcohol Ethoxylates - One Atmosphere
Foaming Studies

Surfactant Designation	Moles EO/ Mole ROH	EO % wt.	Supplier	T (°C)	Surfactant Concentration (% wt.)	Initial Foam Volume (cc)
AE 1213-3	3	40	1	25	0.5	130
AE 1215-7	7.2	61	1	25	0.5	270
AE 1215-12	12	72	1	25	0.5	240
AE 1012-5	5.2	60	2	25	0.5	270
AE 1214-7	7	60	2	25	0.5	240
AE 1218-10.7	10.7	68.5	2	25	0.5	240
AE 911-2.5	2.5	41	1	25	0.5	160
AE 911-8	8.4	70	1	25	0.5	310
AE 1213-3	3	40	1	48.9	0.5	130
AE 1215-7	7.2	61	1	48.9	0.5	220
AE 1215-12	12	72	1	48.9	0.5	300
AE 911-2.5	2.5	41	1	48.9	0.5	130
AE 911-8	8.4	70	1	48.9	0.5	350
AE 1012-5	5.2	60	2	48.9	0.5	220
AE 1214-7	7	60	2	48.9	0.5	200
AE 1218-10.7	10.7	68.5	2	48.9	0.5	190

Continued on next page.

Table 3 (continued)

Foam Volume of Alcohol Ethoxylates - One Atmosphere Foaming Studies[a]

Surfactant Designation	Moles EO/ Mole ROH	EO % wt.	Supplier	T (°C)	Surfactant Concentration (% wt.)	Initial Foam Volume (cc)
AE 1213-3	3	40	1	25	0.1	40
AE 1215-7	7.2	61	1	25	0.1	240
AE 1215-12	12	72	1	25	0.1	240
AE 911-2.5	2.5	41	1	25	0.1	130
AE 911-8	8.4	70	1	25	0.1	250
AE 1012-5	5.2	60	2	25	0.1	230
AE 1214-7	7	60	2	25	0.1	230
AE 1218-10.7	10.7	68.5	2	25	0.1	190
AE 1213-3	3	40	1	48.9	0.1	10
AE 1215-7	7.2	61	1	48.9	0.1	160
AE 1215-12	12	72	1	48.9	0.1	210
AE 911-2.5	2.5	41	1	48.9	0.1	10
AE 911-8	8.4	70	1	48.9	0.1	260
AE 1012-5	5.2	60	2	48.9	0.1	70
AE 1214-7	7	60	2	48.9	0.1	190
AE 1218-10.7	10.7	68.5	2	48.9	0.1	200

Initial Foam Volume = a (HCN) + b (EO, % wt.) +
c (surfactant concentration, % wt.) +
constant

T = 48.9 °C, all surfactants

correlation coefficient =	0.911
constant =	161.917 ± 78.324
a =	-19.428 ± 7.201
b =	27.560 ± 3.854
c =	231.250 ± 46.691

T = 25 °C, surfactants from supplier 1

correlation coefficient =	0.990
constant =	206.324 ± 32.295
a =	-12.500 ± 2.041
b =	2.353 ± 0.832
c =	116.667 ± 8.334

NOTE: Raw data are taken from ref. 13.
SOURCE: Reproduced with permission for ref. 11. Copyright 1987 Society of Petroleum Engineers.

The initial foam volume (IFV) was correlated with the median carbon number of the hydrophobe (HCN), the weight percent EO in the surfactant, and the surfactant concentration (in weight percent) using the following equation:

$$\text{IFV} = a\ (\text{HCN}) + b\ (\text{weight \% EO}) + c\ (\text{surfactant concentration}) + \text{constant} \tag{1}$$

At 48.9°C, use of the data for ten alcohol ethoxylates from supplier 1 and six alcohol ethoxylates from supplier 2 resulted in a correlation coefficient of 0.911. When only supplier 1 data were considered, the correlation coefficient increased slightly to 0.925.

At 25°C, use of data for both sets of alcohol ethoxylates in the above equation resulted in a correlation coefficient less than 0.90. The attempt to add a temperature term to equation 1 and analyze the 48.9°C and 25°C data resulted in a low value for the correlation coefficient. This was not surprising in view of the 25°C results. Consideration of the 25°C data for only the supplier 2 surfactants resulted in a correlation coefficient of 0.990.

The low correlation coefficient observed when combining the data for alcohol ethoxylates from two different manufacturers is probably due to chemical structure variables not incorporated in equation 1. These include different degrees of hydrophobe linearity, different distributions of carbon numbers and ethylene oxide chain lengths around the average values used in the correlation equation, and the presence of differing amounts of other components such as unreacted alcohols in the surfactants.

The data summarized in Table 2 illustrate an important limitation of the correlation analysis approach in predicting surfactant properties; extrapolation outside the range of parameter values used to generate the correlation equation may not be valid. These data indicate the effect of the EO content of alcohol ethoxylates on foam volume observed in the presence of a west Texas stock tank oil. Initial studies utilized alcohol ethoxylates containing up to 20 moles EO per mole of surfactant. Linear equations with r^2 >0.95 were derived.(11) However, additional data for AE surfactants containing >30 moles EO per mole of surfactant indicated that the foam volume passed through a maximum and then declines as the surfactant EO content increased. Thus, a linear correlation equation was unsuitable for predicting foam volume over the entire range of surfactant EO content. Similar trends were observed for alcohol ethoxylates in the presence of other oil phases.

Foaming Properties of Alcohol Ethoxylate Derivatives

Various anionic derivatives of alcohol ethoxylates have been shown to possess improved saline media foaming properties in the presence of oil phases. Alcohol ethoxysulfates (AES) have been used in field tests as CO_2 mobility control agents.(7-9) (Limited hydrolytic stability renders AES surfactants unsuitable for use in many reservoirs.) Equation 2 may be used to relate 25°C foaming volume (FV) obtained in the presence of decane and a mildly saline brine (Table 4) to surfactant chemical structure:

$$FV = a \ (HCN) + b \ (moles \ EO/mole \ surfactant) + constant \qquad (2)$$

The correlation coefficient for the AES surfactants was 0.972. When equation 2 was used to analyze the 25°C foaming properties of a series of alcohol ethoxyacetate (AEA) surfactants, the correlation coefficient was 0.928. The different sign of the "a" coefficient and the different magnitude of "b" for the two classes of surfactants indicated that the effect of the carbon number of the hydrophobe and EO chain length on their foaming properties differed significantly. For instance, the magnitude of the "b" coefficient suggests that AES surfactant foam volumes were more sensitive to changes in EO chain length. These differences could be due to the different polar and steric effects of the sulfate and carboxylate end groups and different unreacted alcohol ethoxylate concentrations in the two classes of surfactants.

Additional AES foaming experiments were performed in the presence of a Gulf Coast stock tank oil (Table 5). The correlation coefficient of equation 2 was 0.995 and 0.983 at pH 7 and pH 2.5 respectively. When the foaming data at both pH values were combined and the data analyzed using equation 3, the correlation coefficient was 0.990.

$$FV = a \ (HCN) + b \ (moles \ EO/mole \ surfactant) + c \ (pH) + constant \qquad (3)$$

The value of "c" was within experimental error of zero.

Critical Micelle Concentration of Alcohol Ethoxylates

Surfactant critical micelle concentration (cmc) may be related to chemical structure using multiple correlation analysis. The cmc value plays an important role in surfactant adsorption, foaming, and interfacial tension properties. The 25°C cmc values of a series of high purity single component highly linear primary alcohol ethoxylates (Table 6) were analyzed using equation 4:

$$log \ (cmc) = a \ (HCN) + b \ (moles \ EO/mole \ surfactant) + constant \qquad (4)$$

The correlation coefficient was 0.992.

A series of "twin-tail" alcohol ethoxylates were then studied (Table 7). The hydrophobe carbon atom bearing the ethylene oxide chain is bonded to two carbon atoms. The chemical structures are listed in Table 7. Although the range of

Table 4

Foaming Properties of Alcohol Ethoxylate Derivatives at $25^{\circ}C$[a)]

Surfactant Designation[b)]	Moles EO/Mole ROH	median C No. of Hydrophobe	Foam Volume (cc)[c)]
AES 911-2.5A	2.5	10	15.8
AES 911-2.5	2.5	10	15.7
AES 1215-3A	3	13.5	11.8
AES 1215-3	3	13.5	10.6
AES 1215-9	9	13.5	14.3
AES 1215-12	12	13.5	15.3
AEA 1213-4.5	4.5	12.5	0.6
AEA 1213-6	6	12.5	5.9
AEA 1215-9	9	13.5	11.8
AEA 1214-10	10	13	12.0
AEA 1214-13	13	13	13.6

FV = a (median carbon number) + b (moles EO/mole ROH) + constant

AES
$$a = -1.355 \pm 0.152$$
$$b = 0.469 \pm 0.069$$
$$constant = 28.131 \pm 1.729$$
$$r^2 = 0.972$$

AEA
$$a = 4.103 \pm 3.081$$
$$b = 1.191 \pm 0.381$$
$$constant = -54.266 \pm 37.811$$
$$r^2 = 0.928$$

a) Initial surfactant concentration in the aqueous phase was 0.50% by weight. The aqueous phase was a brine containing 5.19% NaCl and 0.50% $CaCl_2 \bullet 2H_2O$. The volume of the aqueous phase was 10.0cc and the volume of the decane phase was 3.0cc.
b) AES = alcohol ethoxysulfate (A = ammonium salt)
   AEA = alcohol ethoxyacetate
c) AES foam volume was determined 5.0 minutes after shaking the sample tube.
   AEA foam volume was determined 60.0 minutes after shaking the sample tube.

Table 5

$25°C$ Foaming Properties of Alcohol Ethoxysulfates[a]

Surfactant Designation	Median HCN	Average Moles EO/Mole ROH	Brine pH	Foam Volume (cc)
AES 810-2.6[b]	9	2.6	7	18.9
AES 810-2.6[b]	9	2.6	2.5	17.4
AES 911-2.5	10	2.5	7	17.2
AES 911-2.5	10	2.5	2.5	16.8
AES 1215-3	13.5	3	7	6.8
AES 1215-3	13.5	3	2.5	7.6
AES 1215-12	13.5	12	7	5.0
AES 1215-12	13.5	12	2.5	4.8

Foam Volume = a (HCN) + b (moles EO/mole ROH) + constant

pH 7

correlation coefficient = 0.995

constant = 44.546 ± 2.334
a = -2.738 ± 0.228
b = -0.217 ± 0.115

pH 2.5

correlation coefficient = 0.983

constant = 39.294 ± 3.744
a = -2.256 ± 0.364
b = -0.338 ± 0.183

Foam Volume = a (HCN) + b (moles EO/mole ROH) + c (pH) + constant

correlation coefficient = 0.990
constant = 41.576 ± 2.108
a = -2.497 ± 0.194
b = -0.277 ± 0.098
c = 0.072 ± 0.140

a) The tests were performed in a brine containing 10.38% (by weight) NaCl and 1.01% $CaCl_2 \bullet 2H_2O$. Initial surfactant concentration in the aqueous phase was 0.50% by weight. Volume of the aqueous phase was 10.0 cc and of the oil phase, 3.0 cc. The oil phase was a $42°$ API stock tank oil from the U.S. Gulf Coast. Surfactant counterion was sodium unless otherwise noted. Foam volumes were determined 30.0 minutes after shaking the sample tube.
b) The surfactant counterion was ammonium.

Table 6

Critical Micelle Concentration of Homogeneous Alcohol
Ethoxylates

Surfactant Designation	Hydrophobe Carbon Number	Moles EO/ Mole ROH	cmc (mole/liter)	Reference
AE 6-3	6	3	100,000	14
AE 8-1	8	1	4900	15
AE 8-3	8	3	7500	14
AE 8-6	8	6	9900	14, 16
AE 10-3	10	3	600	14
AE 10-4	10	4	680	17
AE 10-5	10	5	810	17
AE 10-6	10	6	900	14, 18
AE 10-9	10	9	1300	14
AE 12-6	12	6	87	14, 16
AE 13-8	13	8	112	19
AE 13-10	13	10	117	19
AE 13-12	13	12	202	19
AE 14-6	14	6	10.4	18
AE 16-6	16	6	1.0	16
AE 16-6	16	6	1.7	20
AE 16-9	16	9	2.1	21
AE 16-12	16	12	2.3	21
AE 16-15	16	15	3.1	21
AE 16-21	16	21	3.9	21

$$\log (cmc) = a \, (HCN) + b \, (\text{moles EO/mole ROH}) + \text{constant}$$

$$
\begin{aligned}
\text{correlation coefficient} &= 0.992 \\
\text{constant} &= 7.615 \pm 0.162 \\
a &= -0.485 \pm 0.016 \\
b &= -0.479 \pm 0.012 \\
\text{standard error of fit} &= 0.185
\end{aligned}
$$

NOTE: T = 25 °C.
SOURCE: Reproduced with permission for ref. 11. Copyright 1987 Society of Petroleum Engineers.

Table 7

Critical Micelle Concentration of <u>Alpha</u>-Branched Alcohol Ethoxylates[a]

Entry	Surfactant	Temperature °C	Temperature °K	cmc (mole/l)	Hydrophobe Sidechain[b] σ_I	Hydrophobe Sidechain[b] E_s	Reference
1	$CH_3CH_2CH_2CH_2$—$(OCH_2CH_2)_6$—OH	20	293.15	8.0×10^5	0.00	1.24	14
2	CH_3CH_2CH—$(OCH_2CH_2)_6$—OH, CH_3	20	293.15	9.1×10^5	-0.04	0.00	14
3	$CH_3(CH_2)_4$—CH_2—$(OCH_2CH_2)_6$—OH	20	293.15	7.4×10^4	0.00	1.24	14
4	$CH_3CH_2CH_2CH$—$(OCH_2CH_2)_6$—OH, CH_3CH_2	20	293.15	10.0×10^4	-0.03	-0.07	14
5	$CH_3(CH_2)_6$—CH_2—$(OCH_2CH_2)_6$—OH	30	303.15	8.9×10^3	0.00	1.24	18
6	$CH_3(CH_2)_3CH$—$(OCH_2CH_2)_6$—OH, $CH_3CH_2CH_2$	30	303.15	20.0×10^3	-0.02	-0.36	14
7	$CH_3(CH_2)_8CH_2$—$(OCH_2CH_2)_6$—OH	20	293.15	9.2×10^2	0.00	1.24	16
8	$CH_3(CH_2)_4CH$—$(OCH_2CH_2)_6$—OH, $CH_3CH_2CH_2CH_2$	20	293.15	3.1×10^3	-0.04	-0.39	14

Continued on next page

Table 7 (continued)

Critical Micelle Concentration of <u>Alpha</u>-Branched Alcohol Ethoxylates[a]

Entry	Surfactant	Temperature °C	°K	cmc (mole/l)	Hydrophobe Sidechain[b] I	E_s	Reference
9	$CH_3(CH_2)_8CH_2$—$(OCH_2CH_2)_9$—OH	20	293.15	1.4×10^3	0.00	1.24	15
10	$CH_3(CH_2)_4CH$—$(OCH_2CH_2)_9$—OH $CH_3CH_2CH_2CH_2$	20	293.15	3.2×10^3	-0.04	-0.39	14
11	$CH_3(CH_2)_{12}CH_2$—$(OCH_2CH_2)_{10}$—OH	25	298.15	117[c]	0.00	1.24	23
12	$CH_3(CH_2)_5CH$—$(OCH_2CH_2)_{10}$—OH $CH_3(CH_2)_5CH_2$	25	298.15	141[c]	-0.06	--	23
13	$CH_3(CH_2)_{12}CH_2$—$(OCH_2CH_2)_{12}$—OH	25	298.15	202[c]	0.00	1.24	23
14	$CH_3(CH_2)_5CH$—$(OCH_2CH_2)_{12}$—OH $CH_3(CH_2)_5CH_2$	25	298.15	178[c]	-0.06	--	23

Table 7 (continued)
Critical Micelle Concentration of <u>Alpha</u>-Branched Alcohol Ethoxylates[a]

<u>20°C</u>

$$\log(cmc) = a \ (HCN) + b \ (\text{moles EO/mole ROH}) + constant$$

correlation coefficient = 0.985

<u>20°C linear hydrophobe AE's</u>
correlation coefficient = 0.9995

<u>20°C branched hydrophobe AE's</u>
correlation coefficient = 0.996

<u>all data</u>
correlation coefficient = 0.987

$$\log(cmc) = a \ (\text{hydrophobe backbone carbon number}) + b \ (\text{hydrophobe sidechain carbon number})$$
$$+ c \ (\text{moles EO/mole ROH}) + d \ (T, \ ^{\circ}K) + constant$$

correlation coefficient = 0.989

<u>d = 0</u>
correlation coefficient = 0.998

<u>20°C</u>

$$\log(cmc) = a \ (\text{hydrophobe backbone carbon number}) + b \ (\text{moles EO/mole ROH})$$
$$+ c \ (\sigma_I) + d \ (E_s) + constant$$

correlation coefficient = 0.986
$c = -49.908 \pm 23.177$ $d = 2.073 \pm 0.581$

<u>all data</u>
correlation coefficient = 0.919
$c = 35.267 \pm 18.564$ $d = 0.073 \pm 0.506$

a) cmc values determined from surface tension data unless otherwise noted.
Source: Reproduced with permission from Ref. 11. Copyright 1987 Society of Petroleum Engineers.

temperatures studied (20°C and 30°C) was not great, a temperature term was included in the correlation equation:

$$\log (cmc) = a \text{ (hydrophobe backbone carbon no.)} +$$

$$b \text{ (hydrophobe sidechain carbon no.)} + \qquad (5)$$

$$c \text{ (moles EO/mole surfactant)} + d \text{ (T, }^\circ\text{K)} +$$

$$\text{constant}$$

The correlation coefficient was 0.989 and actually increased slightly to 0.998 if the temperature term was omitted from equation 5. (The temperature could be collinear to one or more of the other variables.) The relative values of "a" and "b" indicate that both the hydrophobe backbone and the sidechain have an important influence on the cmc values. The effect of the sidechain is a demonstration of the important differences in properties observed for linear and branched hydrophobe surfactants.

One could investigate the polar and steric effects of the hydrophobe sidechain using the appropriate substituent constants and equation 6:

$$\log (cmc) = a \text{ (hydrophobe backbone carbon no.)} +$$

$$b \text{ (moles EO/mole surfactant)} + c \; (\sigma_I) + d \; (E_s) + \quad (6)$$

$$\text{constant}$$

The relative values of "c" and "d" suggest that polar effects have a greater effect on the cmc value than steric effects of linear alkyl hydrophobe sidechains. Molecular models suggest that the steric effect of the sidechain is limited to the closest 2-5 EO groups of the 6-12 member EO chain. Even for the longest hydrophobe sidechains a substantial portion of the hydrophile chain is uninfluenced by steric effects. However, inductive effects, while weak, could still be transmitted through the chemical bonds of the hydrophile chain for much of its length.

Critical Micelle Concentration of Alcohol Ethoxy Sulfates

Equation 5 was used to analyze the cmc properties for a series of alcohol ethoxysulfates (Table 8). While a "fair" correlation coefficient of 0.946 was obtained, it should be noted that the range of hydrophobe carbon chain lengths studied was limited. AES surfactants are prepared by sulfation of alcohol ethoxylates and can contain unreacted alcohol ethoxylate. Variation in the concentration of unreacted AE, which is not considered in equation 5, could reduce the correlation coefficient.

Table 8

Critical Micelle Concentration of Alcohol Ethoxysulfates[a]

Surfactant	HCN[b]	Moles EO/Mole ROH	cmc, 25°C (10⁴ moles/liter)
$C_{16}H_{33}OSO_3^-Na^+$	16	0	4
$C_{16}H_{33}(OCH_2CH_2)OSO_3^-Na^+$	16	1	2.1
$C_{16}H_{33}(OCH_2CH_2)_2OSO_3^-Na^+$	16	2	1.2
$C_{16}H_{33}(OCH_2CH_2)_3OSO_3^-Na^+$	16	3	0.7
$C_{16}H_{33}(OCH_2CH_2)_4OSO_3^-Na^+$	16	4	0.8
$C_{18}H_{37}(OCH_2CH_2)OSO_3^-Na^+$	18	1	1.9
$C_{18}H_{37}(OCH_2CH_2)_2OSO_3^-Na^+$	18	2	0.8
$C_{18}H_{37}(OCH_2CH_2)_3OSO_3^-Na^+$	18	3	0.5
$C_{18}H_{37}(OCH_2CH_2)_4OSO_3^-Na^+$	18	4	0.4

$$\log(cmc) = a\ (HCN) + b\ (\text{moles EO/mole ROH}) + \text{constant}$$

$$\text{constant} = -2.013 \pm 0.595$$
$$a = -0.090 \pm 0.036$$
$$b = -0.200 \pm 0.027$$

standard error of fit = 0.104
correlation coefficient = 0.946

a) Raw data taken from references 21 and 23.
b) HCN = median hydrophobe carbon number.
Source: Reproduced with permission from Ref. 11. Copyright 1987 Society of Petroleum Engineers.

Cloud Point of Alcohol Ethoxylates

The cloud point (CP) of alcohol ethoxylates is the temperature at which solutions of these surfactants become cloudy. The
turbidity is caused by the separation of the fluid into two
phases: a low surfactant concentration and a high surfactant
concentration phase. This phase separation is a critical property that can effect surfactant performance particularly in
saline waters. Earlier studies indicated that the effect of salt
concentration on alcohol ethoxylate cloud point could be linear,
at least to a first approximation.(11) The following equation
was used to analyze the data for commercial alcohol ethoxylates
summarized in Table 9:

$$\text{CP } (^\circ\text{C}) = a \text{ (HCN)} + b \text{ (moles EO/mole surfactant)} + \\ c \text{ (NaCl concentration, \% weight)} + \text{constant} \tag{7}$$

The correlation coefficient was 0.954.

Interfacial Tension of Alcohol Ethoxylates

Interfacial tension (IFT) between injected fluids and formation
crude oil is a critical parameter in EOR processes. The results
reported herein suggest that the relationship between IFT and
surfactant chemical structure can be quantified. Equation 8 was
used to relate the IFT to alcohol ethoxylate chemical structure:

$$\text{IFT} = \exp \ a \text{ (HCN)} + \exp \ b \text{ (moles EO/mole surfactant)} + \text{constant} \tag{8}$$

The correlation equation obtained using the data summarized in
Table 10 was 0.938. The modest correlation coefficient may be
due to the use of commercial linear primary alcohol ethoxylates
from two different manufacturers. Variations in hydrophobe
linearity, hydrophobe carbon number distribution about the
average value, and EO chain length distribution about the average
value were not considered in equation 8. This interpretation is
supported by the observation that inclusion of data for three
secondary (methyl branched at the alpha position of the hydrophobe) alcohol ethoxylates in the analysis resulted in a decrease
of the correlation coefficient to <0.90.

Discussion

Both the use of one atmosphere foaming experiments and the
technique of multiple correlation analysis have a common purpose:
minimizing the effort required to develop new surfactants for
mobility control and other EOR applications. Proper use of these
techniques with due consideration of their limitations can
substantially reduce the number of experiments required to
develop new surfactants or to understand the effect of surfactant
chemical structure on physical properties and performance parameters.

Table 9. Correlation of Alcohol Ethoxylate Cloud Points

Surfactant Designation	HCN [a]	Moles EO/ Mole ROH	EO, % wt	Molecular Weight	Cloud Point (°C) in			
					Fresh Water	5% NaCl	10% NaCl	15% NaCl
AE 911-6	10	6	62	425	52 [b]	--	--	--
AE 911-8	10	8	70	529	80 [b]	61	49	39
AE 1213-6.5	12.5	13	60	484	45 [b]	--	--	--
AE 1215-7	13.5	7.2	61	519	50 [b]	--	--	--
AE 1215-9	13.5	9	67	610	74 [b]	54	44	35
AE 1215-12	13.5	12	72	729	97 [b]	79	66	54
AE 1415-7	14.5	7	60	539	46 [b]	-- [b] 78 [b]	--	--
AE 1415-13	14.5	13	72	790	--	--	--	--

Cloud Point = a (HCN) + b (moles EO/mole ROH) + c (NaCl % weight) + constant

$$\text{constant} = 39.779 \pm 9.209$$
$$a = -2.591 \pm 0.690$$
$$b = 6.999 \pm 0.567$$
$$c = -2.393 \pm 0.223$$

standard error of fit = 5.477

correlation coefficient = 0.954

Cloud Point = a (EO, % weight) + b (surfactant molecular weight) + constant

$$\text{constant} = 91.474 \pm 8.619$$
$$a = 3.486 \pm 0.178$$
$$b = 0.036 \pm 0.009$$

standard error of fit = 1.514

correlation coefficient = 0.997

a) HCN = median hydrophobe carbon number.
b) Raw data are from ref. 24.
SOURCE: Reproduced with permission from ref. 11. Copyright 1987 Society of Petroleum Engineers.

Table 10

Interfacial Tension of Linear and Branched Alcohol Ethoxylates with Mineral Oil[a]

Surfactant Designation	HCN[b]	Linear Hydrophobe	Moles EO/ Mole ROH	Interfacial Tension (dynes/cm)
Rsec-AE 1115-7	13	No	7	3.1
Rsec-AE 1115-9	13	No	9	3.6
Rsec-AE 1115-12	13	No	12	4.7
AE 1215-7	13.5	Yes	7	4.6
AE 1215-9	13.5	Yes	9	5.0
AE 1215-12	13.5	Yes	12	6.0
AE 1213-6.5[c]	12.5	Yes	6.5	4.4
AE 1215-7[c]	13.5	Yes	7	4.6
AE 1215-9[c]	13.5	Yes	9	5.0

IFT = a (HCN) + b (moles EO/mole ROH) + constant

all surfactants

correlation coefficient < 0.9

linear hydrophobe surfactants only:

correlation coefficient = 0.984
standard error of fit = 0.101

constant = 2.545 ± 1.571
a = 0.005 ± 0.124
b = 0.276 ± 0.025

a) Raw data taken from reference 25. Interfacial tensions were measured at 25°C in fresh water using 0.1% (by weight) surfactant. The surfactant was from supplier 4 unless otherwise noted.
b) Average hydrophobe carbon number.
c) Surfactant obtained from supplier 1.

Source: Reproduced with permission from Ref. 11. Copyright 1987 Society of Petroleum Engineers.

The correlation equations discussed herein were limited to linear relations. Relatively low values of correlation coefficients ($0.90 < r^2 < 0.95$) and relatively large standard deviations in coefficient values and constant terms may be due to two causes. The first is that the correlation may not actually be linear but a linear equation is a good approximation to the true correlation equation. The second is that there may be relatively large uncertainties in the experimental data. A limited number of data points can compound this second difficulty.

The limitation of the use of one atmosphere foaming experiments to rank order the predicted surfactant performance in permeable media rather than in quantitatively or semi-quantitatively predicting the actual performance of the surfactants under realistic use conditions has already been mentioned. Multiple correlation analysis has its greatest value to predicting the rank order of surfactant performance or the relative value of a physical property parameter. Correlation coefficients less than 0.99 generally do not allow the quantitative prediction of the value of a performance parameter for a surfactant yet to be evaluated or even synthesized. Despite these limitations, multiple correlation analysis can be valuable, increasing the understanding of the effect of chemical structure variables on surfactant physical property and performance parameters.

Conclusions

Foaming properties of alcohol ethoxylates and alcohol ethoxylate derivatives are related to chemical structure features such as hydrophobe size and linearity, ethylene oxide chain length, and the terminating group at the end of the ethylene oxide chain. Foaming properties may be mathematically related to chemical structure parameters using multiple correlation analysis.

Other surfactant physical properties and performance parameters such as critical micelle concentration, cloud point, and interfacial tension may be related to surfactant chemical structure using multiple correlation analysis.

When studying commercial surfactants obtained from different suppliers and possibly manufactured by different synthetic routes, variations in hydrophobe carbon number distribution, hydrophobe chain branching, the distribution of ethylene oxide chain lengths about the average value, and the nature and concentration of impurities has to be considered.

Predictions of surfactant performance made based on multiple correlation analysis may not be valid if the surfactants involved have chemical structure parameters outside the range used to define the correlation equation.

References

1. Sibree, T.O. Trans. Faraday Soc., 1943, 325.
2. Marsden, S.S.; Khan, S.A. SUPRI Technical Report 3, 1977.
3. Patton, J.T.; Holbrook, S.T.; Hsu, S. Soc. Pet. Eng. J.,
    1983, 456-460.
4. Craig, Jr., F.F.; Lummus, J.L. U.S. Patent 3 185 634, 1965.
5. Borchardt, J.K.; Bright, D.B.; Dickson, M.K.; Wellington,
    S.T. Proc. 60th Ann. Tech. Conf. and Exhib. of SPE, 1985,
    Paper No. SPE 14394.
6. Wellington, S.T.; Vinegar, H.J. Proc. 60th Ann. Techn. Conf.
    and Exhib. of SPE, 1985, Paper No. SPE 14393.
7. Holm, L.W. J. Pet. Tech., 1970, 22, 1499-1506.
8. Holm, L.W. Proc. SPE/DOE Fifth Joint Symp. on Enhan. Oil
    Recov., 1986, 497-508, Paper No. SPE/DOE 14963.
9. Heller, J.P.; Boone, D.A.; Watts, R.J. Proc. 60th Ann. Tech.
    Conf. and Exhib. of SPE, 1985, Paper No. SPE 14395.
10. Bikerman, J.J. "Foams"; Springer-Verlag; New York, 1973; p.
    85.
11. Borchardt, J.K. Proc. SPE Internat. Symp. Oilfield Chem.,
    1987, p. 395-413, Paper No. SPE 16279.
12. Jaffee, H.H. Chem. Rev., 1953, 53, 191.
13. Patton, J.T.; et. al. "Enhanced Oil Recovery by CO_2 Foam
    Flooding", Final Report; DOE/MC Report No. 03259-15, 1982.
14. Shinoda, K.; Yamanaka, T.; Kinoshita, K. J. Phys. Chem.,
    1959, 63, 648.
15. Corkill, J.M.; Goodman, J.F.; Ottewill, R.M. Trans. Faraday
    Soc., 1961, 57, 1627.
16. Hudson, R.A.; Pethica, B.A. Proc. 4th Intern. Congr. Surface
    Activity, 1964.
17. Balmbra, R.R.; Clunie, J.S.; Corkill, J.M.; Goodman, J.F.
    Trans. Faraday Soc., 1964, 60, 979.
18. Becher, P. Proc. 4th Intern. Congr. Surface Activity, 1964.
19. Elworthy, P.M.; MacFarlane, C.B. J. Chem. Soc., 1963, 987.
20. Schick, M.J.; Gilbert, A.M. J. Colloid Sci., 1965, 20, 464.
21. Weil, J.K.; Bistline, R.G.; Stirton, A.J. J. Phys. Chem.,
    1958, 62, 1083.
22. Chapman, N.B.; Shorter, J. "Correlation Analysis on Organic
    Chemistry", Plenum, New York, 1978, Tables 10.3.5 and 10.3.6.
23. Weil, J.K.; Bistline, R.G.; Maurer, E.W. J. Am. Oil Chem.
    Soc., 1959, 36, 241.
24. "Properties Guide NEODOL® Ethoxylates and Competitive
    Nonionics," Reference No. SC:569-82, Shell Chemical Company,
    1982.
25. Dillan, K.W.; Goddard, E.D.; McKenzie, D.A.; J. Am. Oil Chem.
    Soc., 1979, 57, 59-70.

RECEIVED January 29, 1988

Chapter 10

Adsorption of Binary Anionic Surfactant Mixtures on α-Alumina

Jeffrey J. Lopata, Jeffrey H. Harwell[1], and John F. Scamehorn

School of Chemical Engineering and Materials Science, University of Oklahoma, Norman, OK 73019

The adsorption of binary mixtures of anionic surfactants of a homologous series (sodium octyl sulfate and sodium dodecyl sulfate) on alpha aluminum oxide was measured. A thermodynamic model was developed to describe ideal mixed admicelle (adsorbed surfactant bilayer) formation, for concentrations between the critical admicelle concentration and the critical micelle concentration. Specific homogeneous surface patches were examined by considering constant levels of adsorption. This model was shown to accurately describe the experimental results obtained, as well as previously reported results of another binary anionic/anionic surfactant system. Theoretical predictions of ideal mixture adsorption can be made on an a priori basis if the pure component adsorption isotherms are known.

The adsorption of mixtures of surfactants on mineral oxide surfaces is important in detergency, flotation, and enhanced oil recovery (EOR), among other technologies. If the thermodynamics of mixed surfactant adsorption on mineral surfaces were known, it might be possible to formulate mixtures to either enhance or reduce the total surfactant adsorption, thereby saving reagent costs in flotation (1-3) and enhanced oil recovery (4). Technical problems, such as the selectivity of mineral separation (3,5) and the chromatographic separation of surfactant slugs (6,7), could also be systematically addressed. Surfactants are almost always used as mixtures in

[1]Correspondence should be addressed to this author.

0097–6156/88/0373–0205$06.00/0
© 1988 American Chemical Society

practical applications, since surfactants are usually manufactured as mixtures. These mixtures are often comprised of homologs of a surfactant series that differ only by the alkyl chain length, and they are usually treated theoretically as a single component surfactant, with the mean properties of the mixture. A thermodynamic knowledge of the adsorption of homologous mixtures of surfactants would allow better predictions to be made of the surfactant's performance.

Most oil-field experience with the use of surfactant derives from so-called micellar/polymer or chemical flooding EOR, in which surfactants are used to produce ultra-low interfacial tensions through the formation of a third phase, which coexists with the aqueous and oleic phase and contains most of the surfactant. By the phase rule, in a three-phase system of three components (or pseudocomponents) the compositions of the phases are fixed. So long as the system composition stays within the tietriangle, removal of one of the components from the system does not change the compositions or physical properties of the phases. This stabilizes the properties of the system against loss of surfactant from the fluid phases due to adsorption on the very large surface area of the porous rock (8). Nevertheless, loss of surfactant due to adsorption has long been a major concern in micellar/polymer EOR (9-11).

The surfactant systems used for mobility control in miscible flooding do not form a surfactant rich third phase, and lack its "buffering" action against surfactant adsorption. Furthermore, for obvious economic reasons, it is desirable to keep the surfactant concentration as low as possible, which increases the sensitivity of the dispersion stability to surfactant loss. Hence, surfactant adsorption is necessarily an even greater concern in the use of foams, emulsions, and dispersions for mobility control in miscible-flood EOR. The importance of surfactant adsorption in surfactant-based mobility control is widely recognized by researchers. A decision tree has even been published for selection of a mobility-control surfactant based on adsorption characteristics (12).

Pure component surfactant adsorption has been widely studied (9,13-24). Figure 1 illustrates the four distinct adsorption regions that exist when an anionic surfactant adsorbs on a positively charged surface. Region I is called the Henry's Law region and the surfactant molecules adsorb because of the electrostatic attraction between the surfactant head groups and the oppositely charged surface, and the interaction of the surfactant tail groups with the surface. At a critical concentration, the adsorption is greatly enhanced by the association of the surfactant tail groups. A two-dimensional phase transition is believed to take place on the highest energy patches on the solid surface. The concentration at which the first surfactant aggregate is

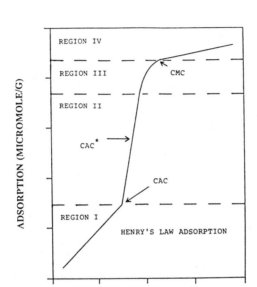

EQUILIBRIUM CONCENTRATION (MICROMOLE/L)

Figure 1. Typical Anionic Surfactant Adsorption Isotherm on a Positively Charged Mineral Oxide Surface.

formed is called the Critical Admicelle Concentration or
CAC (18), or the Hemimicelle Concentration or HMC
(20,23). At the CAC, the surfactant aggregate (or
admicelle) forms on the most energetic patch on the
surface. As the surfactant concentration increases,
successively less energetic patches have admicelles form
on them in Regions II and III. There is no fundamental
significance to the Region II to Region III transition;
Region III exists due to historical precedent as the
region where the adsorption increases less rapidly with
concentration on a log-log plot. With each adsorption
level in Region II, $_*$there is a corresponding equilibrium
concentration (CAC*) that is required to form an
admicelle on a patch of a specific energy level. Region
IV begins with the formation of micelles. Micelles act
as a chemical potential sink for any additional
surfactant added to the solution, thereby keeping the
monomer concentration nearly constant, and the adsorption
level nearly constant. Hence, Region IV is sometimes
called the plateau adsorption region (9).
     The adsorption of mixtures of surfactants has
received comparatively little attention. The adsorption
of mixtures of nonionic and anionic surfactants has been
studied (10,25-27) and strong negative deviations from
ideality were observed (10,27). Attempts to model the
degree of non-ideality using regular solution theory
failed (27). The adsorption of mixtures of anionic and
cationic surfactants would be expected to exibit even
larger deviations from ideality (28).
     Wilson and co-workers developed a statistical
mechanical model for single component surfactant
adsorption (29-31) and expanded it to a binary system
(2,3). Different adsorption curves were generated by
varying the Van der Waals interaction parameters. The
mixed adsorption equations that were developed were very
complex and were not applied to experimental data.
     Scamehorn et. al. expanded a single component
adsorption equation (9) to describe the adsorption of
binary mixtures of anionic surfactants of a homologous
series (11). Ideal solution theory was found to describe
the system fairly well. The mixed adsorption equations
worked very well in predicting the mixture adsorption,
but the equations were complex and would be difficult to
extend beyond a binary system.
     Scamehorn et. al. (32) also developed a reduced
adsorption equation to describe the adsorption of
mixtures of anionic surfactants, which are members of
homologous series. The equations were semi-empirical and
were based on ideal solution theory and the theory of
corresponding states. To apply these equations, a
critical concentration for each pure component in the
mixture is chosen, so that when the equilibrium
concentrations of the pure component adsorption isotherms
are divided by their critical concentrations, the
adsorption isotherms would coincide. The advantage of

these equations is that little adsorption data is required to make approximate mixture predictions.
   The adsorption of binary mixtures of anionic surfactants in the bilayer region has also been modeled by using just the pure component adsorption isotherms and ideal solution theory to describe the formation of mixed admicelles (33). Positive deviation from ideality in the mixed admicelle phase was reported, and the non-ideality was attributed to the planar shape of the admicelle. However, a computational error was made in comparison of the ideal solution theory equations to the experimental data, even though the theoretical equations presented were correct. Thus, the positive deviation from ideal mixed admicelle formation was in error.

Experimental

Materials. Sodium dodecyl sulfate ($C_{12}SO_4$) was purchased from Fisher Scientific with a manufacturer reported purity of at least 95.01%. The $C_{12}SO_4$ was recrystallized one time from reagent grade alcohol and water. Sodium decyl sulfate ($C_{10}SO_4$), from Eastman Kodak Company, was recrystallized twice using the same procedure as for the $C_{12}SO_4$. Sodium octyl sulfate (C_8SO_4), from Eastman Kodak Company, was recrystallized two times from boiling ACS grade 2-propanol and water, and then rinsed three times with ACS grade diethyl ether. The crystals were then dried in a hood for two days, and then under a vacuum for three days.
   The alpha aluminum oxide was purchased from Alpha Products, Thiokol/Ventron Division. The aluminum oxide had a particle size of 40 microns, a surface area of 160 m^2/g, and consisted of 90% Al_2O_3 and 9% H_2O, according to the manufacturer. The aluminum oxide was dried in 50 g batches under a vacuum for two days before use.
   Other materials used were ACS grade sodium chloride and sodium carbonate, HPLC grade methanol, and 0.1 N and 0.01 N hydrochloric acid and sodium hydroxide solutions. All of these chemicals were purchased from Fisher Scientific. The water was distilled and deionized.

Methods. The brine solutions that were used to make the sample solutions were 0.15 M sodium chloride and contained 0.0015 g/l sodium carbonate. Sodium carbonate was added to buffer the solutions against the carbonic acid that formed upon the absorption of carbon dioxide in the solutions. The pH of the brine solutions were adjusted to an initial value of 7.8, so that when the solution was equilibrated with the alumina, the equilibrium pH was 8.4. Ten milliliters of the sample solution were then pipeted to test tubes which contained 0.5 grams of aluminum oxide. The test tubes were then centrifuged for 45 minutes at 1000 rpm and placed in a water bath that was kept at a constant temperature of $30^{\circ}C$. After four to five days, the pH values of the equilibrated samples were measured and the supernatants of the samples were removed from the solids.

The concentrations of the anionic surfactants (C_8SO_4, $C_{10}SO_4$, and $C_{12}SO_4$) in the initial and equilibrated samples were measured using a high performance liquid chromatograph with a conductivity detector. Details of the analytical procedure may be found elsewhere (34).

Theory

The pure component adsorption isotherms of the surfactants used in this study (C_8SO_4, $C_{10}SO_4$, and $C_{12}SO_4$) are shown in Figure 2. All of the adsorption isotherms were continuous in Regions II and III, which suggests that the distribution of energy level patches on the surface of the alpha aluminum oxide was nearly continuous.

The admicelle standard states are defined as the equilibrium monomer concentrations that are required to form the pure admicelles on a specific energy level patch. This particular patch is assumed to correspond to the same adsorption level for either pure surfactants or mixtures in Regions II and III. The equilibrium monomer concentration that is required to form the mixed admicelles on the same energy level patch, is then compared to the monomer concentration predicted from ideal solution theory. By defining the standard states at constant levels of adsorption, we can look at one homogeneous energy level patch on the surface at a time. As we look at increasingly higher adsorption levels, the effects of lower energy level patches and increasing total surface coverage on admicelle formation can be examined. The admicelle standard states were determined from the pure component adsorption isotherms by reading the equilibrium monomer concentrations that corresponded to an arbitrary adsorption level of interest (33).

The approach used to develop the ideal solution theory equations to describe binary mixed admicelle formation, was similar to that used by Roberts et. al. (33). The total monomer equilibrium concentration that it takes to reach a specified level of adsorption was used as the variable which was predicted from the model. The partial fugacities can be written for both the monomer and admicelle phases:

$$f_i^{mon} = Y_i \gamma_i^{mon} CAC_m^* \tag{1}$$

$$f_i^{adm} = Z_i \gamma_i^{adm} CAC_i^* \tag{2}$$

where f_i^{mon} and f_i^{adm} are the partial fugacities of component i in the monomer and admicelle phases, respectively; Y_i is the surfactant-only based mole fraction of component i in the equilibrium solution; Z_i is the surfactant-only based mole fraction of component i

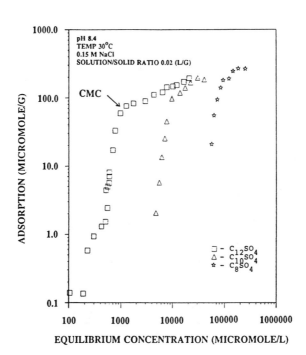

Figure 2.   Adsorption Isotherms of C_8SO_4, $C_{10}SO_4$, and $C_{12}SO_4$ on Alpha Aluminum Oxide.

in the mixed admicelle; γ_i^{mon} and γ_i^{adm} are the activity coefficients of component i in the monomer and admicelle phases, respectively; CAC_m^* is the total equilibrium monomer concentration at the specified total adsorption level; and CAC_i^* is the pure surfactant i monomer concentration at the specified total adsorption level. The asterisks above the critical admicelle concentrations (CAC_m^* and CAC_i^*) designate that these concentrations correspond to admicelle formation on a specific energy level patch, not necessarily the onset of aggregate formation, (which is designated by CAC). The monomer concentration was extremely dilute, and swamping electrolyte was always present in the experiments. Therefore, the monomer was assumed to obey Henry's Law ($\gamma_i^{mon} = 1.0$).

At equilibrium, the partial fugacities of a component are equal in every phase. Therefore, Equations 1 and 2 can be equated to form Equation 3, which is valid for each surfactant component in the mixture.

$$z_i \gamma_i^{adm} CAC_i^* = Y_i CAC_m^* \qquad (3)$$

It is desirable to obtain an a priori prediction of the total equilibrium monomer concentration (CAC_m^*) at set levels of adsorption based on the mixture feed mole fractions, instead of the equilibrium monomer mole fractions (Y_i). The equilibrium monomer mole fractions will differ from the feed mole fractions because of the preferential adsorption of some of the surfactants in the mixture. A mass balance on component i in the feed, equilibrium solution, and adsorbed phase is solved for the equilibrium monomer mole fraction to obtain Equation 4:

$$Y_i = \frac{Q_i (CAC_m^* V + \Gamma W_g) - z_i \Gamma W_g}{CAC_m^* V} \qquad (4)$$

where Q_i is the feed mole fraction of component i, V is the volume of the liquid sample, Γ is the total adsorption, and W_g is the weight of the aluminum sample. Equations can also be written for the surfactant-only based mole fractions in the monomer, admicelle, and feed. For a binary system,

$$Y_1 + Y_2 = 1.0 \qquad (5)$$

$$z_1 + z_2 = 1.0 \qquad (6)$$

$$Q_1 + Q_2 = 1.0 \qquad (7)$$

If ideal solution theory for mixed admicelle formation is assumed ($\gamma_i^{adm} = 1.0$), six independant equations can be written for a binary system. The set of equations consists of Equation 3 written for components 1 and 2, Equation 4 written for component 1, and Equations 5, 6, and 7. The six equations can be solved directly for the six unkowns (Y_1, Y_2, Z_1, Z_2, Q_2, and CAC_m^*) to completely describe the system. Equation 3 written for components 1 and 2 and Equations 4 - 7 can be combined to result in Equation 8.

$$CAC_m^* = \frac{CAC_1^* W_g Q_1 + CAC_2^* (CAC_1^* V + \Gamma W_g (1-Q_1))}{V(1-Q_1)CAC_i^* + VCAC_2^* Q_1 + \Gamma W_g} \quad (8)$$

Once CAC_m^* is known, Y_1 and Z_1 can be calculated by a simultaneous solution of Equations 3 and 4, and then Y_2 and Z_2 can be calculated from Equations 5 and 6, respectively. Equations 3-7 can easily be expanded to more than two components, and can describe non-ideal mixing in the admicelle phase by inserting the correct expressions for the admicellar activity coefficients for each component.

Except at concentrations near the CAC, the amount of surfactant adsorbed in the Henry's law region is small in comparison to the amount of surfactant present in the admicelles. This implies that nearly all of the adsorbed surfactant molecules are associated on the mineral surface in the form of admicelles. It is important to keep in mind that Equation 8 is valid only between the CAC and the CMC. Above the CMC, equations could be included to account for the formation of micelles, by including monomer-micelle equilibrium equations (11).

The only information needed to predict the mixture surfactant concentration to attain a specified adsorption level is the pure component adsorption isotherms measured at the same experimental conditions as the mixture isotherms. These isotherms are needed to obtain the pure component standard states.

The method of predicting the mixture adsorption isotherms is to first select the feed mole fractions of interest and to pick an adsorption level within Region II. The pure component standard states are determined from the total equilibrium concentration that occurs at that set level of adsorption for the pure surfactant component adsorption isotherms. The total equilibrium mixture concentration corresponding to the selected adsorption level is then calculated from Equation 8. This procedure is repeated at different levels of adsorption until enough points are collected to completely descibe the mixture adsorption isotherm curve.

Results and Discussion

The three pure component adsorption isotherms of the homologous series of sodium alkyl sulfates (C_8SO_4, $C_{10}SO_4$, and $C_{12}SO_4$) in Figure 2, were plotted at their reduced concentrations in Figure 3, to extrapolate the C8SO4 data to its CAC. The adsorption data for C8SO4

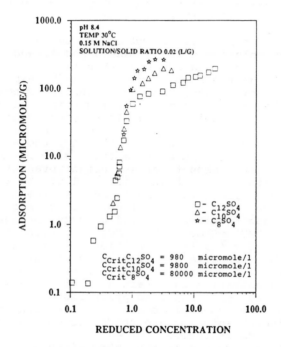

Figure 3.   Reduced Adsorption Isotherms of C_8SO_4, $C_{10}SO_4$, and $C_{12}SO_4$ on Alpha Aluminum Oxide.

could not be measured below a certain concentration and the data was needed to obtain the C_8SO_4 standard state concentrations. The critical concentrations used to calculate the reduced concentrations were chosen for the best coincidence of the reduced adsorption isotherms. As seen in Figure 3, the corresponding states theory (32) worked very well for the pure component surfactants used in this study.

The adsorption isotherms of C_8SO_4 and $C_{12}SO_4$, and mixtures thereof, on alpha aluminum oxide are illustrated in Figure 4. Also shown in Figure 4 is the extrapolation of the C_8SO_4 pure component adsorption isotherm down to its CAC and the ideal solution theory predictions.

The agreement between the mixture adsorption data and ideal solution theory is excellent. It is important to remember that while looking at various constant levels of adsorption in Region II, we are looking at the CAC of the mixed admicelle that has just formed on a particular patch. By looking at different adsorption levels, we are looking at how the two surfactants interact on different energy level patches on the surface.

Equation 8 was also used to predict the mixture adsorption isotherms in Region II of $C_{10}SO_4$ and $C_{12}SO_4$ on gamma aluminum oxide and compared to experimental data from Roberts et. al. (33). Figure 5 is a plot of the total adsorption versus the total equilibrium concentration for two pure component and three mixture mole fractions from that work. The conditions under which the data were obtained were identical to those in this study, except for the substrate (gamma alumina) and the equilibrium pH (6.8). Once again, agreement between the data and ideal solution theory is excellent. It is no surprise that the adsorption of mixtures of $C_{10}SO_4$ and $C_{12}SO_4$ follow ideal solution theory, because the adsorption of mixtures of C_8SO_4 and $C_{12}SO_4$ follow ideal solution theory, and $C_{10}SO_4$ and $C_{12}SO_4$ are closer members of the homologous series. Roberts et. al., reported a positive deviation from ideality for the adsorption of mixtures of $C_{10}SO_4$ and $C_{12}SO_4$ due to a miscalculation, and attributed it to the planer shape of the admicelle. The results presented here clearly shows that the hydrophobic bonding in the admicelle is similar for mixed and single component admicelles.

Micelles and monolayers composed of homologous mixtures of anionic surfactants can be approximately described by ideal solution theory to model the mixed surfactant aggregate (35). Therefore, it is not surprising that mixed admicelles composed of these surfactants also obey ideal solution theory. It is also important to note that this is true at all adsorption levels within Region II, as seen by the excellent agreement between theory and experiment in Figures 4 and 5.

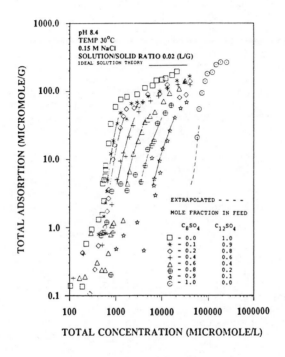

Figure 4. Mixure Adsorption Isotherms of C_8SO_4 and $C_{12}SO_4$ on Alpha Aluminum Oxide.

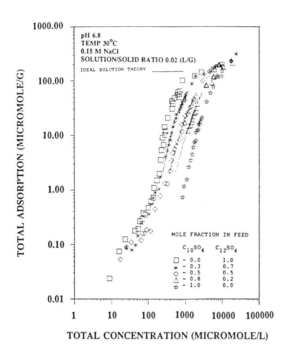

Figure 5. Mixture Adsorption Isotherms of $C_{10}SO_4$ and $C_{12}SO_4$ on Gamma Aluminum Oxide (Data from 33).

Acknowledgments

Financial support for this work was provided by Mobil Research and Development Corp., The Shell Development Co., Arco Oil and Gas Company, DOE Contract No. 19-85BC10845.000, The OU Energy Resources Institute, and The Oklahoma Mining and Minerals Resources Research Institute.

Literature Cited

1. Novich, B. E.; Ring, T. A. Langmuir 1985, 1, 701.
2. Wilson, D. J. Sep. Sci. Technol. 1982, 17, 1219.
3. Wilson, D. J.; Carter, K. N. Sep. Sci. Technol. 1983, 18, 657.
4. Meldrau, R. F., et. al. J. Pet. Technol. 1983, 35, 1279.
5. Aplan, F. F. In Flotation; Kirk-Othmer Encyclopedia of Chemical Technology; Wiley, Inc.: New York, Vol. 10, p 523.
6. Sharma, M. K.; Shah, D. O.; Brigham, W. E. SPE Reservoir Eng. 1986, 1, 253.
7. Harwell, J. H.; Helfferich, F. G.; Schechter, R. S. AIChE J. 1982, 28, 448.
8. Smith, D. H. SPE/DOE Paper No. 14914, Tulsa, April, 1986.
9. Scamehorn, J. F.; Schechter, R. S.; Wade, W. H. J. Colloid Interface Sci. 1982, 85, 463.
10. Scamehorn, J. F.; Schechter, R. S.; Wade, W. H. J. Colloid Interface Sci. 1982, 85, 494.
11. Scamehorn, J. F.; Schechter, R. S.; Wade, W. H. J. Colloid Interface Sci. 1982, 85, 479.
12. Wellington, S. L.; Vinegar, H. J. SPE Paper No. 14393, Las Vegas, September, 1985.
13. Hanna, H. S.; Somasundaran, P. J. Colloid Interface Sci. 1979, 70, 181.
14. Levitz, P.; Damme, H. V. J. Phys. Chem. 1986, 90, 1302.
15. Koopal, L. K.; Keltjens, L. Colloids Surf. 1986, 17, 371.
16. Rosen, M. J.; Nakamura, Y. J. Phys. Chem. 1977, 81, 873.
17. Trogus, F. J.; Sophany, T.; Schechter, R. S.; Wade, W. H. Soc. Pet. Eng. J. 1977, 17, 337.
18. Harwell, J. H.; Hoskins, J. C.; Schechter, R. S.; Wade, W. H. Langmuir 1985, 1, 251.
19. Dick, S. G.; Fuerstenau, D. W.; Healy, T. W. J. Colloid Interface Sci. 1971, 37, 595.
20. Somasundaran, P.; Fuerstenau, D. W. J. Phys. Chem. 1966, 70, 90.
21. Wakamatsu, T.; Fuerstenau, D. W. in Adv. Chem. Ser. 1968, 79, 161.

22. Fuerstenau, D. W.; Wakamatsu, T. Farraday Discuss. Chem. Soc. 1976, 59, 157.
23. Chandar, S.; Fuerstenau, D. W.; Stigter, D. In Adsorption From Solution; Ottewill, R. H.; Rochester, C. H.; Smith, A. L., Eds.; Academic Press: London, 1983; p 197.
24. Dobias, B. Colloid Polym. Sci. 1978, 256, 465.
25. Schwuger, M. J.; Smolka, H. G. Colloid Polym. Sci. 1977, 255, 589.
26. Gao, Y.; Yue, C.; Lu, S.; Gu, W.; Gu T. J. Colloid Interface Sci. 1984, 100, 581.
27. Roberts, B. L.; Harwell, J. H.; Scamehorn, J. F. Colloids Surf., in press.
28. Schwuger, M. J. Kolloid - Z. Z.Polym. 1971, 243, 129.
29. Wilson, D. J. Sep. Sci. 1977, 12, 447.
30. Wilson, D. J.; Kennedy, R. M. Sep. Sci. Technol. 1979, 14, 319.
31. Kieffer, J. E.; Wilson, D. Sep. Sci. Technol. 1980, 15, 57.
32. Scamehorn, J. F.; Schechter, R. S.; Wade, W. H. J. Am. Oil Chem. Soc. 1983, 60, 1345.
33. Roberts, B. L.; Scamehorn, J. F.; Harwell, J. H. In Phenomena in Mixed Surfactant Systems; Scamehorn, J.F., Ed.; ACS Symp. Ser. Vol. 311, American Chemical Society: Washington, D. C., 1986, p 200.
34. Lopata, J. J. M.S. Thesis, University of Oklahoma, Oklahoma, 1987.
35. Scamehorn, J. F. In Phenomena in Mixed Surfactant Systems; Scamehorn, J. F. Ed.; ACS Symp. Ser., Vol. 311, American Chemical Society: Washington, D. C., 1986, p 1.

RECEIVED January 5, 1988

Chapter 11

Adsorption of Nonionic Surfactants on Quartz in the Presence of Ethanol, HCl, or CaCl$_2$

Its Effect on Wettability

A. M. Travalloni Louvisse and Gaspar González

Petrobrás/Cenpes, Cidade Universitária, Q–7, Rio de Janeiro–RJ, Brazil

The effects of HCl, CaCl$_2$, and ethanol on the adsorption
of Triton X-100 on silica, glass, and quartz have been
investigated. Ethanol and HCl reduced the adsorption of
surfactant by the solid; CaCl$_2$ presented the opposite
effect. The results have been attributed to changes in
the solubility behavior of the surfactant in the presence of these compounds. Measurements of the cloud
point and critical micelle concentration (CMC) of the
surfactant solutions under similar conditions confirmed
this view. The adsorption process resulted in a hydro-
phobization of the solid. Measurements of contact
angles on glass and quartz indicate that this transition
takes place at a very low surfactant concentration
($\theta = 26°$ for $C = 10^{-6}$ molar), and that ethanol and HCl
reduce this value of contact angle. The reverse transition to zero contact angle was effective at concentrations well below the critical micelle concentration,
indicating that association of the surfactant at the
solid-solution interface precedes the CMC. The results
are discussed in terms of their relevance to some surfactant mediated (enhanced) oil recovery processes.

Nearly all of the treatment processes in which fluids are injected
into oil wells to increase or restore the levels of production
make use of surface-active agents (surfactant) in some of their
various applications, e.g., surface tension reduction, formation
and stabilization of foam, anti-sludging, prevention of emulsification, and mobility control for gases or steam injection. The
question that sometimes arises is whether the level of surfactant
added to the injection fluids is sufficient to ensure that enough
surfactant reaches the region of treatment. Some of the mechanisms which may reduce the surfactant concentration in the fluid
are precipitation with other components of the fluid, thermally
induced partition into the various coexisting phases in an oil-
well treatment, and adsorption onto the reservoir walls or mineral

particles. Some of these processes may be predicted and prevented by an adequate formulation of the fluids and a proper control of the additives. The effect of adsorption, however, is difficult to evaluate due to the complexity of the rocks forming the oil-bearing geological formations. It is to this aspect of the rock-fluid interaction that this work was directed. We studied the solution behavior of Triton X-100, a nonionic surfactant of structure similar to those used in completion and stimulation fluids, its adsorption at the silica-solution interface, and the effect of these processes on the concentration of surfactant in the fluids injected in porous media.

Experimental

Materials -- The surfactant used in this study was an octyl phenol exthoxylated with 9-10 oxyethylenic units. The product was manufactured by J. T. Baker Chemical Company, and marketed as Triton X-100. Most of the surfactants used as additives for injection fluids present a structure similar to Triton X-100.

For the adsorption tests, a sample of silica (Merk) with a specific surface area of 388 m^2g^{-1} measured by the BET method was used. The solid specimens used to measure the contact angles were microscope glass slides and pieces of quartz polished using a rotating plate covered with a polishing cloth impregnated with 10 μm diamond polishing particles.

These solids were treated with concentrated acid, thoroughly rinsed with double distilled water, and ultrasonicated to remove particles which may have remained deposited on the solid surface.

All the chemicals, including the ethanol, were Analar grade. All water used was twice distilled from an all-Pyrex still.

Experimental Procedure -- The surface tensions of Triton X-100 solutions, with or without other additives, were measured using a du Nouy ring tensiometer (Fisher Scientific Company) at 25°C.

The temperature at which a 1% surfactant solution became turbid ("cloud point") in pure water or in the presence of additives was determined by visual observation of the solution contained in a test tube as the temperature was increased. For this purpose, the test tubes were placed in a transparent water bath equipped with a control unit which permitted temperature increases at the rates of 3°/min and 0.2°/min. A dark background permitted a sharper estimation of cloud points.

The amount of surfactant adsorbed by silica was determined by measuring the difference between the initial and final concentration of 25 ml of surfactant solution after 3 hours of contact with one gram of silica. The surfactant concentration was determined from the adsorbance at 281 nm in 10 mm optical-path quartz cells, using a double-beam Varian 634 spectrophotometer. An alternative procedure used for some cases was based upon the fact that the γ-log C

diagram is roughly a straight line in the range 3.2×10^{-5} to
2×10^{-4} molar. The surfactant solutions were diluted to reach a
value of surface tension corresponding to a solution of composition within this range. Hence, as the dilution factor was known
it was possible to calculate the concentration of the original
solution. Both methods led to similar results.

The contact angles at the quartz- (or glass) water-air interface
were measured using a modified captive-bubble apparatus (1).

Results and Discussion

Solution Behavior of Triton X-100 -- Figures 1 and 2 show the surface tension of aqueous solutions of Triton X-100 and the effect
of various additives at 25°C. The critical micelle concentration
(CMC) for the surfactant in water was 2.8×10^{-4} mol dm^{-3}. This
figure agrees with results reported by other authors. The graph
shows a rather shallow minimum in the CMC which is usually interpreted as evidence of some degree of polydispersity (2). Curves b
and c show the effect of ethanol at 10 and 25 vol%. The reduction
of surface tension for low-surfactant concentration corresponds to
the effect of ethanol. For higher concentrations, the surfactant
is preferentially adsorbed at the liquid surface, displacing the
alcohol from the interface, However, as the slope of the γ-log
C curve in the CMC region is lower in the presence of alcohol, it
seems that this displacement is not complete. Nishikido studied
this effect for various oxyethylenic surfactants and concluded
that the ratio between the maximum adsorption of surfactant in
the presence of alcohols and its corresponding value in pure water
decreased linearly with the mole fraction of the alcohol (3). The
results of Figure 1 also indicate that ethanol does not modify to
a great extent the CMC of Triton X-100. According to the same
author, two opposite effects are responsible for this behavior;
the solvency effect, which would increase the CMC, and a partial
penetration of the alcohol into the micelle structure, which would
result in a reduction of the CMC. Hydrocholoric acid increases
the CMC of Triton X-100, and calcium chloride reduces its value.
The changes shown in Figure 2 may be attributed to changes in the
surfactant solubility in electrolyte solutions.

Nonionic surfactants dissolve in aqueous solutions through hydrogen bonding between the water molecules and the oxyethylenic portion of the surfactant. These interactions are weak but enough
in number to maintain the molecule in solution up to the cloud
point temperature, at which the surfactant separates as a different phase (4). Figure 3 shows that electrolytes like calcium
chloride, potassium chloride, or sodium chloride reduce the cloud
point of Triton X-100. Hydrochloric acid instead promoted a
salting-in effect similar to that observed for ethanol.

According to Schott, et al., H^{+} and all bivalent and trivalent
cations form complexes with the ether linkages of nonionic surfactants, increasing their solubility (5). Among the anions, only
large, polarizable ions like iodide and thiocyanate break the

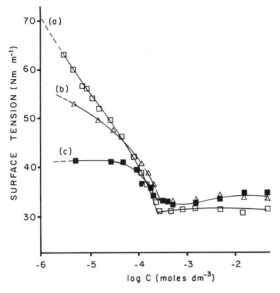

Figure 1.   Plots of the surface tension versus logarithm of the surfactant concentration for Triton X-100 in (a) pure water, (b) 10% ethanol, and (c) 25% ethanol.

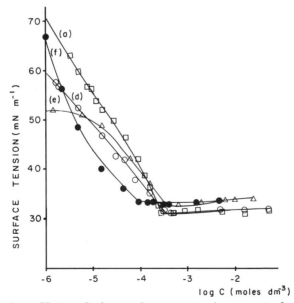

Figure 2.   Plots of the surfactant tension versus logarithm of the surfactant concentration for Triton X-100 in (a) pure water, (d) 1 mole dm^{-3} HCl, (e) 1 mole dm^{-3} HCl, 10% ethanol, and (f) 10 wt% $CaCl_2$.

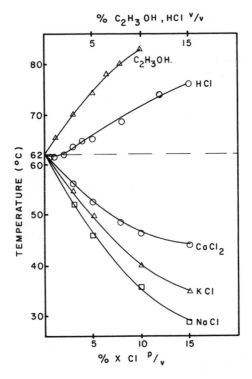

Figure 3. Effect of different additives on the cloud point of Triton X-100 (1 wt%) solutions of ethanol, HCl, CaCl$_2$, KCl, and NaCl.

structure of water to reduce the extent of hydrogen bonding between water molecules to favor the hydration of the ether linkages via hydrogen bonding. Chloride has a very strong salting-out effect, which determines the reduction of the cloud point observed for $CaCl_2$ in spite of the salting-in effect of Ca^{++}.

Adsorption Behavior of Triton X-100 -- The adsorption of Triton X-100 onto silica is shown in Figures 4 and 5. The results indicate that there are three well-defined regions in the isotherm. Initially, for low concentrations of surfactant, the adsorption is rather low. Following this region there is a sudden increment in the adsorption, and finally a saturation plateau is reached for an equilibrium concentration of 6 x 10^{-4} moles dm^{-3} for the case of pure aqueous solutions. The amount adsorbed in the plateau is 1.3 μ moles m^{-2}. This figure is lower than the results reported by Doren, et al., for quarts, but compares well with results of Rouquerol and van den Boomgaard (6-8). This value of adsorption corresponds to a molecular cross section of 1.2 nm^2. The area per molecule in a compact monolayer at the air-water interface, calculated from the results of Figure 1 is 2.22 nm^2. These figures indicates that the adsorption plateau at the solid-solution interface is attained when a bilayer of surfactant is formed on the solid. This association process takes place at the interface before the formation of micelles in solution. Scamehorn and Schechter, et al., have studied in detail the adsorption of ionic surfactants on polar solids, and they denominated as "admicelles" this bilayer structure to distinguish it from the "hemimicelles" described by Gaudin (9,10). The results of Figure 4 indicate that the mechanism of adsorption for nonionics on silica follows a similar overall pattern.

Calcium chloride, hydrochloric acid, or ethanol at low concentrations do not modify the adsorption plateau; nevertheless, the saturation is attained at a different equilibrium concentration due to the changes in the solubility behavior of the surfactant in the presence of these additives. When the concentration of ethanol is 25% (6 molar), the adsorption maximum is reduced to 0.8 μ moles m^{-2}, indicating that for this rather high concentration of ethanol, the surfactant does not completely dislocate the alcohol from the solid surface.

The Contact Angle at the Quartz (Glass) - Solution Interface in the Presence of Triton X-100 -- The values of receding contact angle as a function of the concentration of Triton X-100 in water (Curve a) or in the presence of ethanol (Curves b and c) are shown in Figure 6. The shape of the curves corresponds to a typical wetting isotherm for polar, high-energy solids (11). In this case, the contact angle increases for low-surfactant concentration, reflecting the adsorption of single molecules onto the polar groups of the solid. After reaching a maximum, the contact angle decreases as a consequence of the formation of a second layer of surfactant with their polar groups oriented toward the liquid phase. The addition of ethanol reduces the receding angles, but the shapes of the wetting isotherms are still maintained.

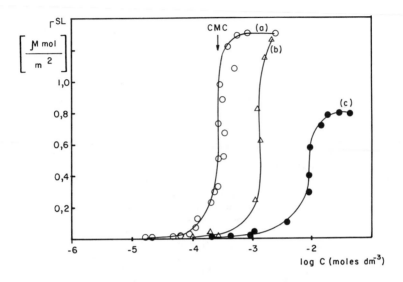

Figure 4.   Adsorption isotherms for Triton X-100 on silica
             (a) pure water, (b) 10% ethanol, and (c) 25% ethanol.

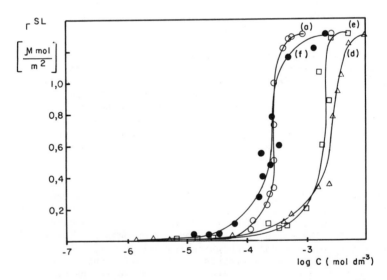

Figure 5.   Adsorption isotherms for Triton X-100 on silica
             (a) pure water, (d) 1 mol dm^{-3} HCl, (e) 1 mol dm^{-3}
             HCl, 10% ethanol, and (f) 10 wt% CaCl$_2$.

Capillary Pressure of Fluids Containing Ethanol and Triton X-100 -- The capillary pressure, P, of a fluid in a porous medium is defined by the relation (12)

$$\Delta P = \frac{2\gamma^{LV} \cos\theta}{r_e} \tag{1}$$

where γ^{LV} represents the surface tension of the fluid, θ is the contact angle, and r_e is the effective radius of the porous medium. The term $2\gamma^{LV} \cos\theta$ represents the product between the capillary pressure and the effective radius of the porous medium. From Figures 1 and 6 it is possible to calculate the product $2\gamma^{LV} \cos\theta$ as a function of the concentration of Triton; the results are shown in Figure 7. It seems clear that the addition of ethanol reduces the capillary pressure for low concentrations of surfactant, and this effect substantiates the use of alcohols in work over fluids when it is necessary to remove aqueous fluids trapped in the formation by capillary forces. Recent field tests indicate that water blockages may be effectively removed by the use of perfluorocarbonate surfactants in connection with ethanol and hydrochloric acid (13).

The Effect of Adsorption on the Levels of Surfactant in Injection Fluids -- From the results of Figures 4 and 5 it is possible to assess the changes in the concentration of surfactant as a consequence of its adsorption on the minerals forming the reservoir. The results reported in this paper are for quartz, silica, or glass; nevertheless, these solids probably represent the worst cases, because the adsorption of nonionics on alumina (and by extension onto alumino silicates) and on calcareous minerals is lower than the adsorption onto quartziferous materials (G. González, A. Travalloni, in progress).

To calculate the reduction in the concentration of surfactant in the fluid by adsorption it is necessary to have an estimation of the inner surface area of the reservoir. This parameter is related to the porosity of the medium and to its permeability. Attempts have been made to correlate these two quantities; but the results have been unsuccessful, because there are parameters characteristic of each particular porous medium involved in the description of the problem (14). For our analysis we adopted the approach of Kozeny and Carman (15). These authors defined a parameter called the "equivalent hydraulic radius of the porous medium" which represents the surface area exposed to the fluid per unit volume of rock. They obtained the following relationship between the permeability, k, and the porosity, θ:

$$k = \frac{\theta^3}{5 \, S_o^2 \, (1-\theta)^2} \tag{2}$$

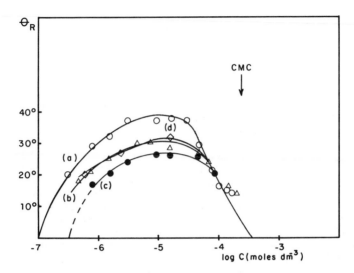

Figure 6.   Receding contact angle at the quartz-aqueous solution-
             air interface versus logarithm of the surfactant con-
             centration in (a) pure water, (b) 10% ethanol,
             (c) 25% ethanol, and (d) 1 mol dm^{-3} HCl.

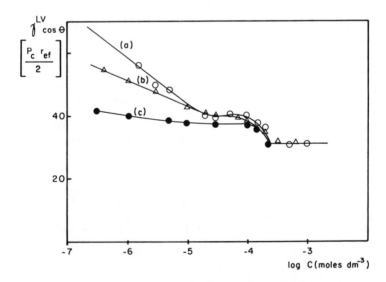

Figure 7.   Calculated capillary pressures for aqueous solutions
             of Triton X-100 in a quartziferous porous media versus
             logarithm of the surfactant concentration in (a) pure
             water, (b) 10% ethanol, and (c) 25% ethanol.

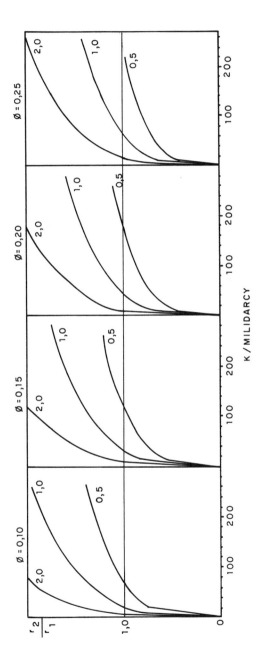

Figure 8. Ratio between the radius of penetration of the surfactant and the radius of penetration of the fluid as a function of the permeability (k) porosity (∅) and concentration of surfactant.

From Equation 2 it is possible to calculate the inner surface area
of the porous medium, S_o. Hence, from the adsorption data of Figure 4 it is possible to calculate the volume of rock necessary to
adsorb all of the surfactant and the total volume of rock attained
by the fluid. Assuming a radial penetration, these two volumes
may be transformed into the radius of penetration of the fluid as
a whole, r_1, and the radius necessary to consume the surfactant,
r_2. The ratio r_2/r_1 will be higher than one if the surfactant is
still available when the fluid reaches its programmed penetration
distance. Figure 8 shows the ratio r_2/r_1 as a function of the
permeability for three different porosities and four different
concentrations of surfactant. For low-porosity solids it is
almost certain that the surfactant will be present in all the
fluid penetration distances, and in this case the use of 0.2, 0.1,
or 0.05 vol% surfactant solution does not result in important
differences. The level of surfactant becomes important for porous
media of high porosity and low permeability. In these cases, due
to the tortuosity of the porous medium, the surfactant will be
depleted unless its concentration in the fluid is rather high.

The addition of ethanol increases the penetration distance of the
surfactant; this is shown for various porosities in Figure 9 where
the ratio r_2/r_1 is plotted as a function of the permeability for a

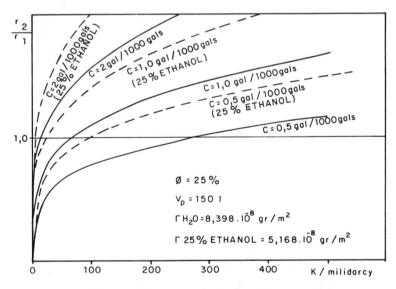

Figure 9. Ratio between the radius of penetration of the surfactant and the radius of
penetration of the fluid as a function of the permeability (k) and concentration of surfactant
(indicated in each figure) in water (____) and in 25% ethanol (--) under the conditions
O = 25%, treatment volume = 150 L, and maximum adsorption as in Figure 1.

porosity of 25% and a solution containing 25% ethanol. It seems reasonable to conclude that for fluids containing alcohols it may be possible, under certain circumstances, to reduce the concentration of surfactant without reducing its penetration distance in porous media. This behavior is related to the cosolvency effect of ethanol and to the reduction of the maximum adsorption shown in Figure 4.

Literature Cited

1. Adamson, A.W. Physical Chemistry of Surfaces. 3rd ed., New York, J. Wiley and Sons, 1976.

2. Crook, E.M.; Fordyce, D.B.; Tregzi, G.F. J. Phys. Chem., 1963, 67, 1987.

3. Nishikido, N. J. Colloid Interface Sci., 1986, 112(1), 87.

4. Shinoda, K.; Saito, H. J. Colloid Interface Sci., 1968, 26, 70.

5. Schott, H.; Royce, A.E.; Suck, K.H. J. Colloid Interface Sci., 1984, 98(1), 196.

6. Doren, A.; Vargas, D.; Goldfarb, J. Trans. IMM, 1975, 84, C33.

7. Requerol, J.; Partyka, S.; Requerol, F. In Adsorption at the Gas-Solid and Liquid-Solid Interface. J. Requerol; K.S.W. Sing, Eds., Elsevier Sci. Pub. Co.: Amsterdam, 1982, p. 69.

8. Van den Boomgaard, Th.; Tadros, Th.F.; Kyklema, J. J. Colloid Interface Sci., 1987, 116(1), 8.

9. Scamehorn, J.F.; Schechter, R.S.; Wade, W.H. J. Colloid Interface Sci., 1982, 85(2), 463.

10. Gaudin, A.M.; Fuerstenau, D.W. Trans. AIME, 1955, 202, 958.

11. Wolfram, E. Adhesion at Solid-Liquid Interface. Lecture Course at University of Bristol, England, 1977/1978.

12. Defay, R.; Prigogine, I. Tension Superficielle et Adsorption. Liege: Desoer, 1951, pp. 6-9.

13. González, G.; Dubai, A. Redutores de tensão superficial para fluidos de injeção, Relatório Técnico CENPES, 1987.

14. Kinghorn, R.R.F. An Introduction to the Physics and Chemistry of Petroleum. J. Wiley and Sons: New York, 1983.

15. Scheidegger, A.E. The Physics of Flow through Porous Media. University of Toronto Press: Toronto, 1973, p. 141.

RECEIVED March 3, 1988

MECHANISMS AND THEORY
OF DISPERSION FLOW

Chapter 12

Foam Formation in Porous Media

A Microscopic Visual Study

Arthur I. Shirley[1]

**Research and Technical Services, ARCO Oil and Gas Company,
2300 West Plano Parkway, Plano, TX 75075**

A microscopic examination of foam flow in model
porous media is presented. Experiments were
conducted examining foam formation and displacement
efficiency in glass micromodels with widely different
pore properties. The micromodels were designed: (1)
by petrographic image analysis to generate complex
pore geometries, and (2) by drafting in order to
obtain more homogeneous pore geometries.

A study of the effect of pore geometry on foam
formation mechanisms shows that "snap-off" bubble
formation is dominant in highly heterogeneous pore
systems. The morphology of the foams formed by the
two mechanisms are quite different. A comparison of
two foam injection schemes, simultaneous
gas/surfactant solution injection (SI) and alternate
gas/surfactant solution injection (GDS), shows that
the SI scheme is more efficient at controlling gas
mobility on a micro-scale during a foam flood.

High gas mobility and low sweep efficiency are typical problems
encountered in oil recovery processes using gas injection.
Improving gas mobility to increase reservoir sweep can add
significantly to recoverable reserves and impact the economics of
these tertiary recovery processes[1]. One technique that has
received much attention as a mobility control aid is foam
flooding. In this process the injected gas is dispersed in a
liquid containing a surface-active agent, forming a foam. The foam
is composed of a stable configuration of microscopic bubbles that
behave as a very viscous material when the foam is
made to flow in a porous medium. This high apparent viscosity

[1]Current address: Group Technical Center, The BOC Group, Inc., 100 Mountain Avenue,
Murray Hill, New Providence, NJ 07974

is responsible for the often sizeable reduction in gas mobility
given by foams, which may have mobilities of less than 1/10000 of
the gas mobility[2]. Foams can improve areal sweep and also limit
gravity override by gases[3,4].

A high-pressure micromodel system has been constructed to visually
investigate foam formation and flow behavior. This system uses
glass plates, or micromodels, with the pattern of a pore network
etched into them, to serve as a transparent porous medium. These
micromodels can be suspended in a confining fluid in a pressure
vessel, allowing them to be operated at high pressure and
temperature. Because of this pressure capability, reservoir fluids
can be used in the micromodel, and any effects of phase behavior or
pressure- and temperature-dependent properties on foam flow can be
examined.

Micromodels have been used in previous studies of foam in porous
media to observe bubble formation and size distribution as well as
the effect of injection scheme or flow heterogeneities (13-17).
These experiments were conducted at low pressure and temperature.
For studying foams with dense hydrocarbon gases, and for examining
the effect of live oils or miscibility, the high-pressure and
temperature system is necessary. Both low- and high-pressure and
temperature micromodel systems have been used in studying other
recovery techniques such as waterflooding[5], immiscible-gas
flooding[6], low-interfacial-tension flooding[7], micellar
flooding[8-9], and first-contact and multicontact miscible
flooding[10-12]. In each case they have provided valuable pictures
of fluid-fluid interactions on the microscopic scale.

The high-pressure and temperature micromodel system has been used
in this study to investigate the formation, flow behavior and
stability of foams. Micromodel etching patterns were made from
binary images of rock thin sections and from other designs for a
comparison of pore effects. These experiments show how
simultaneous injection of gas and surfactant solution can give
better sweep efficiency on a micro-scale in comparison to slug
injection.

Micromodel and High-Pressure System Design

Pressure Vessel and Flow System Design. The operation of glass
micromodels at high internal pressure requires a confining or
external pressure to prevent glass failure. Micromodels have
survived internal-external pressure differences of over 700 PSI,
but prolonged exposure leads to fatigue and rupture. The choice of
confining medium is limited by the additional requirements of
optical transmittability and refractivity needed for viewing the
micromodel from some distance away. The typical solution[10-12] is
to place the micromodel in some kind of pressurizable visual cell
and provide confining pressure with an optically compatible and
transparent fluid.

The pressure vessel design used here is similar to one developed by Campbell and Orr[10]. It consists of a midsection housing the micromodel, sandwiched between two polycarbonate windows, which in turn are covered on each end by doughnut-shaped end plates (Figure 1). The whole assembly is held together by a bolt circle and sealed between mid-section and windows by an o-ring and gasket combination. To achieve a pressure rating of 3000 psi at 150°F, the windows were machined from 3-ply G.E. Lexgard Polycarbonate Laminate, and the metal parts were made of J-45 steel from the E.M. Jorgensen Co.

The micromodel is located in the mid-section of the pressure vessel. It is seated in a slot where the feed ports exit the mid-section and is clamped down by restraining bars. The holes tapped in the micromodel are aligned with the feed port exits, and a seal is maintained by O-rings seated around these exits. This design differs from the Campbell-Orr pressure vessel which uses tubing fitted to C-clamps to carry fluids to and from the micromodel. The design employed here restricts the size of the micromodels to a particular set of specifications, but does provide a more secure way of pressurizing the micromodel.

The flow of fluids to and from the micromodel is accomplished by the flow system shown in Figure 2. Two ISCO 5000 LC pumps are used to meter gas and liquid to the micromodel at high pressure. The pump for the liquid also pressurizes a transfer vessel of glycerol that gives the overburden pressure. When both gas and liquid are flowing into the micromodel a sampling valve breaks the flow into 15 µl slugs to insure intimacy of the fluids without foaming. Downstream a TEMCO static back pressure regulator is used to maintain system pressue with very small fluctuations.

Micromodel floods were recorded on videotape for later analysis. A television camera with a zoom lens allowed magnification levels from 33x to 333x as measured on the monitor's screen. A Fujinon 35mm lens was used for larger fields of view. Recording was made in time lapse for replay in real time, allowing up to 10 days worth of recording on a two-hour tape.

Micromodel Design and Fabrication. In past studies using micrmodels the etching patterns were drafted by hand[5-17]. These patterns were either regular pore networks of identical pore bodies and throats[6,10] or tracings of thin-sections[11,12]. In either case the influence of human hand and mind is great enough to raise the question of whether or not micromodel results are affected by the drafting technique. This may be especially true for foam flow in micromodels, where bubble formation, and breakup and deformation will depend on pore geometry[17]. It may seem odd to be so concerned about having exactly the same pore shapes as are found in nature when the micromodel is already a great approximation; i.e., a three-dimensional problem (flow in porous media) is reduced to two-dimensions, and relatively smooth, clean glass is substituted for rough, dirty rock. Pore geometry is particularly important, however, because the variability in pore dimensions will determine

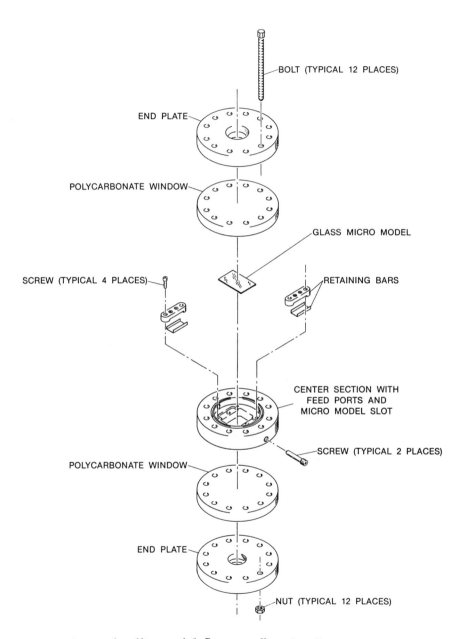

Figure 1.   Micromodel Pressure Vessel: Exploded View

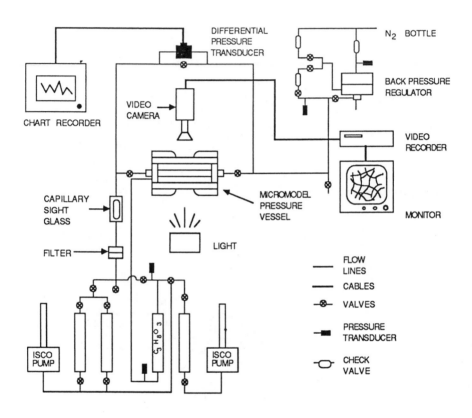

Figure 2.   Micromodel Flow System Schematic

the heterogeneity of the porous medium, and this quality is a
crucial part of many physical processes in flow in porous
media[17]. Hand-drafted patterns cannot duplicate the complexity
of natural pore systems, so it is desirable to copy the natural
pore networks exactly.

To accomplish this a petrographic image analysis (PIA) technique
for generating micromodel patterns was developed by M. Parma and
W.J. Ebanks[18] that can take color images of thin-sections and
convert them to digital maps for making black-and-white etching
patterns. It uses false-color processing to separate the colors in
a thin-section into different grey levels, which can be segmented
by computer into various components of the thin section. Since the
epoxy filling the pore spaces is bright blue, in contrast to the
dull shades of the rock matrix, porosity and non-porosity can be
easily distinguished.

The resulting computer-processed image, as viewed on a video
monitor, is a mixture of green (for porosity) and black (for
non-porosity) pixels. The green pixels can be counted and divided
by the total pixel number to give the porosity of the section under
view. The video binary image can then be fed to a flat-screen TV
for photographing.

Because the thin section has to be highly magnified ($\sim 100x$) to
make the pores show up clearly a single view is insufficient for
making a pattern. To get a large enough area for a pattern, the
field of view is moved over slightly in one direction (to where it
is still partially overlapping the previous view), and another
binary image is generated. This procedure is continued until the
photographs of the overlapping views can be placed together to form
a montage, usually five photos long by two photos wide. The
montage is then reduced (=10x) onto high-contrast black-and-white
photo paper or onto transparencies to get the final etching
pattern. In this way the scaling of the true pore network to that
on the etching pattern can be controlled precisely although an
exact 1:1 scaling would require an impossibly large montage.

Using this technique etching patterns for different rock types have
been generated. One of these rocks is a reservoir sandstone
(Figure 3) showing a fairly regular grain size (black areas) and
highly interconnected pore bodies (whit areas). Porosity for this
sample was calculated to be 30% with very few dead-end pores and
the total magnification or scale from thin-section to etching
pattern (a reduced version of Figure 3) was 10.9x. A second
etching pattern was included in this work to allow for a comparison
between the imaged-analyzed and hand-drafted patterns. The design
is an architectural transfer called "Patio Stone" that is made by
Para-Tone, Inc. It is used to indicate masonry or stone floors on
blueprints, but has been previously used as an etching pattern
because of its similarity to a pore network (Figure 4). As can be
seen from the figure, this pattern gives very regular, large pores
when etched, with a pore body/pore throat aspect ratio of close to
unity. It would be expected that this lack of heterogeneity might

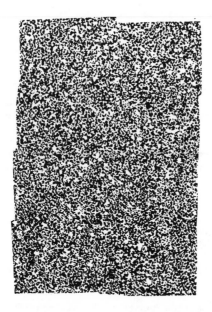

Figure 3. Reservoir Sandstone Binary Image

Figure 4. "Patio Stone" Binary Image

have a profound effect on displacement stability, phase mixing and trapped phase saturations. The etching patterns in Figures 3 and 4 were reduced by roughly the same amount of for the actual etching films. Table I gives the estimated range of pore sizes, scale, and porosity for these etching patterns, along with the pore volume of the etched micromodels.

TABLE I. Permeabilities, Pore Size Ranges, Scale[1],
Pore Volume and Porosity of Micromodels

M-M	k-t, Darcy-μm[1]	Pore Sizes, μm	Scale[2]	P.V., cc	Porosity %
Sandstone	385	50-200	10.9	0.18	30
Patio Stone	473	400-1300[3]	N/A	0.39	--

[1] k-t = $Q\mu L/W\Delta P$ L = 2.75", W = 1.75"

[2] Scale = $\dfrac{\text{Avg. Pore Size of Micromodel}}{\text{Avg. Pore Size of Thin Section}}$

[3] From Ref. 10

Subsequent to the completion of the image analysis work for generating etching patterns, a paper by Trygstad, et al. [19] appeared in the literature which describes essentially the same process. The etching patterns of Trygstad, et al. have scale factors from 10x to 25x, and microscopic examination of their micromodels reveals that the etched pores and the thin-section pores have an almost exact geometric similitude in terms of aspect ratio and pore shapes. Displacement experiments in their micromodels confirm the anticipated flow effects of the real rock heterogeneities, and show that these models allow for a more accurate comparison between pore-level flow behaior in micromodels and core-flood results.

The etching of the micromodels used in this study was performed by M. Graham of Adobe Labs Co. using a technique developed by B.T. Campbell[10] and I. Chatzis[20].

Micromodel Foam Floods

Table II lists the micromodel experiments performed in this study. Only two micromodels, the Reservoir Sandstone (RS) and the "Patio Stone" (PS), were used. In these experiments the performance of different injection schemes and the effects of surfactant on gas displacement of water were investigated.

Each experiment would begin with the micromodel filled with either water or surfactant solution. If only gas were to be injected, the gas pump would be set to meter at the appropriate rate and then the

flow lines to the pressure vessel would be purged until the gas
reached the inlet to the pressure vessel. At this point the pump
would be stopped, the value controlling flow to the micromodel set
open, and the system would settle to the preset back pressure. At
this point injection would begin, and the pressure and video
recording would start.

For simultaneous gas and liquid injection, the procedure was
altered to purge the flow lines with both fluids. A mixing chamber
upstream of the purge outlet allows for a gentle blending of the
fluids. Once the pressure was stabilized injection of the premixed
fluid would begin.

To quantify micromodel flow characteristics, the
mobility-thickness, M-t, and the permeability-thickness, k-t, can
be calculated. These quantities are used here instead of mobility
and permeability because the flow is only two-dimensional.
Mobility-thickness is defined as

$$M\text{-}t = \frac{QL}{W\,\Delta P} \tag{1}$$

where Q is the flow rate, and L and W are the length and width of
the micromodel, respectively. For example, the SAG foam flood gave
M-t=10.7 darcy-um/cp, while the straight surfactant solution flood
gave M-t=55.0 darcy-um/cp. When the flow is single phase at 100%
saturation, then k-t can be calculated from

$$k\text{-}t = \frac{Q\,\mu\,L}{W\,\Delta P} \tag{2}$$

where u is the flowing phase viscosity. Using the surfactant
solution viscosity of 8.6 cp as measured with a Brookfield
viscometer, this gives k-t=473 darcy-um for the PS micromodel.
Similar calculations give k-t=385 darcy-um for the RS micromodel
(Table I).

A. "Patio Stone" (PS) Micromodel

Of the many experiments run in the PS micromodel, only Test 11-19A
is described here (see Table II). It was a gas-drive of surfactant
solution (GDS), in which the pressure drop across the micromodel
was measured and analyzed in terms of the flow behavior recorded
simultaneously on videotape. It was also of interest to examine
bubble formation and breakup processes in the PS model, where the
large and fairly regular pores might give a different behavior than
the smaller, more variable pores of the RS model. The surfactant
used in the PS model was an anionic-nonionic blend in a 10 wt.%
(weight percent active) solution, and nitrogen was the gas used in
the GDS test. Conditions were 1000 psi back pressure and ambient
temperature.

Bubble formation in the GDS process was found to occur by two
separate mechanisms. The first mechanism has been identified

TABLE II. Experimental Conditions, Flow Rates and Fluids
For Micromodel Experiments

Test #	Micromodel	P,psi	T,°F	Inj. Scheme	Q,cc/HR	Gas	Liquid
11-19A	PS	1000	75	GDS	0.3	A	A
11-21A	PS	1000	75	S	0.3	A	A
1-13A	RS	1000	75	S	0.3	B	B
1-14A	RS	1200	75	S	0.3	B	B
1-16A	RS	2000	75	S, GDS	4.0	B	B
1-24A	RS	1000	75	GDS	0.23	B	B
1-27A	RS	500	75	GDS	0.23	B	B
1-27B	RS	2000	75	GDS	0.23	B	B
2-3A	RS	2000	120	GDW	0.23	C	C
2-3B	RS	2000	120	GDW	0.23	C	C
2-3C	RS	2000	120	SI, 3:1	0.23	C	D
2-5A	RS	2000	120	ST, 3:1	0.23	C	D
4-11A	RS	1000	120	GDW	0.3	C	C
4-11B	RS	1000	120	GDW	0.3	C	C
4-14A	RS	1000	120	GDS	0.3	C	E
4-14B	RS	1000	120	GDS	0.3	C	E
4-15A	RS	1000	120	GDS	0.3	C	E
4-15B	RS	1000	120	GDS	0.3	C	E
4-16A	RS	1000	120	GDW	0.3	C	C

KEY

PS	= Patiostone Micromodel
RS	= Sandstone Micromodel
S	= Surfactant Injection Only
GDS	= Gas Drive of Surfactant Solution
GDW	= Gas Drive of Water
SI, 3:1	= Simultaneous Injection, Gas:Liquid Ratio = 3:1

Gases

A = N_2
B = CH_4
C = Reservoir Injection Gas (Mostly C_1)

Liquids

A = 30% Liquinox in H_2O
B = 0.5% Alipal CO-128 in 15% NaCl Brine
C = Reservoir Brine
D = 0.1% AOS-1618 In Reservoir Brine
E = 0.5% Alipal CD-128 In Reservoir Brine

previously as "snap-off"[15],[17], in which a gas finger invading a
liquid-filled pore body will snap off smaller bubbles by forming
lamellae at the pore throat (Figure 5a, b). At first the bubbles
tend to be the same diameter as the throat, but as the liquid is
depleted the bubbles become larger until lamellae formation finally
stops. Snap-off is governed by a set of equations relating pore
dimensions (such as aspect ratio) to pressure drop and surface
tension[21], which for gas-liquid systems typically requires a
large pore body/pore throat aspect ratio ($\sim$3). Because the pore
dimensions were fairly uniform (i.e. constant pore radius) for the
PS model, snap-off was rarely observed.

The second bubble formation mechanism is shown in frames c and d of
Figure 5. It occurred only in those areas where one pore crossed
another. If a long gas finger were moving in one pore, and another
gas bubble were to be forced into the intersection of the two pores
by a prevailing pressure gradient, the long finger would break
towards its tailing end into a long piece and short piece. The gas
bubble in the intersecting pore could then slip between the long
and short pieces, leaving three bubbles where only two had existed
before. This process would continue at different intersections
until the long finger had been broken into several smaller ones.
The majority of the bubbles in the GDS experiment were formed in
this manner. Bubbles formed by this second mechanism are several
times larger than the pore radius, whereas bubbles formed by
snap-off tend to be the same size as the pore throat radius.

As the gas would finger through the surfactant solution during the
GDS sequence, the pressure drop across the micromodel fluctuated as
a slugs of gas and liquid moved through the major flow channels.
This is because the total pressure drop across the model is the sum
of the pressure drops across the individual lamellae moving in the
main flow path. By synchronizing the pressure recording on a strip
chart with the video tape, it was found that the peaks in the
pressure difference fluctuations occurred when no gas movement was
visible. The pressure difference would decrease when a gas finger
would enter a major flow channel, and the gas finger would rapidly
move through the channel. As the flood proceeded, the size of the
liquid slugs separating the gas fingers decreased until liquid
production ceased, at which point an uninterrupted gas path existed
through the micromodel (breakthrough). Simultaneously, the
pressure drop decreased from 1.8 to 0.5 psi. A plot of the volume
of liquid between gas fingers (as measured by the distance between
pressure peaks) against the logarithm of the gas throughput appears
to give a linear relationship (Figure 6). This behavior is
consistent with the observation that as the gas saturation
increases, there is less liquid in the micromodel to be produced;
thus, the size of the lquid slugs separating the gas fingers
decreases. The logarithmic relationship is not as easily
explained, however. As the liquid saturation decreases the
pressure drop must increase to force liquid in other regions of the
model to flow. Thus, the pressure loss gets larger as the flood
progresses until breakthrough, when it must decrease since there is
no longer any resistance by liquid to gas flow. Unless liquid is

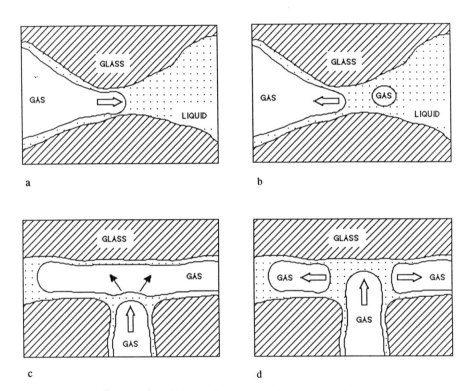

Figure 5.   Bubble Formation During Gas-Drive
a,b.   Front of Finger
c,d.   End of Finger

continually replenished, the mobility control by surfactant will be
only temporary.

The magnitude of the pressure drop fluctuations (the peaks and
valleys on the strip chart recording of ΔP) increased as the
average pressure drop increased during the GDS flood. Similar
results have been found in foam floods in bead packs and core
floods[4]. Figure 7 is a plot of height of a fluctuation, c',
versus the average pressure drop duing that fluctuation, ΔP_{avg}.
The data seem to fall along a line having a slope around unity
(i.e., average pressure drop change proportional to fluctuation
width change) and an intercept ($c' = 0$) of $\Delta P_{avg} \neq 0$. This
intercept may be due to additional pressure losses in the flow
lines outside the micromodel between the pressure transducer taps,
or perhaps due to capillary pressure. The interpretation of Figure
7 would be that the fluctuations, which are correlated with the
movement of the liquid slugs in the main flow channels, are
one-half the amount of pressure loss necessary to initiate movement
of the liquid slugs in the main flow channel. This pressure loss
increases as the volume of liquid decreases in the main channel,
the pressure drop increases until it reaches a level at which the
capillary forces opposing the flow are overcome, it then decreases
until the liquid exits the micromodel. This behavior is
significant because it indicates that the foam pressure drop is not
primarily due to viscous flow but rather capillary resistance, at
least in the two-dimensional model.

Reservoir Sandstone (RS) Micromodel. The RS micromodel was used in
a variety of experiments examining the effects of surfactant, foam
quality, injection scheme and pressure level on foam displacement
efficiency and flow patterns. Various gases and brines or
surfactant solutions were used, primarily field injection gas and
brine, and the surfactant AES (trade name: Alipal CD-128) at a
concentration of 0.5 wt.%.

Figure 8 shows an overall view of the micromodel after simultaneous
injection (SI) of gas and surfactant solution (Frame a) and after
GDS flooding with only one cycle of gas injection (Frame b). The
SI process was filmed during Test 2-5A while the GDS process was
filmed during Test 4-15A. Foam flow in the SI process spreads out
from the major flow channel more than the GDS flood, and the
average size of the bubbles is smaller in the SI case than in the
GDS case. The SI process appears to distribute the gas in a more
uniform fashion throughout the porous medium when compared to the
GDS flood.

When comparing the gas-drive processes GDS and GDW the presence of
surfactant in the displaced liquid has a great effect on the
displacement mechanisms and flow patterns. Figure 9 shows
schematically the final extent of sweep for gas-drive of brine
without surfactant (Frame a) and with surfactant (Frame b). In
each case the gas appears to have preferentially flowed through a
few large channels that zig-zag across the micromodel; however, in

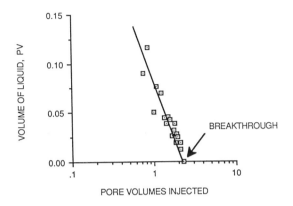

Figure 6.   Volume of Liquid Between Gas Fingers During Drainage
Displacement (GDS Flood) (Test 1-24A)

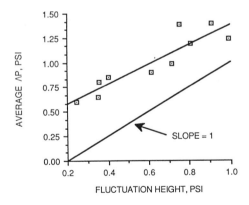

Figure 7.   ΔP Fluctuations vs. Average ΔP (Test 1-24A)

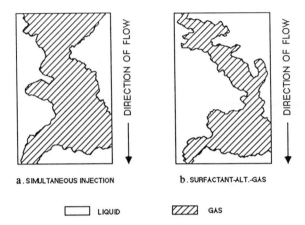

Figure 8. Final Fluid Saturations for Foam Displacement of
           Brine
           a. Simultaneous Injection (Test 2-5A)
           b. Gas-Drive of Surfactant Solution (Test 4-15A)

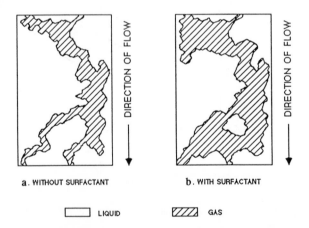

Figure 9. Final Fluid Saturations for Gas-Drive of Brine
           a. Without Surfactant (Test 4-11B)
           b. With Surfactant (Test 4-14B)

the gas-drive where surfactant was used, the gas spread out from
these major channels more than when surfactant was not present,
leaving a larger swept area in the model and delaying
breakthrough. Table III shows this numerically, as the number of
pore volumes injected before breakthrough is nearly twice as large
with surfactant as it is without. Interestingly, pressure drop
measurements at breakthrough as given in Table III show that the
pressure drop for gas-drive is smaller when surfactant is present.
This result is unexpected, and will be examined in the discussion
section of this paper.

TABLE III. Effect of Surfactant on Gas-Drive Pressure Drop

Test	Surfactant?	ΔP Liquid, psi	ΔP Gas-Drive psi	Cum. Flow at B.T., PV
4-11A	N	0.05	0.20	0.13
4-11B	N	0.05	0.25	0.10
4-14A	Y	0.05	0.13	---
4-14B	Y	---	---	0.19
4-15A	Y	0.10	0.18	0.20
4-15B	Y	0.03	0.13	0.18
4-16A	N	0.02	0.24	---

NOTE: See Table II for experimental conditions and fluids.

The SI process is more effective at displacing a liquid than the
GDS process, which in turn is better than the GDW process. The
reasons for this ranking can be understood from a microscopic
examination of the flow behavior. Figures 10 and 11 demonstrate
how intimate mixing of gas and liquid during simultaneous injection
affects the displacement process. In Figure 10a, small bubbles can
be seen flowing into and out of a pore body in the center of the
figure, moving from bottom to top. Stagnant large bubbles fill the
pores below and to the left and right of the center pore. The flow
going through the center pore is very wet, having only a few
entrained bubbles, because a large slug of surfactant is passing
through. The pressure drop measured at this time was 1.65 psi. As
more gas entered the micromodel during simultaneous injection, the
quality of the foam increased and the bubbles got larger (Figure
10b), and the pressure drop increased, reaching 2.1 psi.

As the pressure drop fluctuates and the average bubble size changes
during simultaneous injection, the flow paths also change. Such
changes leave some pores filled with stagnant bubbles while foam
flows by in neighboring pores. Figure 11 shows a larger view of
the area surrounding the center pore in Figure 10 (seen just below
the date and time indicator). A low quality flow of small bubbles
can be seen in the channel that goes diagonally from the bottom
left to the top right of the photograph, while connected to this
channel is a network of other pores filled with larger stationary
bubbles. As time went on, the bubble size increased, filling the
flow path with large bubbles. These became trapped due to their

a

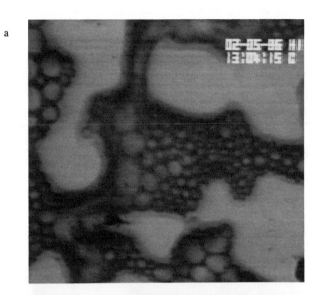

b

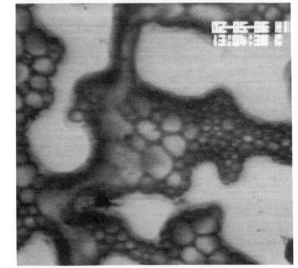

Figure 10. Bubble Size and Flow Resistance (Test 2-5A)
a. Low Quality Foam (ΔP=1.65 psi)
b. High Quality Foam (ΔP=2.10 psi)

size, and the rising pressure drop required to move these bubbles
also dislodged the previously stagnant bubbles. In this way the
flow paths were continually changing and new flow areas were opened
up during SI of gas and surfactant.

This behavior is quite different from that observed during
gas-drive of surfactant solution. In the GDS foaming process the
fingering gas enters the pores along the major flow path and snaps
off the bubbles. Because the liquid filling these pores has been
displaced and no new liquid is entering the micromodel, the foam
that is left in these pores has a very high quality and is almost
always trapped. Figure 12 is a photograph of a highly magnified
view of the area immediately surrounding a major flow channel
running from upper left to lower right. As in previous
photographs, the light areas are the "grains" of the glass matrix,
the darker gray areas are filled with gas, and the thin black lines
are water-films along the pore walls or foam lamellae separating
gas bubbles. This shot was taken shortly before breakthrough, and
it can be seen that the major flow channel is almost entirely
gas-filled with only a few lamellae breaking up the gas. The pores
connected to this channel are filled with a high-quality polyhedral
foam that is stationary. After breakthrough, only gas will flow in
the main channel while the foam in the neighboring pores will be
unmoved.

The back pressure at which a gas-drive is operated will affect the
efficiency of the displacement, as the data in Table IV indicates.
As the back pressure is increased, the ultimate pressure drop also
increases and the number of injected pore volumes before the final
state (the point where essentially only gas is produced)
decreases. A piston-like displacement would require one pore
volume of injected gas before all the liquid was produced, while
increasingly less efficiency displacements would require more
volumes.

TABLE IV. Effect of Pressure on Foam Flow Pressure Drop
and Breakthrough[1]

Test	P, psia	ΔP Final State[2], psi	PVI Final[2], PV
1-27A	500	0.50	1.25
1-24A	1000	0.50	1.10
1-27B	2000	0.60	1.05

[1] See Table II for experimental conditions and fluids

[2] "Final" conditions are defined here as the point at which
only gas is produced from the micromodel, identified by a
constant pressure drop.

Discussion

Effect of Injection Scheme on Foam Displacement. If the main
interest in using foam is for controlling gas mobility, then it is
necessary to have a criterion to judge the effectiveness of a foam

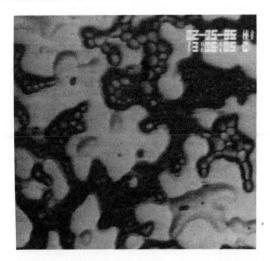

Figure 11. Enlarged View of the Pores in Figure 10, Showing
          Foam Flow in Channels with Neighboring Water-Filled
          Pores (Test 2-5A)

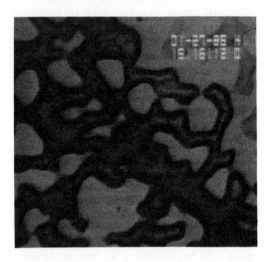

Figure 12.   Gas Finger with High Quality Foam (Test 1-27B)

process. For foam floods in cores or sand packs, the pressure drop
generated during foam injection is commonly the parameter used to
quantify effectiveness, but for floods in micromodels, a visual
estimate of the area swept by foam is the most natural criterion to
choose. A comparison of the areas swept by gas for different
injection schemes gives great insight into the manner in which the
foaming process redistributes and stabilizes the gas flow, as well
as being a measure of effectiveness. Of course, the
two-dimensional nature of the micromodel and its small size make it
difficult to predict behavior in three-dimensional rock. The
effect of the loss of dimensionality and of model size on the
micromodel results is considered at the end of the discussion
section.

When comparing SI and GDS injection processes in the RS model on a
microscopic level, one difference that is immediately apparent is
in the degree of fluid mixing. For the GDS process the
displacement of surfactant solution by gas-drive yields only a
small amount of interfacial area formation at the displacement
front, with little interface generation behind the front. The SI
process, on the other hand, has a continuous mixing of gas and
liquid due to the breaking and reforming of bubbles as they flow,
and this mixing occurs throughout the foam-swept area. Bubble
sizes tend to be much smaller with the SI process, and the foams
are much wetter than those from the GDS process.

Bubble generation during GDS foam flooding occurs by the mechanisms
shown in Figure 5, preferentially the snap-off mechanism. Snap-off
tends to be an inefficient foaming mechanism[15] since most bubbles
are formed at the displacement front, where they are pushed ahead
of the main gas body. Behind this front there are fewer lamellae
because the only operative bubble formation mechanism is the
breakup of gas fingers (Figure 5 c,d). During the SI foam process,
however, the constant flow of surfactant solution allows both
bubble formation mechanisms to operate at anytime and anywhere in
the pore space. The additional liquid, at a flow rate of one-thord
that of the gas, tends to make the foam wet enough that snap-off is
the predominant bubble formation mechanism. Bubbles appear to be
generated more frequently in the SI process.

Another difference between SI and GDS foam lies in the mechanism by
which the foam reduces gas mobility. In each process the foam
flows through the largest channels, creating flow resistance due to
the foam lamellae spanning these pores. In the GDS process, the
foam will invade the pores connected to the main flow channels,
plubbing them off. As the liquid is depleted from the main channel
a continuous gas path is formed, and the microscopic distribution
of gas and liquid becomes permanently fixed. Mobility reduction in
the GDS process is a result of the high trapped gas saturation
which has blocked off some flow paths. In the SI process, foam
will also invade the neighboring pores around the main flow
channels, but this gas is not permanently trapped. Instead, foam
will flow through a channel until the bubbles become so large and
so numerous that the channel is plugged (for example, the

transition from Figure 10a to 10b). The pressure gradient will
increase as a result of this blockage, dislodging bubbles that were
plugging another channel. In this way, the flow will move through
different channels at different times, and more flow paths are
opened up as a consequence (see Figure 8).

The fluctuating pressure gradient that is characteristic of foam
flow in cores was present in the micromodel experiments, and a few
aspects should be noted about the nature of this phenomena. The
fluctuations in the GDS foam flood (Test 11-19A) were found to
correlate with the passage of liquid slugs in the main gas flow
channels, and appear to be a linear function of the average
pressure (Figure 7). By correlating each fluctuation with the
respective liquid slug size (Figure 6), it is found that the
smallest liquid slugs resulted in the largest pressure fluctuations
and average pressures. This relationship might seem
counterintuitive, since it could be expected that when the liquid
fractional flow decreases and gas fractional flow increases that
the gas would be "easier" to push. However, it should be
remembered that the gas is a discontinuous phase and that all the
pressure losses occur in the liquid phase in the thin films and
lamellae. In the SI foam floods the fluctuations were much smaller
than the average pressure drops, presumably because of the wetter
flow, and high pressure drops could be sustained for longer times
than could be done with the GDS scheme.

Mobility Control as a Function of Rock Heterogeneity. Although the
effect of rock heterogeneity on foam formation and propagation was
not studied in a systematic fashion in this investigation, some
observations about it can be made from the experiments described
above. To begin with, a distinction should be made between sweep
efficiency on a macro scale (as in core floods) and on a micro
scale (as in the present experiments). In parallel core tests, a
foam will have roughly the same mobility in a high-permeability
core as in a low-permeability core[4], leading many to believe that
foam can overcome permeability variations in a real reservoir and
give a piston-like displacement. On a microscopic level, however,
the areal sweeps in Figure 8 for the GDS and SI foam processes are
not piston-like. A sizable portion of the micromodel is untouched
by gas, being estimated from 50% in Frame a to 70% in Frame b.
Undoubtedly the results in Figure 8 are influenced by the small
size of the model, but they are still felt to be indicative of the
micro-sweep efficiencies expected in larger, three-dimensional
porous media. Thus, even though the movement of the foam bank may
appear to be piston-like on a macro scale, large amounts of pore
space may be bypassed.

The micro-sweep efficiency will depend on the local heterogeneity
of the porous medium and the foam's flow properties. The RS
micromodel had a few large channels where the flow was heaviest and
other regions with smaller pores that were not invaded by foam.
The PS micromodel, on the other hand, had fewer pore connections,
and the pores were of more uniform size; consequently, it was
almost completely swept of liquid in the GDS foam flood (Test

11-19A). No attempt was made to quantify the heterogeneities of
the two models, so no theoretical interpretation of these data can
be made.

Further evidence of an effect of microscopic heterogeneity on foam
sweep efficiency comes from Figure 9. During gas drive, when no
surfactant is present in the brine, the gas traverses the model
along the larger channels, breaking through at 0.12 pore volumes
injected, as averaged from the data in Table 4. When surfactant is
present, breakthrough is delayed to 0.19 PV injected, a 52%
increase in sweep efficiency. By comparing Frames a and b of
Figure 9, however, it would appear that the gas has followed the
same flow paths in each case, the difference being that the gas has
spread out from the main channels when surfactant is present. This
behavior indicates that the microscopic heterogeneity is
controlling foam propagation in the micromodel, and may mean that
foams cannot give a complete sweep in real porous systems.

One anomalous result from the experiments summarized in Table III
is the lower pressure drop during gas-drive of surfactant
solution. An average of the tabulated data gives $\Delta P=0.15$ psi when
surfactant was in the brine versus $\Delta P=0.23$ psi when it was not.

Typically, the gas-drive of surfactant solution require a higher P
than the gas-drive of brine. One possible explanation is that the
foam formation in the "surfactant-present" gas-drives was poor,
such that very little gas was present in a discontinuous state.
The higher gas saturations in these floods would then allow more
area for gas flow, such that the pressure drop requirements were
lower than when the surfactant was not present. Capillary pressure
might account for the difference, since the presence of surfactant
should decrease the surface tension, resulting in the less
capillary pressure in the GDS flood than in the GDW flood. The
same situation would not necessarily exist in gas-drives in
three-dimensional media, since pressure drop requirements for
breakthrough are less for 3-D than for comparable 2-D networks[22],
and because capillarity is often not correctly scaled in
micromodels[24].

Effect of Micromodel Size and Dimensionality on the Interpretation
of the Results. It is not known how to quantitatively adjust
flooding results in two-dimensional systems to bring them in line
with results in their three-dimensional counterparts, but it is
possible to state how the results might differ. Loss of
dimensionality can lead to lower ultimate recovery, which occurs
for either of two reasons: first, because two-dimensional systems
of the same coordination number have lower macroscopic connectivity
and second, because loss of dimensionality often lowers the
coordination number of the pores[22]. Thus, a lower sweep
efficiency by foam in the micromodels would be expected, although
lower recovery might also result for other reasons. The
pore-shapes and pore size distribution of a two-dimensional thin
section differs greatly from that of the three-dimensional

rock[24], thereby also affecting the displacement. If the
thickness is substantially smaller than the pore—width, then the
thickness would be the controlling length—scale, which could be a
problem if the thickness is very uniform.

For the micromodel experiments described here, it is felt that the
foam sweep efficiency was lower, overall, due to the loss of
dimensionality, but that the relative performance by SI, GDS and
GDW injection are unaffected by dimensionality. It is not likely,
either, that the pore thickness dominated any capillary phenomena
that should instead have depended on the two—dimensional pore size
distribution. If a typical pore thickness between 200 and 300mm is
taken,[10,20] then Table I would seem to indicate that the pore
diameters were at least comparable if not much smaller, such that
the pore diameters would be responsible for any variation in
capillary pressure.

The size of the micromodel, both pore dimensions and the size of
the etched area, can affect micromodel results in a number of
ways. For example, the etched pores in the RS micromodel were
designed to be 10.9x larger than the actual pores in the thin
section. One might therefore expect that the capillary pressure in
the micromodel would be about 1/10th of what it is in the rock.
The larger pore size may also introduce boundary and end effects
into the displacements, since it requires about 40 pore lengths to
remove length effects in two—dimensional simulations of
networks[22]. This requirement is definitely met for the RS model,
but is probably only barely satisfied, or not satisfied at all, for
the PS model. However, from Figures 9 and 10, there is evidence of
both boundary and end effects even in the RS micromodel: incomplete
sweep at the inlet manifold of the model and residual liquid along
the sides.

Conclusions

The conclusions reached from the experiments and analyses presented
here are:

o   Foam flooding by simultaneous injection (SI) of gas and
    surfactant solution gives greater displacement efficiencies
    than gas—drive of surfactant solution injection.

o   With the gas—drive of surfactant solution (GDS) injection
    scheme, the degree of microscopic heterogeneity will determine
    the microscopic sweep efficiency.

o   The dominant mechanism of bubble formation depends intimately
    on the heterogeneity of the porous media and the injection
    scheme. Realistic (i.e. natural) pore structures will give a
    wide bubble size distribution.

o   Pressure drop fluctuations were found to correlate with the
    passage of gas and liquid slugs in the main flow channels of

the micromodel. Although they appear random, the fluctuations are found to depend linearly on the average pressure drop.

Acknowledgments

The author would like to express his appreciation to Tom Lawless for assembling the flow system and performing the experiments. J.P. Heller and I.M. Bahralalom provided very helpful discussions for which the author is also grateful.

Literature Cited

1. NPC Committee on Enhanced Oil Recovery Enhanced Oil Recovery, May 1984, Draft to D.P. Hodel, Secretary of Energy.
2. Bernard, G.G.; Holm, L.W. Soc. Pet. Eng. J., 1964, 267-274.
3. Castanier, L.M.; Brigham, W.E. Chem. Eng. Prog., 1985, 37.
4. Casteel, J.F.; Djabbarah, N.F. SPE 14392, Sept. 1985.
5. Wardlaw, N.C. SPE 8843, April 1980.
6. Chen, J.-D.; Koplik, J. J. Coll. Int. Sci., 1985, 108, 304.
7. Hornof, V.; Morrow, N.R. SPE/DOE 14930, April 1980.
8. Davis, J.A., Jr.; Jones, S.C. SPE 1847, Oct. 1967.
9. Dawe, R.A.; Wright, R.J. Royal Sch. Mines J., 1983, 33, 25.
10. Campbell, B.T.; Orr, F.M., Jr. SPE 11958, Oct. 1983.
11. Bahralolom, I.M.; Bretz, R.E. SPE 14147, Sept. 1985.
12. Bahralolom, I.M.; Orr, F.M., Jr. SPE 15079, April 1986.
13. Holm, L.W. Soc. Pet. Eng. J., 1968, 8, 359.
14. David, A.; Marsden, S.S. SPE 2544, Sept. 1969.
15. Mast, R.F. SPE 3997, Oct. 1972.
16. Kanda, M.; Schecter, R.S. SPE 6200, Oct. 1976.
17. Owete, O.S. Ph.D. Thesis, 1982, Stanford, Univ.
18. Parma, M. Simple Pore Geometry Characterization from Digital Image Analysis Technique, 1985, ARCO Oil & Gas Co.
19. Trygatad, J.C., Ehrlich, R.; Wardlaw, N.C. SPE/DOE 14891, April 1986.
20. Chatzis, I. N.M. Inst. of Mining and Tech Report PRRC 82-12, March 1982.
21. Gardescu, I.I. Trans. AIME, 1930, 86, 351-370.
22. Chatzis, I.; Dullien, F.A.L. J. Can. Pet. Tech, 1977, 16, 97.
23. LeNormand, R., Zarcone, C.; Sarr, A. J. Fluid Mech., 1983, 135, 337.
24. Dullen, F.A.L. J. Microscopy, 1986, 144, 277.

RECEIVED January 5, 1988

Chapter 13

Videomicroscopy of Two-Phase Steady Cocurrent Flow in a Model Porous Medium

Curtis M. Elsik[1] and Clarence A. Miller

Department of Chemical Engineering, Rice University, Houston, TX 77251

A novel flow cell has been developed to observe on a microscopic level the steady state, cocurrent flow of two pre-equilibrated phases in a porous medium. It consists of a rectangular capillary tube packed with a bilayer of monodisperse glass beads 109 microns in diameter. The pore sizes in the model are of the order of magnitude of those in petroleum reservoirs. An enhanced videomicroscopy and digital imaging system is used to record and analyze the flow data.
Several fluid systems covering a wide range of interfacial tensions were studied using a syringe pump capable of producing superficial velocities ranging from 0.1 to 2,000 ft/day. While most of the work employed an injection ratio of 1:1, experiments with various ratios up to 10:1 were performed. At low capillary numbers, the expected stable, continuous, tortuous paths were observed. At a capillary number of about 0.001 the nonwetting phase started to flow freely as large ganglia having lengths of 10-20 pore diameters in the direction of flow. As the capillary number was further increased, these ganglia became shorter, reaching the size of a single pore at a capillary number of about 0.01. When the capillary number was increased above this value, ganglion or drop size continued to decrease to values below the pore throat diameter. Eventually the small drops stretched out, producing a filament type flow.

The success of waterflooding as a petroleum recovery process and the realization that even it leaves behind a substantial amount of residual oil have been major factors stimulating research on two-phase flow in porous media during the past fifty years. Most of the knowledge pertinent to two-phase flow under normal waterflooding conditions is summarized in existing books (1-4). Recently, the

[1]Current address: American Cyanamid Company, Princeton, NJ 08540

0097–6156/88/0373–0258$07.00/0
© 1988 American Chemical Society

development of enhanced oil recovery processes has drawn attention to different mechanisms of two-phase flow which can exist in surfactant systems. For instance, the ultralow interfacial tensions which can be achieved with suitable surfactants have made it possible to conduct displacements in the field at capillary numbers several orders of magnitude larger than those which exist during waterflooding. As a result, changes in the basic mechanism of two-phase flow occur, as shown by our experiments described below.

Achievement of low mobility ratios at the fronts between displacing and displaced fluids is of even greater concern in enhanced oil recovery than in waterflooding owing to the high costs and/or low viscosities of the injected fluids. One response to this concern has been the continuing effort to develop a fundamental understanding of so-called foam flow, which employs aqueous solutions of properly chosen surfactants at relatively low capillary numbers to reduce the effective mobility of low viscosity fluids (see 5,6 and papers on foam flow in this volume).

The mobility of two phases flowing at high capillary numbers is also of interest, however. Such flow occurs in the rear portion of the oil bank formed during tertiary flooding with micellar solutions and is also to be expected in solvent displacement processes when miscibility is just being developed or lost. A two-phase mixture of an aqueous polymer solution and a microemulsion which has an extremely low interfacial tension and would thus progress through a reservoir at a high capillary number has even been proposed as a means of simultaneously displacing oil and maintaining proper mobility control (7). The present work which, as indicated above, identifies the mechanisms of two-phase flow at high capillary numbers, may be viewed as an essential first step in understanding the implications of such flow for the proper mobility design of certain enhanced oil recovery processes.

Macroscopic experiments such as core flooding have been used to obtain relative permeabilities, dispersion coefficients, and other variables relevant to reservoir flow. However, they cannot reveal details of how immiscible phases interact on the pore level. Instead visual experiments have been used to elucidate microscopic flow mechanisms. The latter approach is taken here with experiments using a novel flow cell and state-of-the-art video equipment. The pore level phenomena observed provide a basis for the proper modeling of two-phase flow through porous media at high capillary numbers.

Experimental work describing direct observations of two-phase flow in a pore network was first published by Chatenever and Calhoun (8), and later by Chatenever (9) and Kimbler and Caudle (10). For most of their experiments, which involved cocurrent flow of high tension fluid pairs, these workers observed channel flow, where each phase propagated through the pore network in a continuous manner. The channels or paths did not change with time and were tortuous in nature, being primarily determined by wetting characteristics and small-scale heterogeneities. Such flow can be modeled with conventional relative permeability theory (1-4) since flow in each phase is proportional to its pressure gradient.

It was noted by both groups that when a sufficiently high injection velocity was reached, the nonwetting continuous paths broke up and a new slug flow mechanism developed. This phenomenon was attributed to the extremely high flow rates involved and not investigated further.

The interplay between capillary and viscous forces determines the location of the two phases and the flow mechanism. Hence, their ratio, the capillary number Ca, is an important dimensionless parameter influencing flow. In this work it is defined simply as:

$$Ca = v\mu/\gamma \tag{1}$$

where v is the superficial velocity, μ the viscosity, and γ the interfacial tension. Several different, yet related, definitions have been used by previous workers to correlate their individual data of residual oil recovery performance. A general tabulation of these endeavors has been given elsewhere (11). A popular modification incorporates the porosity ϕ of the porous medium into the denominator of Eq. (1). This procedure amounts to using interstitial velocity, $v_i = v/\phi$, in place of superficial or Darcy velocity v. While this expression suffices for single-phase flow, the interstitial velocity of phase j in two-phase flow involves its saturation S_j, i.e., $v_{ij} = v_j/\phi S_j$. Since fluid saturations are not always known, the simpler definition of Ca was chosen here.

The present work was performed to arrive at a description of two-phase, cocurrent flow through porous media over the entire range of capillary numbers encountered in various processes. Included are low values typical of waterflooding where capillary forces predominate, high values where viscous forces prevail, and intermediate values where both play an important role. The steady state flow mechanism is shown to be dependent on Ca, making it important to consider what range of Ca will be encountered during a particular process of interest. Velocities were limited to values where inertial effects were negligible.

EXPERIMENTAL

Materials

In order to vary interfacial tension over more than four orders of magnitude, several fluid systems were chosen that ranged from high tension surfactant-free formulations to middle phase microemulsions that were at optimal conditions for enhanced oil recovery and had ultralow tensions with the excess brine and oil. Table I lists the specific components used along with their corresponding physical properties. In each case a red water-soluble food coloring dye was added before equilibration to enhance the contrast between phases during microscopy.

The high tension fluid pair chosen was n-decane/water. The interfacial tension of 33.3 dynes/cm (with dye added) was measured using a spinning drop tensiometer at room temperature (22°C). Viscosities were measured with a Brookfield cone and plate viscometer, and densities with a Mettler/Paar digital densitometer. The ultrapure water was prepared using a SYBRON/Barnstead purification system. The local tap water was first pretreated, then distilled, and finally passed through a NANOpure cartridge deionization system. The resulting water conformed to ASTM Type I Reagent Grade Water standards (resistivity > 16.7 megaohm-cm).

A ternary oil-water-alcohol system was chosen that exhibited an

TABLE I.   FLUID SYSTEM PHYSICAL PARAMETERS

Fluid Formulation	Phase Pair	IFT(dyne/cm)	Wetting Phase Viscosity(cp)
50V% Water 50V% Decane T=22°C	L/U aqueous/ oleic	33.3	0.95
28.6W% H_2O 42.8W% NPA 28.6W% DMP T=22°C	L/U aqueous/ oleic	0.14	3.2
28.6W% H_2O 42.8W% NPA 28.6W% DMP T=30.0°C	L/U aqueous/ oleic	0.068	2.1
28.6W% H_2O 42.8W% NPA 28.6W% DMP T=30.0°C	M/U alcohol/ oleic	0.028	2.1
5.67 V% PDM-337 3.33V% TAA 2.5W% NaCl 1:1 vol nC_{13} T=30.0°C	LC/U liq cryst/ ex oil	0.018	10.0 @4.5sec^{-1}
5W% TRS 10-410 3W% IBA 1.5W% NaCl 1:1 vol nC_{12} T=22°C	M/U microemul/ ex oil	0.0015	2.5

L=Lower, M=Middle, and U=Upper Equilibrium Phase

intermediate range of tensions. It consists of 2,3-dimethylpentane (DMP), n-propanol (NPA), and water and has been studied by Giordano and Salter (12) and Raney (13). The small amount of water-soluble dye partitioned predominantly into the aqueous region and did not appear to affect the phase behavior. As Table I indicates, interfacial tensions ranging from 0.028 dyne/cm to 0.14 dyne/cm were measured between phases in equilibrium at room temperature (22°C) and 30°C. Experiments at the latter temperature were performed inside a Hotpack walk-in environmental room where the temperature was held constant to ± 0.1°C.

The next system studied was novel in that the lower phase was a homogeneous lamellar liquid crystal containing a synthetic sulfonate surfactant in equilibrium with an excess oil phase. No previous observations of liquid crystalline flow through porous media have been reported. The initial viscosity of the liquid crystalline phase was a relatively low 10 cp at a shear rate of 4.5 s^{-1}. The interfacial tension between liquid crystal and oleic phase was 0.018 dyne/cm (14).

An ultralow tension system was needed to study mechanisms of high Ca flow at relatively low velocities. A Witco petroleum sulfonate was used that had previously been studied by several researchers including Qutubuddin (15). The tension between the middle phase microemulsion and the excess oil phase was only 0.0015 dyne/cm. Surfactants were used as received, and all other chemicals were reagent grade.

Flow Cell

Numerous arrangements have been used to visualize flow through porous media. Monolayers of beads with varying wettability were the first micromodels developed (8,10). Optical problems caused by internal light reflections and refractions within the spheres made interface tracking difficult. This limitation prompted the development of a two-dimensional model obtained by etching a pore network into a glass plate, and then fusing a second glass plate on top of it (16). While current technology allows etching of a pre-determined pore geometry, limitations on pore throat size and connectivity remain with this type of model. Therefore a novel micromodel was developed, which consists of an optical rectangular capillary tube packed with a bilayer of monodisperse glass beads. The cell is 50mm long, 4mm wide and has an optical path width of 200 μm. Figure 1 shows a schematic cross section of the micromodel. The beads, obtained from Duke Scientific, are 109 μm in diameter with a standard deviation of only 2.8%. This bead size is much smaller than any used previously in flow observation experiments and yields pore throat diameters of about 16 μm between beads and 27 μm along the wall. These dimensions are of the same order of magnitude as the pores in actual reservoir rock. The two layers of beads yield a pore network that has three-dimensional characteristics and is highly interconnected. The ends of the capillary cell are covered with a woven polyester cloth (Gilson Co., 400 mesh) to retain the glass beads.

With proper manipulation large domains of hexagonal close packing can be obtained although all cells also contain small regions of less regular packing between such domains. Figure 2 shows a

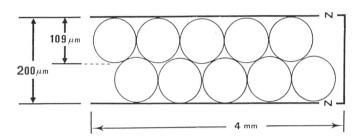

Pore Throat Diameter = **16-27** μm

Porosity = **34.5** %

Permeability = **16** d

Figure 1.   Schematic cross section of micromodel showing physical parameters.

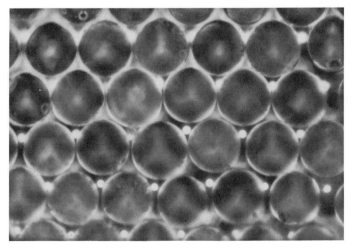

Figure 2.   Bilayer of hexagonal close packed 109 μm glass beads in flow cell with 20x objective using reflected and transmitted light.

photomicrograph of a flow cell taken with a 20x objective using reflected and transmitted illumination. In the top layer the beads are thinly outlined, while the bottom layer can also be seen to be uniformly packed with high intensity light peaks in the center of each bead.

The porosity of the flow cell was found to be 34.5%, and the absolute permeability was 16 darcies. These values are typical of unconsolidated media. All experiments were performed using a completely water-wet glass system. A new flow cell was constructed for each fluid pair studied to eliminate the possibility of contamination. For each individual system the cell was flushed with several pore volumes of alternating water and acetone between experiments, assuring reproducible results.

Flow Experiments

A schematic diagram of the overall experimental arrangement is shown in Figure 3. A gear driven Harvard microliter syringe pump was used to obtain superficial velocities ranging fram 0.1 to 2000 ft/day, which includes representative waterflood velocities of 1 ft/day. This range allows investigation of an overlap of capillary number between fluid pairs, whereby a given Ca can be studied by using various fluids with different interfacial tensions. The pre-equilibrated phases were drawn into Hamilton gas-tight syringes with teflon plungers, which provide nonpulsatile steady state flow rates. The injection ratio between phases was varied by using different syringe sizes of 1, 2.5, 5, and 10 ml. Each phase was delivered through separate microbore tubing all the way to the inlet of the flow cell. The micromodel itself was placed directly on the microscope stage, where a state-of-the-art enhanced videomicroscopy system was used to record the experiments. The basic imaging system has been described briefly by Benton, et al. (17) and Elsik (18).

A Nikon Optiphot-Pol microscope was used with extra long working distance objectives of 4x, 10x, 20x, and 40x magnification. Events within a single pore could be studied with standard projection lenses or by using a Nikon x0.9-2.25 continuously variable zoom lens. The resolving power attained with these optics was sustained using a high resolution Hamamatsu C1000-01 Chalnicon video camera with built-in electronic gain amplification and offset control. Further signal processing includes several techniques detailed by Elsik (18) and described recently in an excellent reference book on videomicroscopy by Inoue (19).

The combination of a blue transmitting filter on the microscope illumination and the red dye in the aqueous phase achieved the contrast necessary to discern interface movement. Without this feature the very thin layer of fluid containing dye did not provide sufficient contrast with the transparent oleic phase. A diffuse filter was positioned directly under the flow cell to smooth out the high intensity light peaks inherent with the glass beads due to internal reflections and refractions. Additional contrast enhancement and data analysis and quantification were performed using a Perceptive Systems Digital Image Processor (DIP).

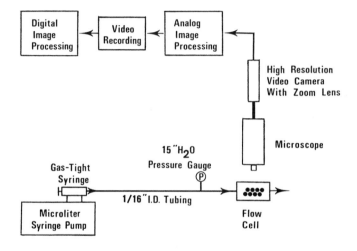

Figure 3.   Experimental apparatus schematic.

RESULTS

The results of the two-phase, steady state, cocurrent flow experiments are summarized in Table II. The various fluid systems from Table I are listed along with selected injection rates and corresponding capillary numbers. Each velocity shown is the sum of the superficial velocities for the two fluids or, equivalently, the ratio of the superficial velocity of either fluid to its fractional flow. The utility of this quantity is discussed later. The viscosities given and used in calculating Ca are those of the wetting phases in the various systems. Most of the data are for 1:1 injection ratios. The observed flow mechanisms are given in the last column of Table II.

With the high tension decane/water system continuous tortuous paths (CTP) of the type reported previously (8-10) were observed at all flow rates studied. Figure 4 shows a video frame taken at a total Darcy velocity of 1490 ft/day, with flow from left to right and $Ca = 1.5 \times 10^{-4}$. The dark phase shown is the aqueous phase containing dye, and the lighter portions of the pore space are filled with oleic phase. Care must be taken not to confuse optical effects caused by the glass beads with fluid interfaces. Some of the small circles shown ($< 20 \mu m$) are actually air inclusions within the glass beads. They do not affect pore structure or the resulting flow mechanism. The beads are packed uniformly and exhibit high intensity light peaks at their centers that leave dark bands around the edges. Although fluid velocities were high, the interfaces between the immiscible phases were completely stable and no movement was detected even at the pore level under high magnification.

The modified Reynolds Number (Re_m) calculated for this flow rate was only 0.93, where $Re_m = vD_p\rho/\mu(1-\phi)$. Therefore inertial effects can be neglected. All other experimental data discussed below have Re_m less than this value. The small bead size of $109 \mu m$ is a key factor in staying well away from the transition region of $Re_m > 10$, where inertial forces become important.

Turning attention to the alcohol system at 22°C and 30.0°C, one finds that continuous paths persist as the flow mechanism at low capillary numbers. However, when Ca reaches about 10^{-3}, the nonwetting oleic phase breaks up into mobile discontinuous ganglia. For example, in the experiment with v = 19.0 ft/day and $Ca = 1.7 \times 10^{-3}$ at 22°C, the oleic phase was observed to flow freely in the form of large ganglia having lengths of approximately ten bead diameters (BD) in the direction of flow. Figure 5 is a video frame of one of these 10 BD ganglia obtained using the real time background subtraction (RTBS) feature of the Digital Image Processor (DIP). The RTBS is used to eliminate optical effects caused by the glass beads. The width of this ganglion is 4 BD. The bead size provides a useful scaling parameter since at any magnification the 109 μm beads can be used as a reference to calibrate distances.

For steady two-phase flow of the M and U phases at 30.0°C the same behavior with ganglia of the same size can be attained with a flow rate of 3.6 ft/day. Ca is comparable at 9.5×10^{-4}. Although no mobile ganglia larger than twenty BD were observed in any of our experiments, their existence in the range of $Ca = 10^{-4} - 10^{-3}$ cannot be ruled out due to the relatively small overall size of the flow cell.

TABLE II. STEADY STATE COCURRENT FLOW OBSERVATIONS

Fluid System	V(ft/day)	Ca	Flow Mechanism
Decane/Water	7.0	7.1 E-7	Stable Continuous
33.3 dyne/cm			Tortuous Paths (CTP)
0.95 cp L/U	1490	1.5 E-4	CTP
H_2O/NPA/DMP	3.0	2.4 E-4	CTP
T=22°C L/U	19	1.7 E-3	Large Ganglia 10 BD
0.14 dyne/cm	37	3.0 E-3	Medium Ganglia 4 BD
3.2 cp	280	2.1 E-2	Subsinglets 0.5 BD
H_2O/NPA/DMP	7.2	7.8 E-4	CTP
T=30.0°C L/U	14.1	1.5 E-3	Large Ganglia 8 BD
0.068 d/cm	27.7	2.9 E-3	3.5 BD Gang. Mobile
2.1 cp	54.2	5.9 E-3	1.5 BD Gang. Mobile
	106	1.2 E-2	0.5 BD Subsinglets
H_2O/NPA/DMP	1.9	5.0 E-4	CTP
T=30.0°C M/U	3.6	9.5 E-4	Large Ganglia 10 BD
0.028 d/cm	54.2	1.4 E-2	0.67 BD Subsinglets
2.1 cp	106	2.8 E-2	Drops Stretched Out
PDM/TAA/C_{13}	6.4	1.2 E-2	Singlet Size Drops
2.5W% NaCl	12.6	2.5 E-2	0.5 BD Drops Mobile
0.018 d/cm	48.4	9.6 E-2	Filament Drops
10 cp LC/U			
TRS/IBA/C_{12}	4.9	2.9 E-2	0.5 BD, Some Elongated
1.5W% NaCl	9.6	5.7 E-2	<.5 BD drops, Snake-Like
0.0015 d/cm	36.0	2.2 E-1	Sev. Filaments In Pore
2.5 cp M/U	141	8.3 E-1	Fast Filament Flow

L= Lower, M=Middle, and U=Upper Equilibrium Phases

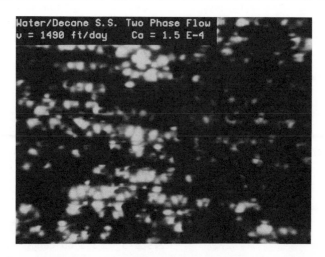

Figure 4.   Continuous tortuous paths, water/decane. Water is the
             dark phase, oil is the clear phase. Ca = 1.5 x 10^{-4}.

Figure 5.   Large mobile ganglion, 10 BD, Ca = 1.7 x 10^{-3}, L/U
             alcohol system at 22°C.   Video frame obtained using
             real time background subtraction.

The steady state values of length listed in Table II are an average of those present in different parts of the flow cell. The lengths observed usually varied by no more than a couple BDs for long ganglia, resulting in a relatively narrow distribution about the reported value. These variations were probably due to subtle permeability fluctuations. In individual areas of the flow cell, however, the ganglia were uniform in size to within experimental limits of measurement.

Transient changes in ganglion length were occasionally observed. They can be understood by recalling that at a given Ca in this range, only ganglia exceeding some critical length in the mean direction of flow are mobile. Isolated, trapped ganglia shorter than the critical length were sometimes observed, but they soon became mobile as moving ganglia coalesced with them. Ganglia longer than the steady state value rapidly broke up into smaller ganglia. These phenomena are discussed further below. Two important points to note in Table II are that the Ca ranges for the different fluid systems overlap and that in the overlap regions the same dependency of flow mechanism on capillary number was observed.

As Table II indicates, ganglion length continues to shrink with increasing Ca until at Ca = 0.01 singlets, i.e., ganglia of one BD, are completely mobile. Included in this range of Ca is the lamellar liquid crystal/excess oil system. Data listed for this system were obtained with a 2.5:1 ratio of the injected fluids, but other injection ratios showed similar results. Although the homogeneous liquid crystalline phase had a somewhat higher viscosity than the other fluid systems studied, it flowed readily through the pore network without plugging. Liquid crystals (LC) of this type exhibit non-Newtonian behavior, with viscosity being both shear rate and time dependent (20). A viscosity measured at 4.5 sec^{-1} was used since this shear rate is representative of those present in the flow cell at moderate velocities. The mechanisms of flow in the LC system were the same as those exhibited by the isotropic phases.

Figure 6 is a 10x video frame of this LC system at v = 12.6 ft/day and Ca = 0.025. Singlets are completely mobile and most ganglia are actually smaller than a single pore body. This section of the flow cell was packed in a body-centered cubic arrangement. The discontinuous ganglia are seen in the pore bodies. They deform upon entering pore throat constrictions. Figure 7 is a high magnification photo of these ganglia in a hexagonal close packed area of the flow cell. The two subsinglets shown are deformed into oval shapes as they pass through the smaller pore throats along the upper layer of beads. Flow is from left to right, and the mobile wetting phase remains continuous.

As Ca is increased further, drop size continues to decrease, falling below the diameter of the pore throats and eventually becoming small enough for several drops to pass through a single pore constriction concurrently. In this last situation the drops are also somewhat elongated in the direction of flow.

At still higher Ca, the drops become even smaller and elongate further along the lines of flow until a filament type flow mechanism is established by Ca = 0.1. The filaments are less than 5 μm in diameter but can sometimes reach over 50 μm in length. The middle phase microemulsion and equilibrium excess oil phase flowing at 141 ft/day provided an opportunity for observation of this phenomenon at

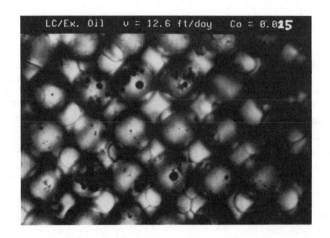

Figure 6.   Mobile singlet and subsinglets, LC/Excess Oil, v =
            12.6 ft/day, Ca = 0.025.   Singlet near center of
            photograph.   10x objective.

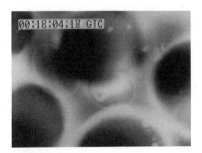

Figure 7.   Subsinglets deforming in pore throats.   LC/Excess
              Oil, 2.5:1 volumetric injection ratio.   v = 12.6
              ft/day, Ca = 0.025. 40x objective.

a very high Ca of 0.83.   This flow mechanism could be important in designing processes that are enhanced by high surface areas.
For Ca > 0.5, no major change in basic flow mechanism is foreseen until the onset of inertial effects.   Care should be taken, however, when dealing with ultralow tension fluids in formations of appreciable thickness since buoyant forces could become important (21).

The observed trend in flow mechanisms is shown in Figure 8. This cocurrent flow capillary number correlation summarizes the steady state experimental observations over the full range of Ca studied.   Injection ratios have been varied between 10:1 and 1:10 with no basic change in the flow mechanism observed.   Some subtle differences were seen at high ratios of wetting to nonwetting phase, including a slightly broader size distribution of the mobile ganglia.   This behavior is a result of nonuniform distribution of the discontinuous phase across the flow cell, causing variation in the pressure gradient within the wetting phase.

At high ratios of nonwetting to wetting phase even larger size distributions developed.   No mobile ganglia smaller than the steady state value were observed, but many larger than the predicted size were seen.   This behavior is a result of the high saturation of the discontinuous phase, which allows for a high frequency of coalescence between mobile ganglia.   The pressure field in the continuous phase also fluctuates as the cross sectional area available for flow changes with local nonwetting phase saturation.

Especially noteworthy in this regard were experiments at a 1:10 ratio of wetting to nonwetting phase which were performed with the L and U phases of the alcohol system at 30°C to determine steady state flow behavior at fluid saturations characteristic of foam flow through porous media.   At this high ratio corresponding to 91% foam quality the wetting phase remained continuous as the oleic phase followed the same ganglia flow mechanisms described previously.   For instance, at v = 39.6 ft/day and Ca = 4.3 x 10^{-3}, 2-3 BD ganglia were observed.   Some ganglia longer than 5 BD were present as a result of the coalescence mentioned above, and singlets were momentarily trapped, as also seen during the 1:1 ratio experiments.   For Ca = 0.0166 singlets were completely mobile.   Temporary pendular rings of aqueous phase were observed as the larger oleic ganglia moved through adjacent pore throats.   The wetting phase retreated to the tight pore space between beads and left the larger pore body filled with nonwetting phase.

DISCUSSION

The mechanisms of steady state, cocurrent, two-phase flow through a model porous medium have been established for the complete range of capillary numbers of interest in petroleum recovery.   A fundamental understanding of the mobile ganglia behavior observed requires a knowledge of how phases break up during flow through porous media.   Several mechanisms have been reported in the literature and two have been observed in this flow cell.

The first mechanism is snap-off, studied by Roof (22), and discussed in more detail recently by Mohanty et al (23), and Falls et al (5).   Basically a thermodynamic instability arises as curvature and hence capillary pressure variations cause the wetting fluid to

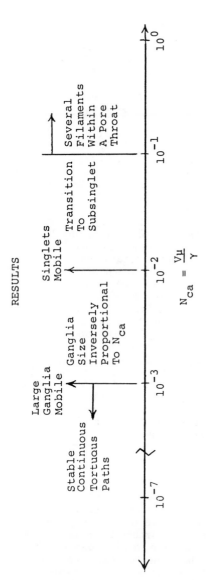

Figure 8.   Cocurrent Flow Capillary Number Correlation.

flow into a pore throat constriction occupied by a ganglion of the nonwetting phase. The result is a local decrease in ganglion diameter and eventually snap-off. This phenomenon has occasionally been observed in the present cocurrent flow experiments. Since it is favored by high aspect ratio, i.e., a high pore body to pore throat ratio, it should be more important in granular media with broader distributions of particle size and shape as well as in consolidated media.

The dominant breakup mechanism seen here is dynamic splitting, which has been observed previously by Payatakes (24) in ganglia displacement experiments. Figure 9 shows how dynamic splitting works in our uniformly packed micromodel. When a mobile ganglion exists that exceeds the minimum length for mobilization for given flow conditions, it can progress into more than one downstream pore throat. As it continues to flow, the trailing interface moves up to the pore bifurcation and the ganglion splits, forming two or more smaller ganglia. In Figure 9 three daughter ganglia are formed from a single larger ganglion as it flows into three downstream pores around beads "A" and "B" from left to right. Figure 9B is a schematic drawing showing the phase boundaries and glass bead locations in Figure 9A. Another drop whose diameter is about equal to that of the pore throats is shown as "C". Dynamic breakup is favored in the present model by the fact that only small geometrical variations typically exist between adjacent pores. Note that in contrast to snap-off, which is a thermodynamic instability, forces produced by flow are essential in the dynamic breakup process.

Figure 10 shows that this mechanism continues to be significant for the breakup of smaller drops. The viscous forces are strong enough to stretch the original subsinglet drop shown into adjacent pore throats, even though it is smaller than a single pore constriction. The first image shows a video frame taken with the 40x objective of a 15 μm drop already deformed into a crescent shape. The next image shows the drop stretched into a "C" shaped blob, and the final image is just prior to breakup. The splitting shown by the sequence of photos takes less than a third of a second, and the subsinglet breaks into two drops approximately the same size. Such dynamic breakup of small drops should be an important aspect of flow at high capillary numbers whatever the pore size distribution of the medium.

At low Ca, capillary forces dictate the tortuous nature of the continuous paths. When the pressure gradient in the flowing phases becomes high enough, viscous forces start to impose a more direct downstream flow. When a sufficiently large Ca is reached, the CTP's of the nonwetting phase break up as they start to enter additional downstream pores. This phenomenon is similar to dynamic breakup, as a CTP splits and flows into one or more pore throats previously occupied by the wetting phase. The wetting phase also invades pore bodies previously occupied by the nonwetting phase.

Once a discontinuous nonwetting phase is formed, the capillary forces opposing ganglia motion must be overcome, or the ganglia will be trapped. The maximum capillary pressure rise exhibited by ganglia of any length in a porous medium can be estimated in terms of throat radius r_t and body radius r_b:

$$\Delta P_{c,max} = 2\gamma(1/r_t - 1/r_b) \tag{2}$$

A

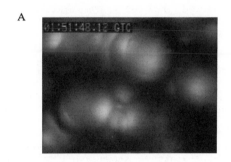

B

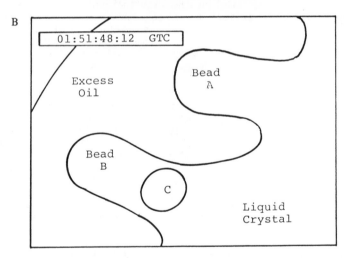

Figure 9.   A) Dynamic splitting of 3 BD ganglion, 20x objective.
             B) Tracing of ganglion interface with bead and
                 subsinglet locations.

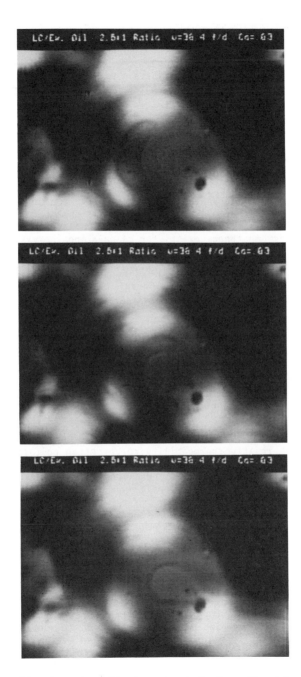

Figure 10.   Dynamic splitting of subsinglet, 40x objective.

This equation assumes a contact angle of zero, a good approximation for the glass micromodel used here where the oleic phase has been observed flowing through pore constrictions without forming contact lines.

The viscous pressure drop in the continuous (wetting) phase provides the force necessary to sustain motion of a ganglion in spite of the capillary forces. Darcy's Law for the wetting phase may be written as:

$$L = \frac{k_{ew} \Delta P_w}{\mu_w} \frac{1}{v_w} \quad\quad\quad (3)$$

where ΔP_w is the pressure drop in this phase over a length L, taken here as ganglion length, and k_{ew} is the effective permeability, i.e., the product of the absolute permeability k and the applicable relative permeability k_{rw}. With μ_w measured independently, and L and v_w known from the steady state flow experiment, the only unknowns are k_{ew} and ΔP_w. The actual pressure drop across a mobile ganglion should be somewhat greater than ΔP_c, since it includes viscous effects. Since no theory exists for calculating the viscous pressure drop within a moving ganglion in this pore geometry, $\Delta P_{c_{max}}$ can be used as a first approximation for ΔP_w in Eq.(3), and the equation becomes basically that employed by previous workers who studied mobilization of trapped drops (25,26). If k_{ew} is approximately constant, not an unreasonable condition if the flow ratio between phases is fixed, a plot of L as a function of $1/v_w$ should be a straight line.

Figure 11 shows the result of a test of this concept, using the data from Table II for the L/U alcohol system at 30.0°C with a 1:1 flow ratio. The data do, in fact, follow a line that has a slope of 1.334 cmft/day, calculated from linear regression analysis. Using r_t = 8 μm and r_b = 18 μm for 109 μm beads (27), k_{ew} becomes 10.6 darcy. For this flow cell k = 16 darcies and k_{rw} is thus 0.66, a reasonable value for high Ca flow with S_w = 0.61. This saturation was determined by the method described below. Basically, $k_{rw} \cong S_w$ within the limits of experimental accuracy, a result consistent with the straight line relative permeability curves often postulated for high Ca. From a physical point of view, this result means that the wetting phase is no longer being excluded from the larger pores, as indeed is observed with the mobile ganglia mechanism of two-phase cocurrent flow.

However, the observed flow mechanism of mobile ganglia raises a fundamental question as to whether relative permeability theory is indeed valid at high Ca for the discontinuous phase since it is unable to communicate a continuous pressure drop over macroscopic distances. This question should be addressed in future research.

The mechanisms of phase breakup discussed above can sometimes produce a ganglion shorter than the minimum mobile size for the experimental conditions. In this case it becomes immobile or trapped, as long as it is larger than pore throat diameter and consequently deforms in pore constrictions to provide capillary trapping forces. Since there is continuous injection of both phases, a mobile ganglion will eventually collide with the trapped ganglion, coalesce, and form a longer ganglion. Thus, trapped ganglia during steady simultaneous flow exist only temporarily as long as each phase is uniformly distributed.

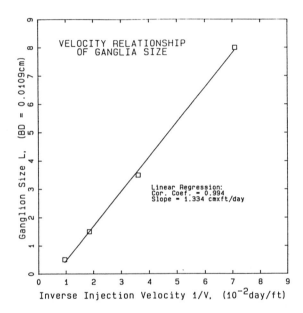

Figure 11.   Dependence of ganglion size on reciprocal of velocity for L/U alcohol system at 30°C.

The mobilization of a trapped phase during displacement has been the subject of extensive research (11,24-26,28-32). These papers all dealt with the concept of viscous forces overcoming capillary trapping forces. Their results can be applied to understand the behavior seen in displacement experiments performed over a range of Ca by several researchers to study residual oil structure and saturation. As Ca increased, the maximum trapped blob size decreased, until eventually singlets were mobile.

This trend is identical to the observation of mobile ganglia presented here. The mobile ganglia size found at a given Ca provides an upper bound on residual ganglia size at that Ca, and should be very close in length to the longest trapped blob possible. If a ganglion should become much longer than that required for mobilization, it would soon break up by one of the observed mechanisms described above.

The digital image processor (DIP) can be used to determine other salient features of ganglia dynamics. Saturations can be estimated and interstitial velocities can be measured from digitized frames of recorded experiments. Papers using DIP methods have shown an ever-increasing variety of applications (33,34). Consider the L/U alcohol system at 30.0°C with v = 14.1 ft/day and 8 BD mobile ganglia. The actual interstitial velocity of ganglia as determined from sequences of video frames was v_{inw} = 48.2 ft/day. A second method for calculating v_{inw} -- and v_{iw} as well -- is to divide the superficial velocity of each phase by the product of its saturation and micromodel porosity. Taking average values over time from the DIP, one obtains S_w = 61% and S_{nw} = 39%. The resultant velocities are v_{iw} = 33.5 ft/day and v_{inw} = 52.5 ft/day. The oleic phase velocities from the two methods agree to within 9%, thus confirming the experimental results obtained using the image processing equipment. Note that the discontinuous ganglia are flowing faster than the continuous wetting phase, a result expected only when the viscosity ratio is less then unity. Here μ_{nw}/μ_w = 0.68 cp/2.1 cp = 0.32 < 1. This phenomenon has also been observed by other researchers in different types of experiments (24,35).

Abrams (36) found an effect on residual oil saturation of the viscosity ratio between phases. While the data in this paper are limited to a small range of viscosity ratios less than unity, Chatenever (9) presented results for systems where the nonwetting phase had the higher viscosity. He observed that the nonwetting fluid continuous tortuous paths changed to a slug flow mechanism at high velocities. Hence, the basic flow mechanism described here likely applies for a wide range of viscosity ratios. Chatenever gave no data on ganglia length, however, and a detailed study of the effect of viscosity ratio on steady flow at high Ca seems desirable.

A heterogeneous pore structure with varying aspect ratio would increase the frequency of breakup and coalescence, which should increase the observed mobile ganglia size distribution. However, the basic flow mechanism should remain unchanged. Also the relative importance of snap-off as a breakup mechanism would be increased relative to dynamic splitting. Here too a detailed study seems desirable.

One interesting consequence of the results described above is that the oil phase should be discontinuous in the portion of an oil bank where Ca is high during an enhanced recovery process such as surfactant flooding. Indeed, the formation of disconnected mobile ganglia was reported by Paterson et al (37) at high Ca in displacement experiments. The implications for mobility control in enhanced recovery processes merits further study.

As mentioned previously, Reed and Carpenter (7) suggested that a two-phase mixture of an aqueous polymer solution and a microemulsion could be used to displace residual oil while simultaneously maintaining mobility control. Drop sizes after shearing were small enough for the emulsions to be translucent, and the interfacial tension in one system was about 4×10^{-5} dyne/cm. This information suggests a very high capillary number in the range of 0.1 where we observed droplets much smaller than pore dimensions to be flowing (see Figure 8).

CONCLUSIONS

1) A novel physical micromodel with a well-defined three-dimensional pore network and with pore sizes comparable to those in petroleum reservoirs has been developed to study multiphase flow through porous media. This cell can be used to visualize pore level mechanisms as well as larger scale phenomena.

2) The mechanism of steady state, two-phase, cocurrent flow through porous media has been determined as a function of capillary number over the full range of conditions pertinent to petroleum recovery. At low capillary numbers stable continuous tortuous paths of both phases exist. When Ca reaches about 10^{-3}, the nonwetting phase breaks up into large mobile ganglia. Increasing Ca reduces ganglia size in an approximately linear manner until Ca = 10^{-2}, where singlets are completely mobile. Further increases in Ca result in subsinglet drops that continue to decrease in size. Eventually two or more drops can pass simultaneously through individual pore throats. At high Ca > 0.1, viscous forces predominate and the drops stretch out into filaments and follow the continuous phase streamlines.

3) A unique steady state mobile ganglion size is established for each Ca when there is a homogeneous pore geometry. Smaller ganglia trap momentarily, and larger ganglia break up by either snap-off or dynamic breakup.

4) Volumetric injection ratios were varied between 10:1 and 1:10 with no basic change in cocurrent flow mechanism.

5) At corresponding capillary numbers the mobile ganglia size will be nearly equal to the upper bound on residual ganglia size determined by displacement experiments.

6) Relative permeability theory may not apply at high capillary number flow for the nonwetting phase since it becomes discontinuous and its flow rate may thus not be proportional to the macroscopic pressure gradient.

7) Certain dilute lamellar liquid crystalline phases having relatively low apparent viscosities can propagate through this porous medium micromodel without plugging it. Their behavior followed the trends established with isotropic phases.

ACKNOWLEDGMENT

This research was supported by grants from Amoco Production

Company, Arco Oil and Gas Company, Chevron Oil Field Research Company, Exxon Production Research Company, the Mobil Foundation, Shell Development Company, and the Standard Oil Company. Discussions with W.J. Benton on various aspects of the experimental techniques were most helpful.

Literature Cited

1.   Craig, F.F. Reservoir Engineering Aspects of Waterflooding. Soc. Petrol. Eng.: Dallas, 1971.
2.   Scheidegger, A.E. The Physics of Flow Through Porous Media, U. of Toronto Press, Toronto, 1974.
3.   Greenkorn, R.A. Flow Phenomena in Porous Media, Marcel Dekker, New York, 1983.
4.   Willhite, P. Waterflooding. Soc. Petrol. Eng.: Dallas, 1986.
5.   Falls, A.H.; Gauglitz, P.A.; Hirasaki, G.J.; Miller, D.D., Patzek, T.W.; Ratulowski, J. SPE/DOE 14961, Presented at Fifth Symp. on EOR, Tulsa, 1986.
6.   Radke, C.J.; Ransohoff, T.C. SPE 15441, Presented at Annual Meeting, New Orleans, 1986.
7.   Reed, R.L.; Carpenter, C.W. U.S. Patent 4 337 159, 1982.
8.   Chatenever, A.; Calhoun, Jr., J.C. Pet. Trans. AIME 1952, 195, 149-156.
9.   Chatenever, A. API Research Project 47-B Final Comprehensive Report, 1957.
10. Kimbler, O.K; Caudle, B.H. Oil and Gas J. 1957, 55, 85-88.
11. Larson, R.G.; Davis, H.T.; Scriven, L.E. Chem. Eng. Sci. 1981, 36, 75-85.
12. Giordano, R.M.; Salter, S.J. SPE/DOE 12697, Presented at Fourth Joint Symp. on EOR, Tulsa, 1984.
13. Raney, K.H. Ph.D. Thesis, Rice U., Houston, 1985.
14. Ghosh, O.; Miller, C.A. J. Colloid Interface Sci. 1987, 116, 593-597.
15. Qutubuddin, S. Ph.D. Thesis, Carnegie-Mellon Univ., Pittsburgh, 1983.
16. Mattax, C.C.; Kyte, J.R. Oil and Gas J. 1961, 59, 115-128.
17. Benton, W.J.; Raney, K.H.; Miller, C.A. J. Colloid Interface Sci. 1986, 110, 363.
18. Elsik, C.M., Ph.D. Thesis, Rice University, Houston, 1987.
19. Inoue, S. Video Microscopy, Plenum Press, New York, 1986.
20. Benton, W.J.; Baijal, S.K.; Ghosh, O.; Qutubuddin, S.; Miller, C.A. SPE Res. Eng. 1987, 2, 664-670.
21. Tham, M.J.; Nelson, R.C.; Hirasaki, G.J. Soc. Petrol. Eng. J. 1983, 23, 746-758.
22. Roof, J.G. Soc. Petrol. Eng. J. 1970, 10, 85-90.
23. Mohanty, K.K.; Davis, H.T.; Scriven, L.E. SPE Res. Eng. 1987, 2, 113-128.
24. Payatakes, A.C. Ann. Rev. Fluid Mech. 1982, 14, 365-393.
25. Melrose, J.C.; Brandner, C.F. J. Can. Petrol. Tech. 1974, 13(4), 54-62.
26. Stegemeier, G. Improved Oil Recovery By Surfactant And Polymer Flooding; Shah, D.O.; Schechter, R.S., Eds.; Academic Press, New York, 1977; pp. 55-91.
27. Graton, L.C.; Fraser, H.J. J. Geology 1935, 43, 785-909.
28. Chatzis, I.; Kuntamukkula, M.S.; Morrow, N.R. SPE 13213, 1984.

29. Chatzis, I.; Morrow, N.R.; Lim, H.T. Soc. Petrol. Eng. J. 1983, 23, 311-326.
30. Wardlaw, N.C.; McKellar, M. Can. J. Chem. Engr. 1985, 63, 525-532.
31. Lenormand, R.; Zarcone, C. SPE/DOE 14882, Presented at Symp. on EOR, Tulsa, 1986.
32. Ng, K.M.; Davis, H.T.; Scriven, L.E. Chem. Eng. Sci. 1978, 33, 1009-1017.
33. Yadav, G.D.; MacDonald, I.F.; Chatzis, I.; Dullien, F.A.L. SPE 15496, Presented at Annual Meeting New Orleans, 1986.
34. Ruzyla, K. SPE 13133, Houston, 1984.
35. Payatakes, A.C.; Dias, M.M. Rev. in Chem. Engr. 1984, 2, 85-174.
36. Abrams, A. Soc. Petrol. Eng. J. 1975, 15, 437-447.
37. Paterson, L.; Hornof, V.; Neale, G. Soc. Petrol. Eng. J. 1984, 24, 325-327.

RECEIVED February 11, 1988

Chapter 14

Snap Off at Strong Constrictions: Effect of Pore Geometry

J. Ratulowski and H.-C. Chang

Department of Chemical Engineering, University of Notre Dame,
Notre Dame, IN 46556

It has been observed that bubbles trapped before a
strong constriction can snap off producing small
bubbles at a high rate. This behavior cannot be
described with usual drainage models. Here, a model
based on a linear stability analysis of a general
evolution equation for an axisymmetric thread in a
channel of arbitrary cross section is developed to
predict snap-off frequency and bubble volume. Two
geometries are investigated--capillaries with square
and circular cross sections. The snap-off rate for
the circular capillary agrees qualitatively with
observed behavior. In the case of the square
capillary, bubble volume was found to be only a weak
function of the liquid flow rate while both the
bubble volume and snap-off frequency are strong
functions of the constriction aspect ratio.

Many enhanced oil recovery processes use gas drives to displace
trapped oil, ie. steam floods and CO_2 floods. The sweep efficiency
of these methods is often low because of the unfavorable mobility
ratio of gas to oil. (The mobility of gas is much greater than the
mobility of oil.) Therefore, gas tends to overide or finger through
oil.
    The mobility ratio may be improved by creating a foam which is a
stable dispersion of gas in a continuous liquid phase (1,2).
Mobility of the gas phase is reduced in two ways when a foam is
used. First, liquid lamallae become trapped and block off a portion
of the porous media. This results in a stationary foam fraction
that can be nearly unity at low flow rates. Also lamellae present
in the flowing foam fraction create an additional resistance to
flow. The apparent viscosity of foam can be several orders of
magnitude larger than the gas viscosity.
    The mobility of foam has been found to be a function of pore
geometry, flow rate, quality, surfactant concentration and bubble

0097–6156/88/0373–0282$06.00/0
© 1988 American Chemical Society

size (3). An additional complication arises when bubble size is
altered within the porous medium. Generation of lamellae through
division at pore branches and snap off at pore constrictions and
destruction through coalescence have been observed in etched plates
and bead packs (4,5). A complete model for foam mobility must
include models for the generation and destruction of lamellae as
functions of the local conditions within the porous medium. Here,
the effects of pore geometry on generation by snap off will be examined.

Snap Off Mechanisms

Lamellae are produced by snap off at constricitons in single capillaries through two mechanisms, drainage and hydrodynamic instability
(see Figure 1). For the drainage mechanism, the bubble front must
pass through the constriction and leave a wetting film of liquid on
the channel walls. When the bubble front is sufficiently far from
the constriction neck, the curvature of the gas liquid interface
results in a pressure gradient in the film which induces flow from
both the upstream and downstream regions of the film into the
constriction neck. The minimum distance of the bubble front from
the pore neck is a function of the pore geometry. Roof (6) found
that for toroidal constrictions, a bubble must extend seven neck
diameters from the constriction before drainage occurs. Once
drainage starts, a liquid collar grows and then snaps off. For this
type of snap off the rate determining step is the growth of the
liquid collar. Gauglitz and Radke (7) have successfully modeled
snap off by the drainage mechanism in circular constricted
capillaries with an approach based on lubrication analysis. Bubbles
formed by the drainage mechanism have a radius larger than the
constriction neck radius.
    Arriola (8) and Ni (5) have observed a second mechanism for snap
off in strongly constricted square capillaries. At low liquid flow
rates, a bubble is trapped in the converging section of the
constriction and liquid flows past the bubble. As liquid flow rate
increases, waves developed in the film profile and at some critical
liquid flow rate these oscillations become unstable and bubbles snap
off. In these experiments, the bubble front is located upstream of
the constriction neck. Therefore, no driving force for the drainage
mechanism exists. Bubbles formed by this mechanism are produced at
a high rate and have a radius on the order of the constriction neck.
No attempt has previously been made to model snap-off rate by this
mechanism in noncircular constrictions.
    Snap off by the instability mechanism may occur in the following
way. A bubble in an angular constriction such as a square channel
will flatten against the walls as shown in Figure 2. The radius of
the circular arcs in the corners for a static bubble is about one
half of the tube half width. At low liquid flow rates, a bubble
trapped behind a constriction has this nonaxisymmetric shape. As
liquid flow rate increases, the bubble moves farther into the
constriction and the fraction of cross sectional area open to liquid
flow at the front of the bubble increases until the thread becomes
axisymmetric at some point near the bubble front.

Ransohoff and Radke (10) have shown with lubication analysis that
the axisymmetric configuration is reached near the bubble front for
a sufficiently high liquid flow rate in a constriction with triangular cross section. The critical liquid flow rate is a function of
the pore geometry. This configuration is linearly unstable for all
liquid flow rates. The bubble volume is then governed by the maximum growing wavelength in the axisymmetric film profile and the snap
off frequency can be estimated from the eigenvalue of the following
linear stability analysis. For constrictions with a low aspect
ratio, the bubble front may pass through the constriction before the
axisymmetric configuration is reached. In this case, a liquid
collar develops by the drainage mechanism. The collar grows until
the unstable axisymmetric configuration is reached and then snaps
off through the growth of instabilities. However, in this case, the
rate determining step is the growth of the liquid collar. This type
of snap off has been observed by Gauglitz et. al. (9) in square
channels.

Evolution Equation

First, an evolution equation for the position of an axisymmetric
thread in a channel of arbitrary cross section is needed. Consider
the axisymmetric gas thread shown in Figure 3. The surface $\partial\Omega_1$ is
the channel wall and the surface $\partial\Omega_2$ is the gas liquid interface.
The region Ω contains a viscous fluid. The distance of the interface from the centerline is the function $C^*(z^*)$. When both the
position of the channel walls and the gas liquid interface vary
slowly in the axial direction, the lubrication approximation of
fully developed unidirectional flow in the axial direction is
appropriate. The liquid flow rate at any axial position, z^*, is
given by the solution of the Stokes equation. The no slip boundary
condition is imposed at the tube wall and the jump momentum balance
is satisfied at the interface. The normal component of the jump
momentum balance provides a relation between the interface shape and
pressure in the film. For gas bubbles in a surfactant free system,
the tangential component of the jump momentum balance requires the
tangential stress to vanish at the interface. When surfactant is
present, the velocity at the surface is some unknown function of z^*
and θ, $u^{\sigma*}(z^*,\theta)$, which may be determined by the simultaneous solution of the surfactant transport equations for both the bulk fluid
and the interfacial regions, the tangencial component of the jump
momentum balance and the film evolution equation. The effects of
finite drop viscosity may be included through the tangential component of the jump momentum balance with the methods similar to
those developed by Teletzki (11) or Jo (12) for the motion of drops
in straight circular tubes. The relation between the local pressure
gradient and the flow rate may be conviently expressed in terms of a
dimensionless flow coefficient which is a function only of the axial
position.

$$Q = \frac{1}{Ca} K (u^\sigma, C, z) \frac{dP}{dz} \qquad (1)$$

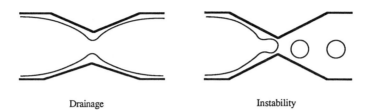

Figure 1.   Snap off mechanisms.

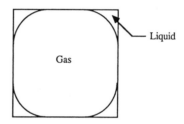

Figure 2.   Nonaxisymmetric thread.

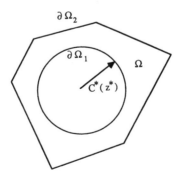

Figure 3.   Nonaxisymmetric channel.

Here the flow rate, Q, and pressure gradient are dimensionless quantities defined as

$$Q = \frac{Q^*}{U\,a^2}, \qquad \frac{dP}{dz} = \frac{a^2}{\sigma}\,\frac{dP^*}{dz^*}$$

The characteristic length scale, a, is the hydraulic radius of the channel at some reference point, the velocity, U, is the gas velocity at that point, and σ is the surface tension. The starred quantities are dimensional. The capillary number, Ca, is based on the reference gas velocity and the liquid phase viscosity. The flow coefficient is defined as

$$K(u^\sigma, C, z) = \int_\Omega u(x,y)\,dx\,dy \qquad (2)$$

Where the quantity u(x,y) is the solution of

$$\left(\frac{\partial}{\partial x^2} + \frac{\partial}{\partial y^2}\right) u(x,y) = 1 \qquad \text{in } \Omega \qquad (3)$$

The no slip boundary condition at is imposed at the tube wall. In a pure gas-liquid system, one with no surfactants, the tangencial stress is set to zero along the interface. However, if surfactants are present the boundary condition at the interface becomes

$$u(x,y) = u^\sigma(z,\theta) \quad \text{on } \partial\Omega_2 \qquad (4)$$

The flow rate can be related to the interface position through the kinematic condition,

$$\frac{1}{Ca}\frac{\partial S}{\partial \tau} = -\frac{\partial Q}{\partial z} \qquad (5)$$

Here $\tau = t\sigma/\mu\,a$ is the dimensionless time and S is the cross sectional area open to liquid flow. For small interface slopes in the axial direction, the pessure gradient is related to the interface location through the normal component of the jump momentum,

$$\frac{dP}{dz} = \frac{C'}{C^2} + C''' \qquad (6)$$

The prime denotes differentiation with respect to z. Combining Equations 1, 5, and 6 gives the general evolution equation for axisymmetric threads,

$$\frac{\partial C}{\partial \tau} = \frac{1}{2\pi C}\frac{\partial}{\partial z}\left[K\left(u^\sigma, C, z\right)\left(\frac{C'}{C^2} + C'''\right)\right] \qquad (7)$$

The constriction shape is characterized by two quantities, the aspect ratio and the constriction length. The aspect ratio is the ratio of the straight tube or pore body hydraulic radius to the hydraulic radius at the constriction neck. A strong constriction is one with a large aspect ratio and a slender constriction is one with a length much larger than the change in hydraulic radius from the pore body to the neck. The above evolution equation is then valid for strong constrictions as long as they remain slender.

Linear Stability

Equation 7 may be linearized by introducing the following deviation variables.

$$\phi(\tau,t) = C_0(z) - C(\tau,z) \tag{8}$$

$$\eta(\tau) = u^\sigma(\tau,z,\theta) - u^\sigma_0(z,\theta) \tag{9}$$

The quantities with the 0 subscript are the steady state solutions to the transport and film profile equations. Substituting equations 8 and 9 into 7 and neglecting terms second order and higher in ϕ and η, gives a linear approximation to equation 7.

$$\frac{\partial\phi}{\partial\tau} = \frac{1}{2\pi C_0} [K(u^\sigma_0,C_0,z)(\frac{\phi''}{C_0^2}+\phi'''') + (\frac{\partial K}{\partial z})_{u^\sigma_0,C_0}(\frac{\phi'}{C_0^2}+\phi''')] \tag{10}$$

The deviation ϕ can be represented as a sum of normal modes of the form

$$\phi = \exp(\omega\tau + ikz) \tag{11}$$

Where ω is a complex frequency and k is a real wavenumber. Because the geometry is infinite in the axial direction, k can vary continuously. Substituting the normal mode into equation 10 gives the dispersion relation for ω.

$$\omega = \alpha[x^4 - x^2] + i\ \beta(x-x^3) \tag{12}$$

where

$$x = k\ C_0 \ , \quad \alpha = \frac{K\ (u^\sigma_0,C_0,z)}{2\pi C_0^5} \ , \quad \beta = \frac{(\frac{\partial K}{\partial\tau})_{u^\sigma_0,C_0}}{2\pi C_0^4}$$

Disturbance will grow with time when the real part of the complex frequency is positive. This is true at all capillary numbers for wavenumbers less than $1/C_0$. The growth rate reaches its maximum positive value when x is equal to $1/\sqrt{2}$. The maximum growth rate is then

$$\omega^m_r = \frac{K(u^\sigma_0,C_0,z)}{8\pi C_0^5} \tag{13}$$

and the maximum growing wavelength is

$$\lambda^m = 2\ \sqrt{2}\ \pi C_0$$

If the disturbances are assumed to grow at the linear rate, the dimensionless frequency of lamellae generation is

$$f = \ln(\frac{B}{C_0}) \frac{K(u_0^\sigma, C_0, z)}{8\pi C_0^5} \tag{14}$$

where $f = f^* \mu a / \sigma$ and the dimensionless volume of the bubbles formed is

$$V = 2 \sqrt{2} \pi^2 C_0^3 \tag{15}$$

where $V = V^*/a^3$. The constant B is related to the magnitude of the initial disturbance. The linear growth rate will persist only as long as sufficient liquid reaches the region of the growing instability. In the case of thin circular films this condition is not always satisfied. The nonlinear analysis of Hammond (13) for circular thin films predicts a decrease in the growth rate with time and the eventual breakup of the film into rings when there is not sufficient liquid to form lamellae. Because of the larger flow channels available to the liquid in the corners of channels with angular cross section, one would not expect this phenomenon to occur in square capillaries.

The presence of surfactant only affects the snap off frequency and the volume of the bubbles formed through the base state interface location and the expression for the flow coefficient. The axisymmetric configuration is linearly unstable at all flow rates even when surfactant is present. However, additional nonlinear effects may slow the disturbance growth. As the instability grows, the interface is stretched and the surface concentration of surfactant in the interfacial region is reduced. As a result the surface tension is higher in the region of the growing instability than in the surrounding regions. This induces a liquid flow out of the collar and thus, reduces the growth rate. This effect is second order.

Circular and Square Constrictions

The relations presented in the previous section are valid for an axisymmetric thread in a channel with an arbitrary cross sectional shape. Now the analysis will be applied to two cases that have been studied extensively experimentally, circular and square capillaries. The circular channel is the simplest geometry in which snap off occurs. In this case, the flow coefficient can be calculated analytically. Although the linear stability analysis is not valid in general for the drainage mechanism typically observed at circular constrictions, it still is a good approximation of the initial growth rate of disturbances in axisymmetric films. As mentioned previously, this growth rate could continue only as long as sufficient liquid is present in the region of the growing instability. However, because the bubble front has passed through the constriction, the maximum growing wavelength is not related to the bubble volume.

Consider a long bubble which travels at a constant velocity. The bubble passes through a constriction and lays down a film with a thickness that is a function of the bubble speed and the surfactant concentration. In this case, the flow coefficient for the pure system is

$$K(C_0) = \frac{-2\pi R(R-C_0)^3}{3} \qquad (16)$$

where R is the constriction aspect ratio.
When this result is substituted into equation 7, one obtains the extended evolution equation of Radke and Gauglitz (7) for a straight channel of radius r, the constriction neck radius. If surfactant is present the flow coefficient is

$$K(C_0) = \frac{-2\pi R(R-C_0)^3}{12} + \pi R \, u^\sigma \, (R-C_0) \, Ca \, (\frac{dP}{dz})^{-1} \qquad (17)$$

Substituting the Ca dependence of C_0 predicted by Bretherton (14) for the pure system and by Chang and Ratulowski (15) for the low capillary limit of a surfactant system, one obtains for both cases that the snap-off frequency varies as Ca^{-2}. This is in good agreement with experimental data (5,16). The film thickness, $1-C_0$, is much less than one. As a result, the ratio of the snap off frequencies of the pure and surfactant systems is approximately equal to the ratio of their flow coefficients. This ratio is a complex function of the properties and concentration of the surfactant. However, a few general observations can be made. As surfactant concentration is increased, both the thickness of the wetting film and the rigidity of the bubble interface increase. These two effects tend to cancel one another and the flow coefficient remains relatively unchanged. In fact, as demonstrated by Gauglitz et. al. (16), if one assumes a completely rigid interface, the snap-off frequency of the two systems are identical.
      The flow coefficient of the square constriction cannot be solved analytically. It was determined numerically by the solution of equation 3 for the geometry shown in Figure 4. The geometry in figure 4 is mapped into a rectangle and the problem is solved with a Galerkin spectral technique. The relation between the flow coefficient and interface position is shown in Figure 5 for the surfactant free system.
      If C_0 is known as a function of the capillary number and the surfactant properties, the functional form of the frequency and bubble volume can be approximated from the linear results. However, a model for C_0 in constricted angular tubes does not exist. If one assumes that snap off occurs as soon as the thread becomes axisymmetric, then the base state thread radius is approximately the half width of the channel at the point snap off occurs. The experimental observations of Arriola and Ni along with the theoretical predictions of Ransohoff and Radke indicate that snap off takes place very near the constriction neck. Therefore, the radius of the bubbles formed should be slightly larger than the half width of the constriction neck. In fact, approximating C_0 by the constriction half width, one observes from equations 14 and 15, that the snap off frequency and bubble volume are independent of the liquid flow rate once the critical liquid flow rate has been exceeded. Ni measured the dependence of snap off on the bubble velocity, the velocity of

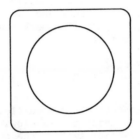

Figure 4. Square tube.

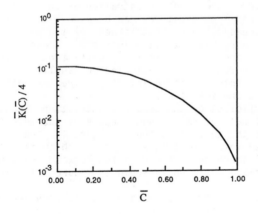

Figure 5. Square channel flow coefficient.

the rear of the trapped bubble. From the above theory the volume of the bubbles formed at the constriction can be approximated from equation 15 with C_0 taken as the neck half width. The volume should be independent of the measured capillary number. On the other hand, the frequency is related to the measured capillary number based on the average gas phase velocity through a mass balance on the gas phase.

$$f\star = \frac{v_b A_g}{V\star} = \frac{\sqrt{2}\, a^2 \sigma}{\pi^2 r\star^3 \mu}\; \tilde{C}a \qquad (18)$$

where $f\star$ is the dimensional frequency, v_b is bubble velocity, A_g is the area open to gas flow, a is the half width of the straight section, $r\star$ is the neck half width, μ is the liquid viscosity, and $\tilde{C}a$ is the capillary number based on the velocity of the gas bubble. The area occupied by the liquid in the corners, Figure 2, is much smaller than the total cross sectional area, and therefore was neglected in the above expression.

Care must be taken in the interpretation of equation 18. It is the result of a simple mass balance on the gas phase which stipulates the bubble velocity at the constriction is governed by the snap-off frequency. The snap-off frequency, on the other hand is given by equation 14 as a function of the bubble position and the pore geometry. Equations 15 and 18 are compared to the experimental data of Ni in Figures 6 and 7. The magnitude of the bubble volume is in good agreement with the data. However, experimentally there does seem to be a slight increase in bubble volume as the capillary number increases. The actual position of the front interface in the converging section of the constriction is determined by a balance of viscous and capillary forces. From equation 15, it appears that C_0 is a weakly increasing function of Ca for the constriction used by Ni. If this is the case then from equation 18 the frequency should be proportional to the measured capillary number raised to a power that is slightly less than one. This shifts the curve in Figure 7 closer to the experimental data.

From the linear stability analysis one can deduce the effect of the aspect ratio on the snap-off frequency. Let $\tilde{C}_0$ be a locally scaled through radius. Here the characteristic length is the hydraulic radius of the channel at the point snap off occurs. If one assumes that this radius is nearly the constriction neck radius than the flow coefficient may be expressed in terms of the locally scaled thread radius as

$$K(C_0) = R^4\, \overline{K}\,(\overline{C_0}) \qquad (19)$$

The snap off frequency in terms of the locally scaled functions is then

$$f = \ln\left\{\frac{B}{R\overline{C_0}}\right\}\; R\; \left[\frac{\overline{K}(\overline{C_0})}{8\pi\overline{C_0}^5}\right] \qquad (20)$$

If snap off occurs at the point the thead just becomes axisymmetric

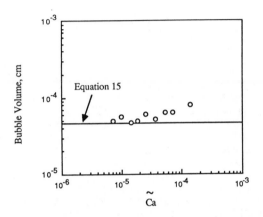

Figure 6. Volume of bubbles formed from a 1 wt.% solution of Amphosol at a square constriction (5), a=500 μm, r*=118 μm, μ=1cp, σ=33 dynes/cm.

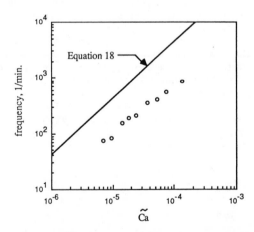

Figure 7. Snap off frequency for the system shown in Figure 6.

then $\overline{C}_0$ is always nearly one. Therefore the frequency should be approximately proportional to the aspect ratio. Unfortunately no data is available for snap off frequency of gas bubbles through the instability mechanism as a function of aspect ratio.

Conclusion

A linear stability analysis of a general evolution equation for axisymmetric threads can provide information on the snap-off frequency and volume of bubbles formed by growing instabilities at strong constrictions. The addition of surfactant is found to affect the frequency and bubble volume only through the base state interface location and the form of the flow coefficient. It is argued that these two effects nearly cancel one another leaving the frequency and volume relations unchanged. For strong square constrictions, it is found that the bubble volume is strongly dependent on the constriction aspect ratio and only a weak function of the liquid flow rate. In addition, the snap off frequency was approximately proportional to the constriction aspect ratio.

Acknowledgments

This work was carried out at the University of Houston where John Ratulowski is a graduate student. We are grateful to the UH-EOR consortium and NSF grant no. ENG-845116 for their support.

Legend of Symbols

a	Reference hydraulic radius
B	Disturbance magnitude
C	Interface location
$\underset{\sim}{C}a$	Capillary number
Ca	Measured capillary number
f	Snap off frequency
k	Wavenumber
K	Flow coefficient
P	Liquid phase pressure
Q	Liquid flow rate
r	Constriction neck radius
R	Aspect ratio
S	Area open to liquid flow
u	Axial liquid velocity
u^σ	Surface velocity
U	Reference gas phase velocity
v_b	Measured bubble velocity
V	Bubble volume
η	Surface velocity deviation variable
λ^m	Maximum growing wavelength
τ	Dimensionless time
ϕ	Interface location deviation variable
ω	Complex frequency
*	Superscript indicates deminsional quantities
0	Subscript refers to the base state values
	Overbar indicates locally scaled quantities

Literature Cited

1. Fried, A. N., The Foam Drive Process for Increasing the Recovery
    of Oil, U.S. Bureau of Mines Report of Investigation 5866, Dept.
    of the Interior, Washington, DC, 1961.
2. Dilgren, R. E.; Deemer, A. R.; Owens, K. B., California Regional
    Meeting, 1980, SPE Paper 10774.
3. Hirasaki, G. J.; Lawson, J. B., Soc. Petr. Engr. J. 1985, 25,
    170-190.
4. Falls, A. H.; Gauglitz, P. A.; Hirasaki, G. J.; Miller, D. D.;
    Patzek, T. W.; Ratulowski, J., SPE/DOE 5th Symposium on EOR,
    1986, SPE Paper 14961.
5. Ni, S.-S. MS.S Thesis, University of Houston, 1986.
6. Roof, J. G., Soc. Petr. Engr. J. 1970, 10, 85-90.
7. Gauglitz, P. A.; Radke, C. J., submitted to J. of Coll. and
    Inter. Sci., 1986.
8. Arriola, A. T. Ph.D. Thesis, University of Kansas, 1980.
9. Gauglitz, C. M.; St. Laurent; Radke, C. J., submitted to SPE
    Res. Engr., 1986.
10. Ransohoff, T. C. Ph.D. Thesis, University of California,
    Berkeley, 1986.
11. Teletzke, G. F. Ph.D. Thesis, University of Minnesota, 1983.
12. Jo, E.-J. Ph.D. Thesis, University of Houston, 1984.
13. Hammond, P. S., J. of Fluid Mech. 1983, 137, 363-334.
14. Bretherton, F. P., J. of Fluid Mech., 1961, 10, 160-188.
15. Chang, H.-C.; Ratulowski, J., in preparation, 1987.
16. Gauglitz, P. A.; C. M. St. Laurent; Radke, C. J., submitted to
    Ind. Engr. Chem. Fun., 1986.

RECEIVED January 5, 1988

Chapter 15

Mobility of Foam in Porous Media

Raymond W. Flumerfelt[1] and John Prieditis[2]

[1]Department of Chemical Engineering, Texas A&M University,
College Station, TX 77843
[2]Department of Chemical Engineering, University of Houston,
Houston, TX 77001

A constricted tube model is used to analyze viscous and
capillary effects associated with foam flow in porous
media. The foam moves through the pore structure as
single layer, continuous bubble trains. Capillary
resistance stems from the drainage and imbibition
surfaces as well as the internal lamellae structure.
This gives rise to a mobilization pressure which is
higher than gas-liquid flow with no surfactant.
Viscous effects of the Bretherton type are included in
a network model to derive permeability expressions cor
responding to different interfacial mobilities. The
significant reduction in gas permeability of foams is
attributed to (1) the significant decrease in the
fraction of channels containing flowing gas (compared
to gas-liquid flow with no surfactant), and (2) the
increase in viscous and capillary effects associated
with bubble train lamellae.

A basic problem associated with many secondary and tertiary
recovery processes is the high mobility of the displacing phase
relative to the displaced phase. Such situations are particularly
critical when low viscosity, low density drive fluids are used to
displace higher viscosity, higher density fluids, and/or when the
displacement takes place in a highly heterogeneous reservoir. The
existence of such unfavorable displacement conditions gives rise to
viscous fingering and gravity segregation with the result that
large sections of the reservoir can be bypassed. Even though the
local displacement in the swept zone may be high, because of poor
sweep, the overall recovery efficiency can be quite low.

In recent years there has been considerable interest in the
use of foams in chemical steam flood, CO_2, and low tension processes. To date, principal applications have been as diverting
agents where the foam has been used to block high permeability, low
oil saturation zones and hence force drive fluids through lower
permeability, higher oil saturation zones. The utility of foams in
more general mobility control roles has not been extensively

0097–6156/88/0373–0295$08.75/0
© 1988 American Chemical Society

tested, even though in principle they offer a spectrum of fluid
mobility behavior depending on the in situ foam phase stability;
weakly stable foams exhibit very high mobilities (approaching non-
dispersed gas phase mobilities) and highly stable foams exhibit
very low mobilities.

Extensive mobility control applications of foams are limited
by inadequate knowledge of foam displacement in porous media, plus
uncertainties in the control of foam injection. Because of the
importance of in situ foam texture (bubble size, bubble size
distribution, bubble train length, etc.), conventional fractional
flow approaches where the phase mobilities are represented in terms
of phase saturations are not sufficient. As yet, an adequate
description of foam displacement mechanisms and behavior is
lacking, as well as a basis for understanding the various, often
contradictory, macroscopic core flood observations.

In the present paper, pore level descriptions of bubble and
bubble train displacement in simple constricted geometries are used
in developing mobility expressions for foam flow in porous media.
Such expressions provide a basis for understanding many of the
previous core flood observations and for evaluating the importance
of foam texture and interfacial mobility. Inclusion of the effects
of pore constrictions represents an extension of the earlier
efforts of Hirasaki and Lawson (1).

Macroscopic Observations

Numerous investigators have conducted foam displacement tests in
bead packs and rock cores. From these studies, the macroscopic
features which characterize foam flow and make it different from
conventional gas-liquid flow in porous media have been identified.
These are summarized as follows:

a) The gas phase permeability is reduced by several orders of
    magnitude over that which would be associated with a non-
    dispersed phase system, i.e., gas-liquid system without foaming
    agent (1-7). Such reductions have led some investigators to
    state that stable foam will not flow through porous media under
    practical reservoir pressure gradients (5,8,9).

b) The gas permeability increases with flow rate and applied
    pressure gradient (3,7). Also, a minimum pressure must be
    applied before foam flows in porous media (7).

c) The gas saturation in foam flow is high relative to non-
    dispersed gas flow for equivalent gas flow rates (2-5,10,11)

d) The effective liquid permeability is reduced relative to non-
    dispersed systems (6,10); however, the liquid relative
    permeability dependence on saturation remains nearly unchanged
    (10) or only slightly changed (6).

e) A high trapped gas saturation occurs and water imbibition does
    not displace significant amounts of foam unless a large
    pressure gradient is applied. The highest trapped gas
    saturations occur in the most permeable cores (2,10).

f) Some investigators report little change in gas and liquid
    saturations with changes in steady state liquid and gas flow
    rates (4). Others report that the relative gas permeability is
    a weak function of saturation (6,10,11).

g) Complete gas blockage can occur at high gas saturations and large injection pressures (12).

h) Large injection pressures and pressure gradients are required to maintain flow of low quality foam. For fixed injection gas pressure, continued increase in liquid rate will eventually cause complete gas flow stoppage (3,4,7).

i) Less gas permeability reduction occurs in cores where absolute permeabilities are small (2).

j) The liquid saturation can be reduced by foam below the irreducible liquid saturations associated with non-dispersed gas-liquid displacement (3).

These observations, coupled with the effects of bubble texture (1,13-15) and various history dependent phenomena, clearly demonstrate the inadequacy of conventional fractional flow approaches to describe foam flow in porous media. Also, early approaches which treated the foam simply as a fluid of modified viscosity are also inadequate in explaining the above characteristics. To achieve a fuller understanding of such phenomena, a detailed description of the pore level events is required. In what follows, a simple pore level model is utilized to explain some of the above macroscopic features and to identify some of the key pore level mechanisms.

Nature of Foam Displacement

In most foam displacement applications, the gas and liquid phases (the latter containing the foaming agent) are injected simultaneously or intermittently with the foam being formed in situ. In particular, as gas passes through the porous medium in the presence of foaming agent, several pore level events contribute to the formation of the foam bubbles. Such processes are similar to those observed with oil-water systems (16).

First, if lubricated fingers of the gas phase become sufficiently long and the displacement rate sufficiently small, the fingers can breakup into a set of small bubbles in the same way a long, stationary fluid thread breaks when suspended in a second immiscible phase (17-19). In tubes, such thread breakup produces bubbles on the order of 4 times the undisturbed thread diameter.

In addition to this thread breakup mechanism, gas fingers and large bubbles can also experience bubble snap-off when passing through narrow pore constrictions (20,21). Although snap-off phenomena can be quite complex (21-24), the static analysis of Roof (20) indicates that the resulting bubble diameters are at least twice the pore constriction diameter.

A third bubble generation mechanism arises from bubble division at pore junctions (25). Here, bubbles with diameters smaller than the respective pore channel diameters would not be expected.

In addition to these bubble generation events, bubble-bubble coalescence events can increase the average bubble size. As a result, unless injected as a fine foam with bubble sizes below those of the pore channels, or at rates above those possible in actual reservoirs, foam is expected to advance through the pore channels as single layer bubble trains with the bubbles separated

by thin liquid lamellae as illustrated in Figure 1. Such an
interpretation is supported by the etched plate observations of
Mast (25) as well as similar observations in this laboratory.
Whereas the foam flow channels are filled with continuous or
nearly continuous bubble trains, the adjacent pore channels are
filled with liquid or with trapped gas bubbles. Even though the
gas saturation in the pore space can be considerably below that
associated with foam systems outside porous media, gas bubble and
bubble train assemblages are referred to as foams because they
satisfy the definition of Bikerman (26), that is, an assemblage of
bubbles separated by thin films.

Capillary Resistance

When a foam moves through a porous medium, capillary and viscous
effects can resist its advance. If the foam moved as single
bubbles through the constrictions, the pressure gradients required
to overcome the capillary resistance would be prohibitive. (Note:
The pressure drop required to push a single bubble with σ = 50
dyn/cm through a 10 μm constriction would be about 10^5 dyn/cm².
For a bubble length of 100 μm, this would correspond to a pressure
gradient in the continuous phase of 10^7 dyn/cm³ or 4350 psi/ft. To
prevent fracturing, reservoir pressure gradients must be below 10
psi/ft.)
    The question then arises as to how foam moves through porous
media. Foam injection and displacement is achieved in both the
field and the laboratory. Obviously, a displacement mechanism other
than single bubble displacement is operative.
    In laboratory experiments and field applications the gas is
delivered to the rock face either as continuous gas or as a course
gas-liquid dispersion. In both cases, for the gas to move into the
porous medium, the gas pressure at the rock face must be higher
than the capillary entry pressure. For a gas finger or a bubble
train to advance through the porous medium, the face pressure must
be maintained at a level above the maximum capillary pressure that
the gas finger or bubble train will experience along its path
through the medium.
    Once in the medium, bubbles can be displaced through pore
constrictions only by the concerted action of long, continuous
bubble trains. As illustrated in Figure 1(a), bubble 1 will not
move through constriction E until the bubble train behind it
catches up and pushes it through the constriction. The latter is
possible since the pressure drop across the long continuous train
is much larger than across the individual bubble.
    Similarly, in Figure 1(b) bubble 2 will not advance until the
train behind it advances through constriction E. Once a continuous
train is formed, bubble 2 will then experience a Haines jump
through F and press against bubble 1. Following this jump, the
train will break at the smallest constrictions (E and F) with
bubble 3 being stopped at F and bubble 5 at E. As bubbles continue
to move in at A, the train will close up again and the entire
process will be repeated.
    During this cascading displacement process, the maximum
capillary resistance will generally occur when the drainage surface

of the train is at a minimum constriction along the bubble train path. The magnitude of this capillary resistance will depend not only on the capillary effects at the drainage and imbibition surfaces, but also the capillary effects associated with all of the internal lamellae along the train.

This is illustrated in Figures 2(a) and 2(b) where the two-bubble foam body requires less applied pressure to push it through the constriction than that for a single bubble. The difference is the contribution of the internal lamella in reducing the overall capillary resistance. In more general cases, the contributions of the internal lamellae to the overall capillary resistance of a given train may be positive or negative depending on the size distribution and respective positions of the bubbles in the pore channel (see Figure 3).

In order to obtain an estimate of such effects, the pressure associated with the capillary resistance of the internal lamellae structure was calculated for different sets of bubbles, with different numbers and sizes. The specific pore geometry chosen was that shown in Figure 3. The pressure $\hat{P}_0$ is the pressure in the front bubble and pressure $\hat{P}_N$ is that of the Nth bubble; $\hat{P}_N - \hat{P}_0$ then gives the capillary resistance associated with the internal lamellae structure. The pressure difference $\Delta \hat{P}_\ell$ across each lamellae in Figure 3 is given by

$$\Delta \hat{P}_\ell = \frac{4\sigma}{R} \sin\theta \qquad (1)$$

where R is the pore radius at the lamellae position and θ is the slope of the pore wall at this position. When the pressure differences across all lamella in a train are summed, one obtains the total internal lamella capillary resistance, $\hat{P}_N - \hat{P}_0$. If the lamellae train is moved along the channel and pressure calculations are made at each position, the maximum pressure difference can be determined; it is the latter values that are reported in Figure 4.

Specifically, Figure 4(a) shows the frequency distribution for 1000 calculations (trials) with 100 bubbles in the bubble train. For this case, uniform distributions of R_t and R_b were used, with $0.5 < R_t/R_0 < 1.5$ and $3.0 < R_b/R_0 < 5.0$. The segment lengths were fixed at $5R_0$. The bubble volumes were distributed uniformly between 1.2 and 2.0 times the mean pore size. The pressures $\hat{P}_0^*$ and P_N^* have been made dimensionless using the capillary pressure associated with the front bubble. Also, no values of $\hat{P}_N^*$ below $\hat{P}_N^* = 1$ occur because these cases would correspond to discontinuous trains. During the calculation, if the P^* associated with any internal bubble dropped below 1 (the dimensionless capillary pressure), it was reset to 1 and the calulation continued; the idea being that in an actual porous medium, bubble trains in side channels would adjust to maintain $P^* > 1$ along any continuous train. (Note added in final proof: More recent calculations allowing the train to break show similar qualitative results, but with mobilization pressures about 30% of those reported here).

The results for bubble trains with 100 bubbles (Figure 4(a)) show that the capillary contributions associated with the internal lamellae structure are significant and result in mobilization pressures ($\hat{P}_N$) as high as 9 times the capillary pressure of the

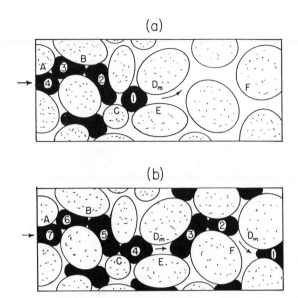

Figure 1.   Illustration of foam displacement: (a) For bubble 1
to be displaced through constriction E,   the bubble train behind
it must first advance through constriction C,   form a continuous
train,   and   then push bubble 1 through constriction E.   (b) The
displacement of bubble 2 first requires the advancement of
bubble 4 through E,   bubble 5 through C,   etc., to form a continuous   train.   Once this train pushes bubble 2 through F,   the
train momentarily breaks with 3 trapped at F and 5 trapped at E.

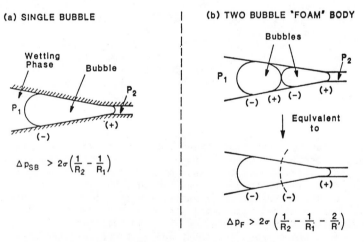

Figure 2.   Mobilization pressures required to mobilize a   single
bubble   and   a two bubble "foam" body.   In (b),   the capillary
effects associated with the internal lamella results in a   lower
mobilization pressure than found in (a).

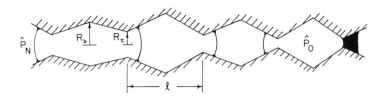

Figure 3. Constricted pore geometry used in mobilization pressure calculations.

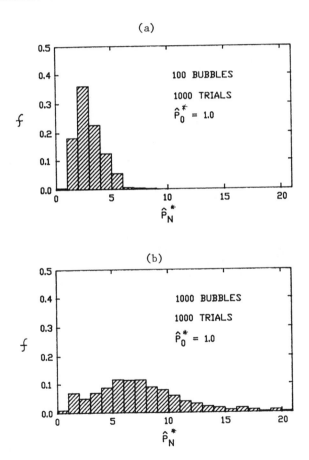

Figure   4.   Number fraction distributions of calculated mobilization pressures.   Case (a) corresponds to bubble trains containing 100 bubbles; case (b) to 1000 bubbles.

drainage surface, with the expected value being about 3. For
longer trains, the maximum and expected values shift to higher and
higher values as illustrated in Figure 4(b), which represents
results for trains of 1000 bubbles. In the latter case, the
maximum dimensionless mobilization pressure is 20 and the mean
value is about 7.

Additional trials suggest that the expected value of the
pressure difference over a series of lamellae due to capillary
resistance can be approximated by

$$\overline{(\hat{P}_N^* - \hat{P}_0^*)} = C_1 N_L^{1/2} L^{1/2} \Delta\hat{P}_{\ell m}^* \tag{2}$$

where N_L is the number of lamellae per unit length, L is the length
of the continuous train, and $\Delta P_{\ell m}$ is a dimensionless average
pressure difference across a single lamella given by

$$\Delta\hat{P}_{\ell m}^* = \frac{2}{R_m^*} \sin\theta_m \tag{3}$$

Here, $\Delta\hat{P}_{\ell m}^*$ has been made dimensionless using the capillary pressure
associated with the front bubble, $2\sigma/R_0$. Also, R_m^* represents the
mean radius in the flow channel made dimensionless using R_0, and
θ_m represents the mean slope of the channel and is indicative of the
aspect ratio. The mean pressure difference is larger for channels
of smaller radius or larger aspect ratio.

It should be noted that an upper bound for the capillary
resistance occurs when all lamellae are simultaneously at the
minimum constrictions (exit side of constricted pore bodies). For
this case,

$$\hat{P}_N^* - \hat{P}_0^* = C_2 N_L L \Delta\hat{P}_{\ell m}^* \tag{4}$$

In this relationship the capillary resistance is directly
proportional to the number of lamellae in the flow path and the
average pressure drop across a single lamella.

Clearly, internal lamella structure effects can produce
significant increases in the capillary resistance associated with
bubble trains. The magnitude of this effect will vary with the
bubble size, the pore geometry, and the length of the train.
Extremely high values are not expected because the bubble train
length will be limited by the breaks occurring at the minimum pore
diameters along the bubble train path (see Figure 1(b)). Further
limitations on the capillary resistance may exist because
coalescence or film rupture phenomena is enhanced by associated
increases in capillary pressure (15). Just as in non-dispersed
two-phase flow, it is the capillary pressure which dictates the
pores which are penetrated by the foam. However, because of the
effects of the internal lamellae structure, a pressure larger than
the normal capillary pressure is required and this larger pressure
is termed the mobilization pressure. In the work here, the
mobilization pressure is represented by

$$P_{c,mob} = K_m (2\sigma/R_{t,min}) \tag{5}$$

where K_m is greater than 1 and is a factor which, at this point, must be determined experimentally. The quantity $2\sigma/R_{t,min}$ is the capillary pressure associated with mobilization of a continuous gas through the pore structure.

Viscous Resistance

When bubbles advance through channels, viscous effects can arise which can seriously alter the rate of advance. For large bubbles in tubes the situation is similar to that shown in Figure 5(a). The bubble is pushed through the tube by the liquid phase moving at an average velocity V. If the liquid phase wets the tube surface (e.g., an aqueous phase on quartz), a thin film is formed between the bubble and the tube wall with the thickness depending on the bubble velocity U, the liquid viscosity μ_L, the interfacial tension σ, the tube radius R, the gas-liquid interfacial mobility, and the disjoining pressure of the film.

Depending on the presence and magnitude of dynamic interfacial effects (surface tension gradients and surface viscous and elastic effects), the gas-liquid interfacial mobility can range from perfectly mobile with complete slip and zero tangential shear stress at the interface to completely immobile where the no slip condition applies and the velocity of the liquid at the interface is the same as the gas velocity U. Such dynamic interfacial phenomena occur in many systems involving surface active agents and are particularly important when the characteristic length scales of flow system are small.

The disjoining pressure effects embody the contributions of van der Waals and electric double layer interactions across the thin film. A positive disjoining pressure acts to open the film and a negative disjoining pressure to collapse it. In general, the film will be stable if the disjoining pressure is positive. If the disjoining pressure is negative, the film stability will depend on the level of the film pressure arising from lubrication flow effects, the latter increasing with capillary number. For aqueous foam systems on quartz, a positive disjoining pressure is predicted (27) and the bubble displacement occurs as in Figure 5(a) with the existence of a stable thin film. In all following discussions, it is assumed that such a concentric film displacement mechanism is operative.

Over the years, a number of investigators have considered the displacement of long bubbles in tubes (1,27-31). The early work of Bretherton (28) set the stage for these studies and hence the problem is often termed the "Bretherton problem". The principal result of the various analyses of this problem is the prediction of the film thickness b in terms of the capillary number:

$$b = BRCa^{2/3} \tag{6}$$

where
$$Ca \equiv \frac{\mu_L U}{\sigma} \tag{7}$$

For "thick films" where disjoining pressure effects are negligible, the constant B in Equation 6 ranges from 1.337 for the perfectly mobile interface case to 2.123 for the immobile interface case.

For "thin films" where the disjoining pressure is important, the quantity B must be considered a function which varies with the disjoining pressure parameters, the interfacial mobility, and the capillary number (27,29). In general, with all other things being fixed, B increases with disjoining pressure and decreases with interfacial mobility.

Of direct interest in the study here is the pressure drop across the bubble in the liquid phase. For the mobile interface case this is given in dimensionless form by

Mobile
Interface: $$\Delta P^\star \equiv \frac{\Delta P}{(2\sigma/R)} = KCa^{2/3}$$ (8)

where K is a constant equal to 4.72 for "thick films" and is a function of Ca and disjoining pressure in the case of "thin films" (27).

For the case of immobile interfaces and long bubbles $(L \sim O(2R)$ or greater), the principal contribution to the pressure drop is the viscous drag in the uniform film region. The pressure drop relation is then:

Immobile
Interface: $$\Delta P^\star = \tilde{K} \left(\frac{L}{R}\right) Ca^{1/3}$$ (9)

with $$\tilde{K} = 1/B$$ (10)

the B being the same as that in Equation 6.

The pressure drop results of Equations 8 and 9 are different in two important respects. First, the functional dependence of $\Delta P^\star$ on Ca is different in the two cases, being 2/3 in the mobile case and 1/3 in the immobile case. Obviously, the displacement rate has a stronger influence on ΔP in the mobile case than in the immobile case; the same being true for μ_L. On the other hand, the effect of σ on ΔP (dimensional) is greatest in the immobile case.

The second major difference in Equations 8 and 9 is that ΔP is independent of bubble length in the mobile case and increases in direct proportion to L in the immobile case.

The above discussion is concerned with single bubble displacement. To obtain results analogous to Equations 6, 8, and 9 for bubble trains, it is necessary to account for changes in curvature at the bubble ends (in the Plateau border regions) due to the compression between adjacent bubbles. Referring to Figure 5(b), the contact radius R_c can be related to the capillary pressure $P_c = \hat{P}_1 - P_1 = \hat{P} - P$ by

$$R_c = \begin{cases} R\left(1 - \frac{2\sigma}{RP_c}\right)^{1/2} & \text{for } P_c > \frac{2\sigma}{R} \qquad (11) \\[3mm] 0 & \text{for } P_c \leq \frac{2\sigma}{R} \qquad (12) \end{cases}$$

where $P_c > 2\sigma/R$ denotes the necessary condition for compression between adjacent bubbles, and hence, the existence of a continuous bubble train.

The film thickness is found to be similar to Equation 6 except that the coefficient is replaced by

$$B' = B \left(\frac{1 - R_c^2/R^2}{1 + R_c^2/R^2} \right) \tag{13}$$

These results indicate that the larger the capillary pressure, the larger the contact radius and the smaller the film thickness.

It also turns out that for the mobile interface case the pressure in the gas phase across the lamella of Figure 5(b) is given by

$$\Delta \hat{P}^* \equiv \frac{\hat{P}_2 - \hat{P}_1}{2\sigma/R} = KCa^{2/3} \tag{14}$$

This result is obtained in the same way as Equation 8 (see Bretherton (28)), but because of the bubble contact and flattening, the curvature in the Plateau border regions is $2R/(R^2-R_c^2)$ instead of $2/R$ in the free bubble case. It is interesting to note that the results in both cases are the same.

For the immobile case, if the pressure drop in the liquid phase across a bubble is largely due to the drag force in the film region, then

$$\Delta \hat{P}^* = \tilde{K} \frac{Ca^{1/3}L}{R} \left(\frac{1 + (R_c^2/R^2)}{1 - (R_c^2/R^2)} \right) \tag{15}$$

This equation follows directly from Equation 9 with b being replaced by the form appropriate for the bubble train case (Equations 6 and 13).

Using the results of Equations 14 and 15 for a bubble train of $N + 1$ bubbles, the total pressure difference in the gas phase can be determined by a simple summation of the contributions across each lamella:

$$\Delta \hat{P}^* \equiv \frac{\hat{P}_{N+1} - \hat{P}_o}{(2\sigma/R)} = \begin{cases} N \, K \, Ca^{2/3} & (16) \\[2em] \tilde{K} \dfrac{Ca^{1/3}L_f}{R} \left(\dfrac{1 + (R_c^2/R^2)}{1 - (R_c^2/R^2)} \right) & (17) \end{cases}$$

where L_f is equal to the sum of the wall contact lengths of the N bubbles, that is, the total length of the thin film regions in the train.

Equation 16 indicates that for bubble trains with mobile interfaces, the pressure drop is related to the number of lamellae in the train (or number of bubbles per unit length). This important observation, first noted by Hirasaki and Lawson (1),

partially explains why some foam systems exhibit higher apparent
viscosities in porous media as the bubble size is reduced.
In comparison, Equation 17 indicates that the total contact
length L_f of the films is the important parameter for immobile
interface systems. Foaming systems of this type would show no
dependence on foam texture (bubble size and bubble size
distribution), but would exhibit very large apparent viscosities in
porous media if the bubble trains were sufficiently long.

Permeability Model

The above results for bubble and bubble train displacement are now
utilized in developing a simple permeability model for foam
displacement. The displacement process envisioned is that of
simultaneous injection of gas and liquid (with surfactant) into a
porous medium under steady conditions. The gas phase is injected
either as a high quality foam or as a non-dispersed gas phase which
experiences rapid breakdown to form the bubbles and bubble trains
which make up the "foam phase". Just as in conventional multiphase
flow, the liquid and gas phases are assumed to move along separate,
but intersecting channels in the porous medium. In general, the
non-wetting gas phase exists as bubbles and bubble trains in three
states: As an isolated and disconnected fraction, as a flowing
fraction (as continuous or nearly continuous bubble trains), and as
a dendritic non-flowing fraction connected to the flowing fraction.
The wetting liquid phase exists and flows in separate channels as
well as in the "nooks and crannies" and thin films of pores filled
primarily by the non-wetting gas phase.
This complex two phase flow is now modeled as a network of
interconnected channels consisting of constricted tube segments
which contain the flowing and non-flowing fractions of the gas and
liquid phases. Although overly simplistic in many respects, this
representation provides a basis for obtaining qualitative and
semi-quantitative permeability results which are useful in
interpreting various macroscopic phenomena.
The specific characteristics of this model can be summarized
as follows:
a) The channels in the porous medium are made up of connected tube
segments of the type illustrated in Figure 6. Each channel is
characterized by its minimum pore (or tube) diameter D_m and its
pore size distribution $\eta(D,D_m)$. Here, $\eta\,dD$ represents the
number of tube segments with diameters between D and D + dD in
a channel with minimum pore diameter D_m. Following other
investigators (32, pp. 60-66), the lengths of the tube segments
are related directly to their diameters, i.e., $\ell = \ell(D)$.
Hence, the total length of any given channel D_m is given by

$$L = J \int_{D_m}^{\infty} \eta \ell\, dD \qquad (18)$$

where $J(D_m)$ is the number of D segments in the D_m channel.
Although in any real porous medium the lengths of the channels
are also distributed, uniform channel lengths are used here.

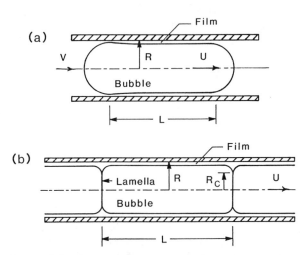

Figure 5.   (a) Single bubble displacement in a tube;   (b) displacement of a continuous train of bubbles.

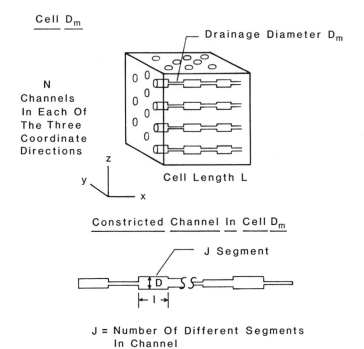

Figure 6.   Geometric nature of D_m cells used in   permeability model development.

b) The D_m channels making up the porous medium are distributed
    according to the distribution $\gamma(D_m)$. Here, γdD_m represents
    the number fraction of D_m channels with diameters between D_m
    and $D_m + dD_m$. It then follows that the total pore volume V_p is
    given by

$$V_p = 3N_c \int_0^\infty V_{pm} \gamma dD_m \qquad (19)$$

where N_c is the number of channels in each of the 3 coordinate
directions and V_{pm} is the pore volume of the D_m channel. The
latter can be written as

$$V_{pm} = \frac{\pi}{4} J \int_{D_m}^\infty D^2 \ell \eta dD \qquad (20)$$

c) The foam and liquid phases are assumed to move in separate but
    communicating channels. The non-wetting foam phase is
    restricted by capillary forces from flowing in channels with
    diameters below the foam drainage diameter D_{fd}. For a given
    capillary pressure P_c, the drainage diameter for continuous gas
    is less than that for foam ($D_{gd} = 4\sigma/P_c < D_{fd}$).

d) In the set of channels defined by $D_m > D_{fd}$, the fraction
    containing flowing foam phase is denoted by $f_G(D_m)$. The other
    channels in this set ($D_m > D_{fd}$) are filled with flowing liquid
    and trapped gas. For channels with $D_m < D_{fd}$, only trapped gas
    and flowing liquid are found and $f_G = 0$.

e) The flow of the foam phase occurs as a result of bubble train
    movement through free-flowing gas channels. The total pressure
    drop in the flowing foam phase is composed of a mobilization
    pressure and a dynamic pressure. The first arises from static
    capillary effects along the respective bubble train paths and
    the second from dynamic capillary and viscous effects
    associated with bubble train movement. Bretherton type
    relations (Equations 16 and 17) are used to describe the
    viscous contributions to the dynamic pressure changes in each
    constant diameter tube segment of the constricted channels.
    Viscous effects associated with moving from one tube diameter
    to another (expansion and contraction effects) are neglected as
    are inertial effects and transverse flow effects.

f) The texture of the foam phase is invariant during the
    displacement process and the texture in a given channel D_m is
    characterized by an average bubble volume $V_{bm} = V_{bm}(D_m)$. This
    situation would correspond to very stable foam systems where
    the foam texture is dictated by pore structure characteristics.

g) The capillary pressure along any continuous bubble train is
    treated as uniform and the pressure gradients in the associated
    gas (foam) phase and adjacent liquid phases are equal; the
    latter arising from the assumed interconnectivity of the
    channels.

Following approaches similar to Scheidegger (33) and Dullien
(32,34), single tube pressure drop-flow rate relations are now used
to obtain permeability relations for cells of the type just des-

cribed.   In particular,   permeability relations are developed   for continuous bubble trains moving through such channels.   The analysis   focuses   first on the case of foam systems with mobile   interfaces; later, results for rigid interface systems will be given.

Geometric Relations. To express the results in terms of pore distributions based upon volume fractions instead of number fractions, it is first noted that in channel D_m the volume associated with the segments with diameters between D and $D + dD$ is given by

$$V_{pm} \alpha dD \ = \ \frac{\pi}{4} \ D^2 \ell J\eta dD \tag{21}$$

where αdD represents the volume fraction associated with tube segments in the range D to $D + dD$ in channel D_m. Dividing each side by $\pi D^2/4$ and integrating over the range of D segments in the D_m channel, and then using Equation 18, it follows that

$$V_{pm} \ = \ \frac{\pi}{4} \ \mathbb{D}_m^2 \ L \tag{22}$$

where
$$\mathbb{D}_m^2 \ \equiv \ \left[\int_{D_m}^{\infty} \frac{\alpha}{D^2} \ dD \right]^{-1} \tag{23}$$

     If βdD_m represents the volume fraction of channels with D_m values between D_m and $D_m + dD_m$, the pore volume associated with the channels in this range is given by

$$V_p \beta dD_m \ = \ 3N_c V_{pm} \gamma dD_m \tag{24}$$

Dividing by $3N_c V_{pm}$ and integrating over the range of D_m values

$$\frac{V_p}{3N_c} \int_0^{\infty} \frac{\beta}{V_{pm}} \ dD_m \ = \ \int_0^{\infty} \gamma dD_m \ = \ 1 \tag{25}$$

or using Equation 22, the total pore volume takes the form

$$V_p \ = \ \frac{3}{4} \ \pi N_c \ \mathbb{D}^2 \ L \tag{26}$$

where
$$\mathbb{D}^2 \ \equiv \ \left[\int_{D_{fd}}^{\infty} \frac{\beta}{\mathbb{D}_m^2} \ dD_m \right]^{-1} \tag{27}$$

Viscous Pressure Drop. For continuous bubble trains with perfectly mobile interfaces moving through a given D_m channel, the dynamic pressure drop in the gas (foam) phase over a single tube segment D follows from Equation 16:

$$(\Delta \hat{\mathcal{P}})_D \ = \ \frac{4\sigma}{D} \ KNCa^{2/3} \tag{28}$$

The total number of bubbles N in the D tube segment can be given in terms of the average bubble volume V_{bm} in the D_m channel by

$$N = \frac{\pi}{4} D^2 \ell / V_{bm} \qquad (29)$$

Using this result in Equation 28,

$$(\Delta \hat{\mathscr{P}})_D = \pi K \left(\frac{\sigma D \ell}{V_{bm}} \right) Ca^{2/3} \qquad (30)$$

where

$$Ca \equiv \frac{4(Q_G)_m \mu_L}{\pi D^2 \sigma} \qquad (31)$$

and $(Q_G)_m$ is the volumetric gas (foam) flow rate through the D_m channel.

Summing up the pressure drops across each of the D tube segments, the total pressure drop across the porous medium is obtained, i.e.,

$$\Delta \hat{\mathscr{P}} = \int_{D_m}^{\infty} (\Delta \hat{\mathscr{P}})_D J \eta \, dD \qquad (32)$$

or using Equations 21 and 29,

$$\Delta \hat{\mathscr{P}} = 4 K \sigma \left(\frac{V_{pm}}{V_{bm}} \right) \left(\frac{4(Q_G)_m \mu_L}{\pi \sigma} \right)^{2/3} \frac{1}{\tilde{D}_m^{7/3}} \qquad (33)$$

where

$$\tilde{D}_m^{7/3} \equiv \left[\int_{D_m}^{\infty} \frac{\alpha}{D^{7/3}} \, dD \right]^{-1} \qquad (34)$$

The total gas flow rate through the medium Q_G is obtained from

$$Q_G = N_c \int_{D_{fd}}^{\infty} (Q_G)_m \, f_G \, \gamma \, dD_m \qquad (35)$$

where f_G is the fraction of D_m channels containing flowing foam. This equation can be expressed in terms of the distribution using Equation 24; then

$$Q_G = \frac{1}{3} V_p \int_{D_{fd}}^{\infty} \frac{(Q_G)_m \, f_G \, \beta}{V_{pm}} \, dD_m \qquad (36)$$

If Equation 33 is solved for $(Q_G)_m$ and substituted into Equation 36 it is possible to show after some reduction that:

$$\frac{\Delta \hat{\mathscr{P}}}{L} = K_1 \left(\frac{\sigma}{\overline{V}_b} \right) \overline{Ca}^{2/3} \, \overline{D} \qquad (37)$$

where
$$K_1 \equiv 3^{2/3}\pi K \tag{38}$$

$$\overline{C}a \equiv \frac{Q_G \mu_L L}{\phi V \sigma \overline{f}_G} \tag{39}$$

$$\overline{D} \equiv \left[\int_{D_{fd}}^{\infty} \left(\frac{V_{bm}}{\overline{V}_b}\right)^{3/2} \left(\frac{f_G}{\overline{f}_G}\right) \frac{\beta}{\overline{D}_m^{3/2}} dD_m \right]^{-2/3} \tag{40}$$

with
$$\overline{D}_m \equiv \left[\int_{D_m}^{\infty} \frac{\alpha}{D^{7/3}} dD \right] \left[\int_{D_m}^{\infty} \frac{\alpha}{D^2} dD \right]^{-5/3} \tag{41}$$

Also, the characteristic bubble volume $\overline{V}_b$ corresponds to the average bubble size in the porous medium, and $\overline{f}_G$ represents the fraction of channels in the porous medium containing flowing foam.

The foam (or gas) permeability is now defined in the conventional way:

$$k_G = \frac{\mu_G Q_G}{(V/L)(\Delta \hat{\mathscr{P}}/L)} \tag{42}$$

or, from Equation 39,

$$k_G = \phi \left(\frac{\mu_G}{\mu_L}\right) \left(\frac{\sigma}{\Delta \hat{\mathscr{P}}/L}\right) \overline{f}_G \overline{C}a \tag{43}$$

where μ_G is the gas viscosity. Using Equation 37, it then follows that the gas (or foam) permeability corresponding to the mobile interface case is given by

$$k_G = \frac{\phi}{K_1} \left(\frac{\mu_G}{\mu_L}\right) \left(\frac{\overline{V}_b}{\overline{D}}\right) \overline{f}_G \overline{C}a^{1/3} \tag{44}$$

This result indicates that the gas (foam) permeability increases with capillary number and average bubble size. Also, k_G varies with $\phi^{2/3}$, $\mu_L^{-2/3}$, and $\sigma^{-1/3}$.

A similar analysis can be carried out for rigid interfaces and the dynamic pressure drop is found to be

$$\frac{\Delta \hat{\mathscr{P}}}{L} = K_2 \left(\frac{\sigma}{\overline{\overline{D}}^2}\right) \overline{C}a^{1/3} \tag{45}$$

Here
$$K_2 \equiv 8 \sqrt[3]{3} \, \tilde{K} \tag{46}$$

$$\overline{\overline{D}} \equiv \left[\int_{D_{fd}}^{\infty} \left(\frac{f_G}{\overline{f}_G}\right) \overline{\overline{D}}_m^6 \, \beta dD_m \right]^{1/6} \tag{47}$$

$$\overline{\overline{D}}_m \equiv \left[\int_{D_{fd}}^{\infty} \kappa \left(\frac{2}{D_{fd} D^{11/3}} - \frac{1}{D^{14/3}}\right) \alpha dD \right]^{-1/2} \left[\int_{D_m}^{\infty} \frac{\alpha}{D^2} dD \right]^{2/3} \tag{48}$$

In obtaining Equation 45 we have let the bubble contact length L_f in each tube segment be $\kappa \ell$, where ℓ is the tube length.

Using Equation 45 in Equation 43, the permeability for the rigid interface case can be obtained as

$$k_G = \frac{\phi}{K_2} \left(\frac{\mu_G}{\mu_L}\right) \bar{\bar{D}}^2 \ \bar{f}_G \ \bar{Ca}^{2/3} \tag{49}$$

Here, the dependence on displacement rate (through Ca) is stronger than that for the mobile interface case. Also, bubble size does not appear in this relation as in Equation 44. Further, it is noted that k_G varies with $\phi^{1/3}$, $\mu_L^{1/3}$, and $\sigma^{-2/3}$, the latter being different from those found in the mobile interface case.

The above results are based upon the pressure drop-flow rate relations of Equations 16 and 17. Experimental studies for bubble displacement in round and square tubes (35) suggest a generalization of these relations. In particular, relations of the form

$$(\Delta \hat{\mathcal{O}})_D = K_3 \frac{\sigma \ell}{D^2} Ca^n \tag{50}$$

can be used to describe the experimental observations. For example, bubble train displacement in round tubes (186 bubbles, L = 26.7 cm, and D = 0.75 mm) over the range $4 \times 10^{-6} <$ Ca $< 2 \times 10^{-4}$ gave

$$K_3 = 111, \quad n \cong 0.5 \tag{51}$$

For square tube tests (148 bubbles, L = 41.3 cm, and D = 0.5 mm) in the range $1.2 \times 10^{-5} <$ Ca $< 2 \times 10^{-4}$, the values were

$$K_3 = 154, \quad n \cong 0.85 \tag{52}$$

If Equation 50 is used to determine the permeability, we obtain

$$k_G = \hat{K}\phi \left(\frac{\mu_G}{\mu_L}\right) \hat{D}^2 \ \bar{f}_G \ \bar{Ca}^{1-n} \tag{53}$$

where

$$\hat{K} \equiv 1/(3^n K_3) \tag{54}$$

and

$$\hat{D}^2 \equiv \left[\int_{D_{fd}}^{\infty} \frac{[\tilde{D}_m^{2n+4})/\boldsymbol{D}_m^2]^{1/n}}{\boldsymbol{D}_m^2} \left(\frac{f_G}{\bar{\bar{f}}_G}\right) \beta dD_m \right]^n \tag{55}$$

Equation 53 has the same general form as Equations 44 and 49. At this point, K_3 and n must be considered as empirical parameters which will be dependent on the in situ foam texture, gas saturation, and the porous medium structure.

For comparison purposes, the permeability associated with non-dispersed phase gas flow (no surfactant) is also given:

$$k_G^o = \frac{\phi}{96} \ \bar{f}_G^o \ \tilde{\tilde{D}}^2 \tag{56}$$

where $\widetilde{f}_G^o$ is the fraction of pores with flowing gas and $\widetilde{D}^2$ is a characteristic diameter quantity defined by

$$\widetilde{D}^2 \equiv \int_{D_{gd}}^{\infty} \frac{\widetilde{D}_m^{\,6}}{D_m^{\,4}} \left(\frac{f_G^o}{\overline{f}_G^o}\right) \beta\, dD_m \tag{57}$$

where
$$\widetilde{D}_m^{\,6} \equiv \left[\int_{-D_m}^{\infty} \frac{\alpha\, dD}{D^6}\right]^{-1} \tag{58}$$

It then follows that

$$\frac{k_G}{k_G^o} = 96\widehat{K} \left(\frac{\mu_G}{\mu_L}\right)\left(\frac{\widehat{D}^2}{\widetilde{D}^2}\right)\left(\frac{\overline{f}_G}{\overline{f}_G^o}\right) \overline{Ca}^{\,1-n} \tag{59}$$

This ratio of dispersed phase (foam) and non-dispersed phase permeabilities indicates that the decrease in permeability of a foam relative to a non-dispersed phase is due largely to (1) the decrease in the fraction of pores occupied by flowing gas, (f_G/f_G^o), and (2) the low value of μ_G/μ_L. In addition, we find the ratio of permeabilities increases with displacement rate when $n < 1$.

A simplification which may have practical utility follows from Equation 59 if one assumes that for a fixed injection quality, $\widehat{K}$, $\overline{f}_G/\overline{f}_G^o$, and $\widehat{D}^2/\widetilde{D}^2$ are independent of gas injection rate. Under these conditions, Equation 59 reduces to

$$\frac{k_G}{k_G^o} = K_4\, Q_G^{1-n} \tag{60}$$

where K_4 is a constant. Hence, if k_G^o is known as a function of gas saturation, Equation 60 might represent a useful correlating equation for tests with a specific fluid system in a specific porous rock.

Experimental Observations

A complete evaluation of the permeability model just described would require displacement tests where the gas-liquid interfacial properties (including dynamic interfacial effects) and pore space geometry (capillary pressure and thin section determinations) were fully characterized. This goes beyond the scope of the present work; however, we do provide observations relating to the effects of the macroscopic variables Q_G and Q_L.

In this connection, a series of steady state displacement tests were conducted in a 170-200 mesh bead pack with a porosity of 0.4 and an absolute permeability of 7000 md. The experimental apparatus is shown in Figure 7. Two types of gas injection were employed - constant pressure injection and constant rate injection; air was used in the former, nitrogen in the latter. The liquid phase was water plus surfactant (0.25% by volume Amphosal CA, Stepan Chemical Company) and was always injected under constant

rate conditions.   In all tests, pressure was measured at the inlet
and along the bead pack.

Figure 8 shows typical data of gas rate versus pressure drop;
the liquid rate in this case was just sufficient to avoid liquid
depletion in the cell.   Two important features should be noted:
First, in the limit of zero gas flow, a residual or mobilization
pressure is observed.   Second, at pressures above the mobilization
pressure, the Q_G versus ΔP behavior is non-linear with $\Delta P \sim (Q_G)^n$
and n being less that 1.   A value of n = 0.5 allows a reasonable
fit of the data, which is consistent with Equation 53 and intermediate to the rate dependence predicted by Equations 44 and 49.

The results of Figure 8 were obtained in tests where the
pressure drop was successively increased from one test to another.
Other displacement histories were also investigated where ΔP was
increased and decreased one or more times with data being obtained
during both the up and down parts of the cycles.   In general, the
mobilization pressure ΔP_{mob} varied considerably in the different
tests, but the dynamic pressure gradient (total pressure drop minus
mobilization pressure, all divided by cell length) was relatively
constant at fixed gas and liquid rates, regardless of the
displacement history.   This is illustrated in Figure 9 which gives
gas permeabilities based upon the dynamic pressure gradient over a
range of gas rates and for different displacement histories.   The
liquid rate Q_L in these tests was < 0.3 µl/s.   Even though there
are some non-random deviations to suggest some displacement history
dependence, it is surprisingly weak.   If the foam texture varied
significantly with displacement history, these results would
indicate a weaker dependence of k_G on foam texture than predicted
by Equation 44.

In Figure 10 axial pressure profiles are shown for initial and
final drainage tests.   The curve labelled "initial drainage"
corresponds to the steady state profile observed in a fresh
displacement test with no prior imbibition.   The "final drainage"
curve corresponds to the profile observed after successive
imbibition and drainage experiments.   The significant increase in
the mobilization pressure appears to be a foam texture effect.
Also, there is only a slight increase in the dynamic pressure
gradient with successive drainage and imbibition experiments.

Another point which should be made with respect to Figure 10
is that it is the dynamic pressure gradient which should be used in
calculating the permeability and not simply the total pressure drop
divided by the cell length.   If the latter is used, excessively low
values of k_G will result.   Such effects will be particularly
exaggerated in laboratory tests in short cores where the
mobilization pressure is the dominant contribution to the total
applied pressure.   In the field, the mobilization pressure will
represent only a very small fraction of the total ΔP.

To access the importance of $\bar{f}_G$ in Equation 53 and to determine
how this quantity changes from non-dispersed flow to foam flow, a
transient gas permeability test was performed and the results
presented in Figure 11.   In particular, drainage and imbibition
tests were first conducted to reach a final drainage state with Q_L
being fixed at 0.53 µl/s.   The measured permeability corresponded
to the lower curve in Figure 11.   Here again, the dependence on gas

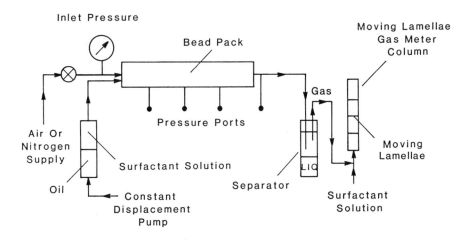

Figure 7.   Bead pack apparatus for foam displacement tests.

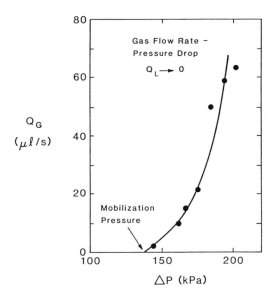

Figure 8.   Typical pressure drop-gas rate curve obtained at steady state in bead pack tests.

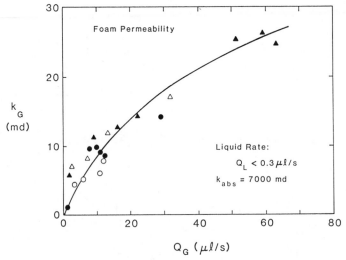

Figure 9.   Observed dependence of gas (foam) permeability on
gas flow rate for low liquid rates.   The data shown   correspond
to different   displacement histories: ● and Δ for gas   only
displacements with the pressure being increased between each
test; ▲ for gas only displacement,   but with the pressure being
lowered; and   ○   to Q_L = 0.3 μl/s and the pressure being
increased.

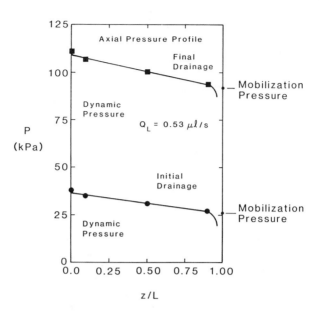

Figure 10. Axial pressure profiles observed at steady state for initial drainage experiment (no prior imbibition) and final drainage experiment (after a number of successive drainage and imbibition experiments).

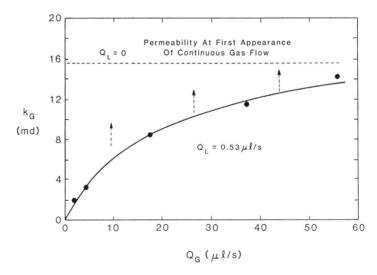

Figure 11. Permeability changes with gas rate. If liquid injection is stopped, the permeability increases with time until a constant value is reached (the dotted line).

rate was non-linear with n = 0.5. The liquid injection was then
stopped, while the gas rate was maintained. With time, because of
bubble-bubble coalescence (15), the gas permeability increased
until continuous gas flow was achieved where there was no mobilization pressure and a constant gas permeability (dotted line in
Figure 11). When this permeability is compared with that for the
non-dispersed phase system under the same conditions (k_G = 5000 md),
one concludes that a significant fraction of the gas flow channels
which are open for continuous gas flow in the non-dispersed case
are blocked during foam displacement. In fact, based upon the
ratio of permeabilities in the two cases, the number of channels
open in the non-dispersed phase case is a factor of 100 or more
greater than in the foam case. Clearly, the fraction of pores
containing flowing gas in foam displacement ($\bar{f}_G$ in Equations 44, 49,
and 53) is quite small and its decrease from the non-dispersed case
is a primary factor in the large permeability reductions observed
with foams.

Figure 12 shows the effect of liquid rate on foam
permeability. The curves shown reflect the results obtained with
different displacement histories and at two different liquid flow
rates (Q_L = 0.53 µl/s and Q_L = 1.56 µl/s). Increasing the liquid
rate clearly decreases the permeability of the foam. This decrease
is likely a reflection of the creation of a finer foam texture. In
particular, increases in liquid rate could increase liquid
saturation and bubble snap-off and decrease bubble-bubble
coalescence. This would result in a finer foam texture and an
increase in the number of bubbles per unit train length.

Another effect which arises as the liquid rate is increased is
an increase in the mobilization pressure. The latter could arise
from an increase in the number of bubbles per unit length in the
continuous trains (refer to Figures 4(a) and 4(b)), and/or from a
decrease in the train lengths. The decrease in train length
without a proportionate decrease in mobilization pressure would
result in a higher pressure gradient in the liquid phase to
maintain gas flow.

To understand the latter effect in a more formal way, consider
the displacement of a continuous train as illustrated in Figure
13(a). The constricted pore represented is an "open pore" in that
it is in direct communication with adjacent channels containing
foam as well as continuous liquid.

The capillary pressure at the position of the kth lamella can
be given by

$$\hat{P}_k - P_k = \frac{2\sigma K_{m,k}}{R_o} + \hat{K}_1 L_k Q_G^n - \frac{Q_L L_k}{\lambda_L} \tag{61}$$

Here, $K_{m,k}$ is the mobilization factor (see Equation 5) for the
bubble train segment in front of the kth lamella. Also, R_o is the
radius of curvature of the drainage surface at the front of the
train, L_k is the length of the train from the kth lamella to the
front drainage surface, K_1 is a resistance factor for the gas
displacement, and $\lambda_L = k_L A/\mu_L$ is a liquid mobility parameter.

Now, a necessary and sufficient condition that the bubbles at
the kth position are in contact is that the capillary pressure must
exceed 2 σ/R_k, where R_k is the pore radius at that point. It then

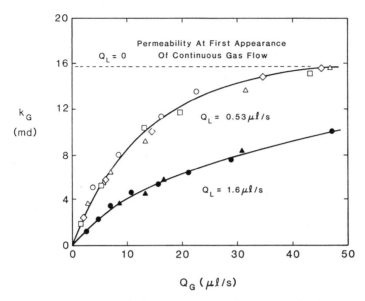

$$Q_G \ (\mu\ell/s)$$

Figure 12.   Permeability curves at different liquid rates.   In both cases, shut-off of liquid injection to cell results in a constant permeability (dotted line) after long periods of time. Data correspond to different displacement histories: □ , second imbibition;   O, third drainage;   △ , fourth drainage; ◇ , third imbibition;   ●, drainage with succeeding tests being conducted   at increasing Q_G's;   ▲ ,   imbibition with Q_G decreasing in succeeding tests.

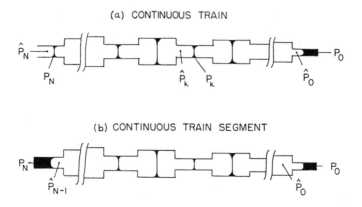

Figure 13.   Illustration of long continuous train and shorter continuous train segment.

follows from Equation 61 that for the bubble train to be continuous over all k positions in the train,

$$\frac{2\sigma K_{m,k}}{R_o} + \hat{K}_1 L_k Q_G^n - \frac{Q_L L_k}{\lambda_L} > \frac{2\sigma}{R_k} \tag{62}$$

For a fixed gas rate Q_G, if the liquid rate is increased sufficiently high, the criterion of Equation 62 will be violated and the bubble train will break at the smallest constriction along the train. As the liquid rate is increased higher and higher, the train will break at other points (successively higher R_k's) with the result that the number of continuous trains will increase, but their lengths will decrease.

The opposite behavior arises for a fixed liquid rate and increasing gas rate. In particular, the number of trains along the path will decrease but their length will increase. For sufficiently high Q_G, a single continuous bubble train would arise.

As noted previously, the longer the bubble train, the smaller the pressure gradient required to maintain gas flow. In particular, for the bubble train shown in Figure 13(b),

$$\frac{\Delta P^*}{L/R_o} = \frac{1}{(L/R_o)} \frac{\left(P_N - P_o\right)}{2\sigma/R_o} = \hat{K}_1 Q_G^n + \frac{1}{(L/R_o)} (K_m - \frac{R_o}{R_N}) \tag{63}$$

where the left hand side represents a dimensionless pressure gradient and L is the total length of the train. In light of the arguments above, the bubble train length varies with Q_G, Q_L, plus the other parameters in Equation 63, i.e.,

$$\frac{L}{R_o} = \frac{L_o}{R_o} [1 - \psi(Q_L, Q_G, \ldots)] \tag{64}$$

where L_o is the train length when $Q_L \to 0$ (e.g., high quality injection). Hence, any parameter such as Q_G (or Q_L) which will affect L will also directly affect the pressure gradient required to maintain gas flow. Obviously, increases in Q_L will result in higher and higher pressure gradients to maintain the gas flow at a given rate.

As a final point in this section, we should mention that as the bubble trains advance through the constricted channels, the capillary resistance will assume its maximum value (the mobilization pressure) only when the lamellae in the train assume their most unfavorable positions with respect to displacement. At other times, the capillary resistance will be below this maximum value with the result that the actual work required to maintain foam flow at a given rate will be below that which would be required if the mobilization pressure was operative at all times. This is easily understood if one pushes a bubble through a single constriction in a tube and notes that the pressure in the train builds up to the mobilization pressure as the drainage surface advances into the constriction and then rapidly falls as the front bubble experiences a Haines jump. To account for such effects in the present model, the K_m term in Equation 63 would vary with time as the bubble train moved through the constricted channels.

A somewhat different interpretation of such phenomena has been given recently by Falls et al. (14). These investigators use energy arguments to account for the work associated with moving lamellae into the constrictions along the bubble train path. Such an approach is attractive from a mechanistic standpoint; however, it predicts that the capillary resistance increases with the length of the porous medium, which has not been observed experimentally. Obviously, more extensive analysis and experimentation are required before such pore constriction effects are fully understood and accurately represented.

Phase Saturations

If the foam phase is thought of as a pseudo continuous fluid with an apparent viscosity $\mu_{app} = \mu_G(k_G^0/k_G)$, it follows that μ_{app} is greater than that of the aqueous liquid phase. (For the tests here, values of μ_{app} were on the order of 1 to 50 times that of water). Because of this, when foam and liquid move through a porous medium under an applied pressure drop, the foam, being the most viscous phase, must occupy a larger region of the pore space. Consequently, as observed, the gas saturation is increased over that of non-dispersed phase flow and the liquid permeability is correspondingly decreased.

Once such high gas saturations have been established, it is difficult to decrease these through liquid imbibition because of the breakup and trapping of the bubble trains once gas flow is stopped. If the gas flow is reinitiated, higher gas saturations can be realized; however, the possible saturation changes in such dispersed flows are considerably smaller than those possible in non-dispersed flows. Further, the effects of flow rate or applied pressure drop on saturation are also small.

The liquid saturations in foam flow are typically close to irreducible liquid saturation. As a result, the liquid saturation in a foam filled medium is generally not a good measure of the quality of the in situ foam, but rather the fraction of pore segments completely filled with liquid. More permeable media, such as unconsolidated media, generally have smaller residual liquid saturations (32,33) and thus tend to have higher gas saturations when foam is flowing.

When gas alone is injected into a porous medium where foam had been flowing, the liquid saturation can be reduced below the irreducible liquid saturation. It would be expected that the liquid phase becomes discontinuous at this point and the further reduction of the liquid saturation occurs as a result of liquid flowing from the core as bubble train lamellae. This does not occur in conventional gas-liquid displacement, and the lower limit of the liquid saturation corresponds to the irreducible value.

Concluding Remarks

The present paper has provided a basis on which foam flow in porous media can be analyzed and described. The gas mobility was found to be dictated by static capillary effects (mobilization pressure), dynamic capillary effects (dynamic effects altering bubble train

length and internal lamellae structure), and dynamic viscous
effects. Although important in the laboratory, static capillary
effects would be of little importance in actual field applications.
The significant permeability reductions which are commonly observed
with foams are principally tied to (1) decreases in the fraction of
pore space containing flowing foam, and (2) increases in viscous
and capillary effects associated with bubble train lamellae.
     Although the current permeability model properly reflects many
of the important features of foam displacement, the authors
acknowledge its limitations in several respects.   First, the open
pore, constricted tube, network model is an oversimplification of
true 3-D porous structures.   Even though communication was allowed
between adjacent pore channels, the dissipation associated with
transverse motions was not considered.   Further, the actual local
displacement events are highly transient with the bubble trains
moving in channels considerably more complex than those used here.
Also, the foam texture has been taken as fixed; the important
effects of gas and liquid rates, displacement history, pore
structure, and foam stability on in situ foam texture were not
considered.   Finally, the use of the permeability model for
quantitative predictions would require the apriori specification of
f_G, the fraction of D_m channels containing flowing foam, which at
present is not possible.   Obviously, such limitations and factors
must be addressed in future studies if a more complete description
of foam flow and displacement is to be realized.

Acknowledgments

     The support of the Shell Company Foundation and University of
Houston EOR Laboratory is acknowledged.

Legend of Symbols

A	cross sectional area of porous medium
b	film thickness
B, B'	film constants or functions, Equations 6 and 13
C_1, C_2	constants
Ca	capillary number, Equations 7 and 31
Ca	macroscopic capillary number, Equation 39
D	diameter of tube or pore segment (Figure 6)
D_{fd}, D_{gd}	drainage diameters corresponding to P_c, foam and continuous gas
D_m	minimum pore diameter in channel, see Figure 6
$\bar{D}$	diameter defined by Equation 40
$\bar{D}_m$	diameter defined by Equation 41
$\hat{D}$	diameter defined by Equation 55
$\hat{D}_m$	diameter defined by Equation 34
$\bar{\bar{D}}$	diameter defined by Equation 47
$\bar{\bar{D}}_m$	diameter defined by Equation 48
$\hat{\hat{D}}$	diameter defined by Equation 57
$\hat{\hat{D}}_m$	diameter defined by Equation 58
D	diameter defined by Equation 27
D_m	diameter defined by Equation 23

$f_G, \bar{f}_G$ fraction of channels with flowing foam, in cell D_m and in total porous medium

$f_G, \bar{f}_G^0$ fraction of channels with flowing continuous gas, in cell D_m and in total porous medium

J number of D segments in D_m channel

k_G gas (or foam) permeability; defined by Equation 42

k_G^0 permeability of non-dispersed gas

K_m mobilization factor, Equation 5

$K_{m.k}$ mobilization factor, Equation 61

$K, K_1, K_2,$
$\hat{K}_3, \hat{K}_4$ constants

$\hat{K}, \hat{K}_1$ constants

ℓ length of pore or tube segment

L length of bubble or continuous bubble train; length of cell

L_f total film length of bubble train

L_k length of train from kth lamella to front drainage surface

L_o train length when $Q_L \rightarrow 0$

n parameter in Equation 50

N number of bubbles in bubble train

N_c number of channels in each coordinate direction

N_L number of lamella per bubble train length

$\Delta\hat{\mathscr{P}}$ dynamic pressure drop in gas phase

$(\Delta\hat{\mathscr{P}})_D$ dynamic pressure drop (in gas) across tube segment D

$\Delta\hat{P}_\ell$ pressure drop across lamella

$\Delta\hat{P}^*, \Delta P^*$ dimensionless pressure drops in gas and liquid (Equation 8)

$\Delta\hat{P}_{\ell m}$ dimensionless mean pressure drop across lamella (Equation 3)

ΔP_{mob} mobilization pressure drop

P_c capillary pressure

$\hat{P}, P$ gas and liquid pressures

$\hat{P}^*, P^*$ dimensionless gas and liquid pressures (PR/2)

$\hat{P}_k, P_k$ gas and liquid pressures, kth lamella

$\hat{P}_0, \hat{P}_N$ pressure in bubbles (see Figure 3)

$(Q_G)_m$ volumetric gas (foam) flow rate through the D_m channels

Q_G gas flow rate

Q_L liquid flow rate

R radius of tube or pore

R_b, R_t radii of pore body and pore throat (Figure 3)

R_c contact radius, Figure 5(b)

R_k radius of pore at kth lamella, Figure 13(a)

R_m^* dimensionless mean radius of flow channel (R_m/R_o)

R_o radius, front pore constriction of bubble train

$R_{t,min}$ pore radius associated with mobilization of continuous gas through a porous medium

U bubble velocity, Figure 5(a)

V porous medium volume; average velocity of liquid, Figure 5

$\bar{V}_b, V_{bm}$ average bubble volume, in porous medium and in cell D_m

V_P, V_{Pm} total pore volume; pore volume in D_m cell

α volume distribution of D segments in D_m channel

β volume distribution of D_m channels

γ number distribution of D_m channels

η number distribution of tubes with diameter D in channel D_m

θ angle of pore wall (Equation 1)

θ_m mean angle of pore wall

κ   defined below Equation 48
λ_L liquid mobility parameter, defined below Equation 61
μ_{app} apparent foam viscosity
μ_G, μ_L viscosities of gas and liquid
σ   surface tension
φ   porosity

Literature Cited

1. Hirasaki, G. J.; Lawson, J. B. Soc. Pet. Eng. J. 1985, 25, 176.
2. Bernard, G. G.; Holm, L. W. Soc. Pet. Eng. J. 1964, 4, 267.
3. Holm, L. W. Soc. Pet. Eng. J. 1968, 8, 359.
4. Raza, S H. Soc. Pet. Eng. J. 1970, 10, 328.
5. Minssieux, L. J. Pet. Tech. 1974, 25, 100.
6. Dellinger, S. E.; Holbrook, S. T.; Patton, J. T. Proceedings, 2nd Joint SPE/DOE Symposium on Enhanced Oil Recovery, 1981, SPE/DOE Paper 9808, 503-508.
7. Ali, J.; Burley, R. W.; Nutt, C. W. Chem. Eng. Res. Des. 1985, 63, .
8. Slattery, J. C. AIChEJ 1979, 25, 283.
9. Hahn, P-S.; Ramamohan, T. R.; Slattery, J. C. AIChEJ 1985, 31, 1029.
10. Bernard, G. G.; Holm, L. W.; Jacobs, W. L. Soc. Pet. Eng. J. 1965, 5, 295.
11. Lawson, J. B.; Reisberg, J. Proceedings, 1st Joint SPE/DOE Symposium on Enhanced Oil Recovery, 1980, SPE Paper 8839, 289-304.
12. Albrecht, R. A.; Marsden, S. S., Jr. Soc. Pet. Eng. J. 1970, 10, 51.
13. Falls, A. H.; Gauglitz, P. A.; Hirasaki, G. J.; Miller, D. D.; Patzek, T. W.; Ratulowski, J. Proceedings, SPE/DOE Enhanced Oil Recovery Symposium, 1986, SPE Paper 14961.
14. Falls, A.H.; Musters, J. J.; Ratulowski, J. Soc. Pet. Eng. J. 1986, SPE Paper 16048, submitted for publication.
15. Khatib, Z. I.; Hirasaki, G. J.; Falls, A. H. Proceedings, 61st Annual SPE Meeting, 1986, SPE Paper 15442.
16. Payatakes, A. C.; Dias, M. M. Rev. Chem. Eng. 1984, 2(2), 85.
17. Tomotika, S. Proc. Roy. Soc. London 1935, A150, 322.
18. Lee, W. K.; Flumerfelt, R. W. Int. J. Multiphase Flow 1981, 7, 363.
19. Lee, W. K.; Yu, K. L.; Flumerfelt, R. W. Int. J. Multiphase Flow 1981, 7, 385.
20. Roof, J. G. Soc. Pet. Eng. J. 1970, 10, 85.
21. Mohanty, K. K. Ph.D. Thesis, University of Minnesota, Minneapolis, 1981.
22. Arriola, A.; Willhite, G. P.; Green, D. W. Soc. Pet. Eng. J. 1983, 23, 99.
23. St. Laurent, C. M.; Gauglitz, P. A.; Radke, C. J. Lawrence Berkeley Laboratory Report LBID-1165, 80 pgs, 1985.
24. Hammond, P. S.; "Non-linear Adjustment of an Annular Film of Viscous Fluid Within an Axisymmetric Constricted Tube," Research Note, Schlumberger-Doll Research, Ridgefield, Conn., 58 pgs, 1985.

25. Mast, R. F. Proceedings, 47th Annual Meeting SPE-AIME, 1972, SPE Paper 3997.
26. Bikerman, J. J. Foams; Springer-Verlag, New York, 1973.
27. Jo, E. J. Ph.D. Thesis, University of Houston, Houston, 1984.
28. Bretherton, F. P. J. Fluid Mechanics 1961, 10, 166.
29. Teletzke, G. F. Ph.D. Thesis, University of Minnesota, Minneapolis, 1983.
30. Park, C.-W.; Homsy, G. M. J. Fluid Mech. 1984, 139, 291.
31. Chen, Jing-Den J. Colloid Int. Sci. 1986, 109, 341.
32. Dullien, F. A. L. Porous Media: Fluid Transport and Pore Structure; Academic Press, New York, 1979.
33. Scheidegger, A. E. The Physics of Flow Through Porous Media; The MacMillian Company, New York, 1960.
34. Dullien, F. A. L. AIChEJ 1975, 21, 299.
35. Ni, S. M. S. Thesis, University of Houston, Houston, 1986.

RECEIVED January 29, 1988

Chapter 16

Description of Foam Flow in Porous Media by the Population Balance Method

T. W. Patzek

Bellaire Research Center, Shell Development Company, P.O. Box 481, Houston, TX 77001

The mobility of foams in porous media is dominated by foam texture (i.e., foam bubble size). The effect of bubble size can be incorporated into a reservoir simulator by adding balances on the number density of bubbles in flowing and stationary foam to the equations of mass, momentum, and energy conservation. These balances account for changes in foam texture caused by mechanisms which create, destroy, or transfer bubbles, e.g., capillary snap-off, bubble division, coalescence, diffusion, evaporation or condensation, as well as bubble trapping and mobilization. The population balance of the number bubble density is volume-averaged and the resulting equations are simplified by a series expansion in the moments of bubble mass. The zeroth order moment equations have a form directly applicable to reservoir simulations, but they cannot be solved without some knowledge or assumptions about the higher order (≥ 2) moments.

All enhanced oil recovery processes of practical significance utilize gases. In 1983 (1), 10 percent of the U.S. oil production came from steam floods and 1 percent from miscible floods. Gas displacement processes are efficient, but they inevitably suffer from gravity override and channeling through oil because gases have lesser densities and higher mobilities than displaced liquids. Volumetric sweep efficiency of a gas displacement process can be improved if the injected gas is dispersed as a foam in a continuous liquid phase and the gas mobility is reduced (2-12). Foam flood technology is not commercial yet, but steam foams have been successfully field tested in heavy oil reservoirs in California. For example, Shell has conducted two steam foam pilots in Kern River (13-14), and Unocal a pilot in Guadelupe River (15) and Midway

0097–6156/88/0373–0326$06.00/0
o 1988 American Chemical Society

Sunset (16). CO_2 foams have not been field tested at the time of this writing.

The scope of possible foam applications in the field warrants extensive theoretical and experimental research on foam flow in porous media. A lot of good work has been done to explain various aspects of the microscopic foam behavior, such as apparent foam viscosity, bubble generation by capillary snap-off, etc.. However, none of this work has provided a general framework for modeling of foam flow in porous media. This paper attempts to describe such a flow with a balance on the foam bubbles.

Numerous studies (2-12) have shown that the behavior of foams flowing in porous media depends on a host of variables, e.g., capillary pressure, capillary number, foam quality, presence of oil, and composition of oil, but is dominated by foam texture (5,6) (i.e., foam bubble size). Experiments (4,5) provide ample evidence that foam mobility is sensitive to foam texture and show that texture itself can be drastically altered by porous media, i.e., foams can be collapsed or refined depending on the flow conditions. Hence, foam texture appears to be the most important variable governing foam mobility.

The purpose of this paper is to derive equations governing the transport of bubbles during flow of foam through porous media. To be readily incorporated into an appropriate reservoir simulator, these equations should have a form similar to those of the transport equations of mass and energy. The new equations can be coupled to the existing ones via the gas mobility in Darcy's Law (4-6). The scope of this paper is limited to the description of foam flow by a balance on the number density of bubbles. This approach falls within a broader class of population balance methods (18-23). The general, volume-averaged (24,25) transport equations of bubbles in flowing and stationary foam are rendered tractable by a series expansion of the bubble density distribution in the moments of bubble mass. The principal mechanisms altering texture of foams in porous media are considered.

A simplified one-dimensional transient solution of the bubble population balance equations, verified by experiments, has been presented elsewhere (5) for a special case of bubble generation by capillary snap-off.

Model of Foam Flow

Foam (5) is a collection of gas bubbles with sizes ranging from microscopic to infinite for a continuous gas path. These bubbles are dispersed in a connected liquid phase and separated either by lamellae, thin liquid films, or by liquid slugs. The average bubble density, related to foam texture, most strongly influences gas mobility. Bubbles can be created or divided in pore necks by capillary snap-off, and they can also divide upon entering pore branchings (5). Moreover, the bubbles can coalesce due to instability of lamellae or change size because of diffusion, evaporation, or condensation (5,8). Often, only a fraction of foam flows as some gas flow is blocked by stationary lamellae (4). Changes in flow conditions can result in trapping or mobilization of bubbles.

To arrive at an approximate description of foam flow in porous media, it is recognized that each bubble and its environment are characterized by the following variables and parameters:
1. Internal variables - peculiar to each bubble, its mass m, and velocity, $\underset{\sim}{v}$. (Note that bubble volume is a poor choice as an internal variable because foam is compressible.)
2. External variables - the bubble position, $\underset{\sim}{x}$, as well as parameters, $\underset{\sim}{\pi}(\underset{\sim}{x},t)$, computed from the conservation equations and relationships external to the population balance, e.g., pressure, temperature, capillary pressure, etc.
3. Interaction parameters, $\underset{\sim}{\sigma}(\underset{\sim}{x},t)$, calculated from the bubble population balance itself, e.g., the total bubble density in flowing and stationary foam, the higher moments of the bubble number density distribution, etc.

To simplify further the description of foam flow, bubble velocity is removed from the set of internal variables. It is assumed that a fraction, X_f, of all bubbles is carried with velocity, $\underset{\sim}{v}(\underset{\sim}{x},t;\underset{\sim}{\pi},\underset{\sim}{\sigma})$ as flowing foam (f). This assumption is justified because flowing bubbles remain in contact most of the time, and their average velocity, $<\underset{\sim}{v}>$, can be related to Darcy's velocity of the gas phase, $<\underset{\sim}{v}> = \underset{\sim}{u}_g/(\phi S_g X_f)$. The remaining fraction, $X_t = 1 - X_f$, constitutes stationary foam (t) which is connected to the flowing part but does not flow itself.

The simplest **number density** of bubbles, $F(\underset{\sim}{x},m,t)$, can be defined in a four-dimensional phase space (19,20) $(0,\infty) \times R^3$, as a distribution over the internal variable, m, with the external variables, $\underset{\sim}{x}$, held fixed. Then, $F(\underset{\sim}{x},m,t)d^3xdm$ is the number of bubbles with masses between m and m+dm, present at time t in an infinitesimal neighborhood, d^3x, of point $\underset{\sim}{x}$.

The following balance equation can be derived for the bubble number density in this space (19,20)

$$\frac{\partial F}{\partial t} + \qquad \nabla \cdot (\underset{\sim}{v}\ F) + \qquad \frac{\partial}{\partial m}(\dot{m}F) - \qquad\qquad H = \qquad 0 \qquad (1)$$

accumulation + convection + $\begin{array}{c}\text{growth without}\\\text{change of identity}\end{array}$ - net generation = 0

As shown in Appendix A, Equation (1) can be averaged over the volume of the porous medium to yield the population balances of bubbles in flowing foam

$$\frac{\partial}{\partial t}\left(\phi S_g X_f <F>^f\right) + \nabla \cdot \left(\underset{\sim}{u}_g <F>^f\right) + \nabla \cdot <(\underset{\sim}{v}'F')_f> + \phi S_g X_f \frac{\partial}{\partial m} <\dot{m}F>^f \qquad (2)$$

$$+ \tilde{\gamma} + \tilde{\delta} - \phi S_g X_f <H>^f = 0$$

and in stationary foam

$$\frac{\partial}{\partial t}\left(\phi S_g X_t <F>^t\right) + \phi S_g X_t \frac{\partial}{\partial m} <\dot{m}F>^t - \tilde{\gamma} + \tilde{\delta} - \phi S_g X_t <H>^t = 0 \qquad (3)$$

where

$<(\cdot)>^i$ is the intrinsic domain average of quantity $(\cdot)$ over flowing and stationary foam, i=f,t;

$<(\underset{\sim}{v}'F')_f>$ is the rate of change of bubble number density due to a dispersion flux that arises from the difference between the local and average velocity;

$\overset{\bullet}{m} \equiv \dfrac{dm}{dt}$ is the rate of change of bubble mass due to diffusion;

$\underset{\sim}{\gamma}$ is the rate of change of bubble number density due to trapping of flowing bubbles;

$\underset{\sim}{\delta}$ is the rate of change of bubble number density due to condensation $(\underset{\sim}{\delta} > 0)$ or evaporation $(\underset{\sim}{\delta} < 0)$.

The net rate of bubble generation, H, describes redistribution of mass in bubble-bubble interactions. Thus, H is a nonlinear functional of $F(\underset{\sim}{x},m,t)$ and Equations (2) and (3) are a pair of coupled, nonlinear, integro-differential equations in the bubble number density, similar to Boltzmann's equation in the kinetic theory of gases (26,27) or to Payatakes et al (22) equations of oil ganglia dynamics.

    Equations (2) and (3) are very difficult to solve and to be of practical use must be approximated by, say, a series expansion in the moments of bubble mass:

$$\mu_i^{(n)} (\underset{\sim}{x},t) \equiv \int_o^\infty m^n <F(x,m,t)>^i \, dm; \quad n=0,1,2,\ldots; \quad i=f,t \qquad (4)$$

The zeroth order moments, $\mu_f^{(o)} \equiv n_f$, $\mu_t^{(o)} \equiv n_t$, represent the total number of bubbles per volume of flowing and stationary foam, respectively, and are measures of **foam texture**. Thus, the zeroth moments of bubble population balances (2) and (3) are the transport equations of average bubbles in flowing and stationary foam. As such, the zeroth moments are necessary to model flow of foams in porous media, but they are not sufficient and must be solved in conjunction with the higher moments of Equations (2) and (3).

    The zeroth moments of Equations (2) and (3) are

$$\frac{\partial}{\partial t} \left(\phi X_f S_g n_f \right) + \nabla \cdot \left(\underset{\sim}{u}_g n_f \right) - \nabla \cdot \underset{\sim}{j}'_D + d_f + \gamma + \delta - h_f = 0 \qquad (5)$$

$$\frac{\partial}{\partial t} \left(\phi X_t S_g n_t \right) \qquad\qquad\qquad\qquad + d_t - \gamma + \delta - h_t = 0 \qquad (6)$$

where

$\underset{\sim}{j}'_D \equiv -\int_o^\infty <(\underset{\sim}{v}'F')_f> \, dm$ is the total bubble dispersion flux   (7)

$d_i \equiv -\phi S_g X_i \; \underset{m \to 0}{\lim} \; <\overset{\bullet}{m}F(\underset{\sim}{x},m,t)>^i$ i=f,t   , is the total rate of bubble-bubble diffusion   (8)

$\gamma \equiv -\int_o^\infty \underset{\sim}{\gamma} \, dm$ is the total rate of bubble trapping   (9)

$\delta \equiv \int_o^\infty \underset{\sim}{\delta} \, dm$ is the total rate of bubble evaporation due to external factors   (10)

$h_i \equiv \phi S_g X_i \int_o^\infty <H>^i \, dm$;   i = f,t is the total net rate of generation   (11)

Relations (7)-(11) are known functions of $\underset{\sim}{g}(x,t)$ and $\underset{\sim}{\pi}(x,t)$. Since the gas density $\rho_g = \int_0^\infty m <F>^f dm = \int_0^\infty m <F>^t dm$, the sum of the first moments of Equation set (2) and (3) is a mass balance of the gaseous phase. Hence, Equations (5) and (6), together with the usual gas mass balance, are a first order approximation of the bubble population balance.

As mentioned before, Equations (5) and (6) are the differential transport equations of average bubbles and could be written from scratch without the convoluted derivations invoked here. Unfortunately, modeling of foam flow in porous media is a lot more complicated than Equations (5) and (6) lead us to believe. Having started from a general bubble population balance, we discovered that flow of foams in porous media is governed by Equations (2) and (3), and that Equations (5) and (6) are but the first terms in an infinite series that approximates solutions of (2) and (3). Moreover, Equations (5) and (6) cannot be solved alone; they are coupled to the higher order moments that define shape of the bubble size distribution. For example, to model bubble trapping and mobilization, we need some knowledge of the bubble size distribution (usually smaller bubbles get trapped because they offer higher resistance to flow). Diffusional mass transfer between bubbles d_i, $i=f,t$, can be modeled only if we account for bigger bubbles growing at the expense of smaller ones. In addition, any model that relates the flowing fraction of foam X_f to flow conditions and foam texture will have to include bubble trapping and mobilization (as bubbles are mobilized, X_f increases, and vice versa). In short, any realistic description of foam flow in porous media will make use of at least the second order moments of the population balances (2) and (3). The second order moments are related to variance (average width) of the bubble size distribution. If the second and higher order moments are neglected, only bubble generation by capillary snap-off (5) and/or bubble coalescence at high capillary pressure (8) can be retained in Equations (5) and (6), provided that the flowing fraction of foam is known and constant. With these rather severe restrictions, Equations (5) and (6) have been solved (5) for a one-dimensional flow of foam in a glass bead pack with a tight section in the middle.

Finally, all the bubble generation and destruction functionals in Equations (2) and (3) must be known in advance to calculate their moments. It is vastly more difficult to find these functionals than to construct approximate generation and destruction functions in Equations (5) and (6). If we start from the zeroth moment equations, however, we forfeit the ability to calculate the higher order moments of the generation and destruction functionals that in turn are necessary to solve (5) and (6). To break this vicious circle without solving the full-blown population balances (2) and (3), we need to make guesses about shape of the bubble size distribution, and then iteratively solve Equations (5) and (6) until some specified criteria are met.

Conclusions

The following conclusions have been reached in this paper:

1. Mechanistic prediction of foam flow in porous media seems to be impossible without a transport equation governing foam texture, i.e., foam bubble size.
2. The transport equation for the bubble number density distribution can be derived by use of the population balance method.
3. The continuum form of the bubble population balance, applicable to flow of foams in porous media, can be obtained by volume averaging. Bubble generation, coalescence, mobilization, trapping, condensation, and evaporation are accounted for in the volume averaged transport equations of the flowing and stationary foam texture.
4. A meaningful simplification of the bubble population balance can be achieved by expanding the bubble number density into the moments of bubble mass.
5. The zeroth order moments of the volume averaged bubble population equations, i.e., the balances on the total bubble density in flowing and stationary foam, have the form of the usual transport equations and can be readily incorporated into a suitable reservoir simulator.
6. Some knowledge of the higher (≥ 2) order moments is required to solve the zeroth moments of the bubble population balances.

Suggestions for Future Work

The functional forms and relative importance of mechanisms which change the density of bubbles in flowing and stationary foam still are not well known. In particular, the functional form of d_i, h_i, $i=f,t$, and that of γ and δ in Equations (5) and (6) needs to be investigated more thoroughly. Also, a model linking the flowing fraction of foam, X_f, to the gas flux and predicting the conditions of total bubble mobilization should be developed.

Nomenclature

$A_i(\underline{x},t)$ mathematical surface separating region $i=f,t,r$ in averaging volume.

d_i rate of change of total bubble density due to mass transfer from small to bigger bubbles in flowing ($i=f$), and stationary foam ($i=t$); defined in Equation (7).

$d^3x, d^3\zeta$ volume element in Euclidean space.

F distribution of bubble number density; defined in Equation (1).

$\langle F_i \rangle$ domain average of F; defined in Equation (A-5).

$\langle F \rangle^i$ intrinsic domain average of F; defined in Equation (A-8).

F' fluctuation bubble number density distribution; defined in Equation (A-26).

h_i total net generation rate of bubbles in flowing ($i=f$), and stationary ($i=t$) foam; defined in Equation (11).

H   net generation rate of bubbles of mass m.

$j'_{\sim D}$ dispersion flux of the total bubble number density; defined in Equation (7).

m   bubble mass.

$\dot{m}$ rate of change of bubble mass due to the environment.

n_i total bubble number density in flowing (i=f) and in stationary (i=t) foam.

$\underset{\sim}{n}_i$ unit outward normal from region i=f,t,r in averaging volume.

R^3 three-dimensional Euclidean space.

S_g gas saturation.

$\underset{\sim}{u}_g$ Darcy's velocity of gas.

$\underset{\sim}{v}$ instantaneous bubble velocity.

$\langle \underset{\sim}{v} \rangle$ average bubble velocity; defined in Equation (A-27).

$\underset{\sim}{v}'$ fluctuation bubble velocity; defined in Equation (A-25).

V   constant averaging volume.

$V_i(\underset{\sim}{x},t)$ part of the averaging volume occupied by region i=f,t,r.

$\underset{\sim}{w}$ velocity of surface $A_i(\underset{\sim}{x},t)$.

$\underset{\sim}{x}$ position vector.

$\underset{\sim}{x}_{ik}$ portion of the surface dividing region with property i from region with property k.

X_f fraction of foam that is flowing.

$X_t = 1 - X_f$ fraction of foam that is stationary.

$\tilde{\gamma}$ rate of change of bubble number density distribution due to trapping; defined in Equation (A-33).

γ rate of change of total bubble number density due to trapping; defined in Equation (9).

$\tilde{\delta}$ rate of change of bubble number density distribution due to condensation ($\tilde{\delta} > 0$), or evaporation ($\tilde{\delta} < 0$); defined in Equation (A-33).

δ rate of change of total bubble number density due to condensation or evaporation; defined in Equation (10).

δ Dirac's delta function.

ϕ porosity.

$\mu_i^{(n)}$ moments of bubble number density distribution function; defined in Equation (4).

$\underset{\sim}{\sigma}(x,t)$ array of parameters describing bubble and bubble-pore space interactions.

$\underset{\sim}{\pi}(x,t)$ array of parameters external to the bubble population balance.

χ_i characteristic functions defined in Equations (A-2)-(A-4).

ϵ_i fraction of averaging volume occupied by region i=f,t,r; defined in Equation (A-10).

Subscripts

f   flowing foam.

r   remaining phases.

t   stationary foam.

Appendix A

Volume-Averaging the Bubble Population Balance

There are at least three different approaches to modeling flow in porous media. Since a porous medium in general is not an ordered system, the idea of developing statistical models seems appealing. Another approach is the method of geometric modeling in which a postulated geometry of pore space allows the governing equations to be solved. A third approach, which falls between the other two, is the development of correct, averaged forms of the governing differential equations. These continuum forms hold for a broad class of porous media, and are more general than the geometric models.

   The purpose of this Appendix is to volume-average the population balance of bubble number density

$$\frac{\partial F}{\partial t} + \nabla \cdot (\underset{\sim}{v} F) + \frac{\partial}{\partial m} (\dot{m} F) - H = 0 \qquad (A-1)$$

The derivation presented here follows with some changes volume-averaging methods developed by Gray and Lee(24) and Whitaker (25). Volume-averaging is considerably simplified with the use of characteristic functions which are nonzero and constant only in a region of porous medium occupied by specified components and zero otherwise. The characteristic function can have an arbitrarily complex domain, i.e., it can be defined in a region of arbitrary shape and connectiveness (see Figure A-1).

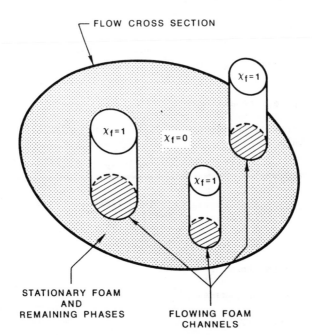

Figure A-1. Two-dimensional characteristic function of flowing foam, χ_f. This function is defined to be one in regions of the porous medium occupied by flowing foam, and zero otherwise. Note that the gradient of χ_f is nonzero only on the surfaces separating the flowing foam from other regions; there it is equal to $\pm \infty$.

The following **characteristic functions** are defined here

Flowing foam (f)

$$\chi_f(\underset{\sim}{x},t) \equiv \begin{cases} 1, \text{ if } \underset{\sim}{x} \; \epsilon \text{ (domain occupied by flowing foam)} \\ 0, \text{ otherwise} \end{cases} \quad \text{(A-2)}$$

Stationary foam (t)

$$\chi_t(\underset{\sim}{x},t) \equiv \begin{cases} 1, \text{ if } \underset{\sim}{x} \; \epsilon \text{ (domain occupied by stationary foam)} \\ 0, \text{ otherwise} \end{cases} \quad \text{(A-3)}$$

The remaining phases (r), i.e. oil, water, solids, etc.

$$\chi_r(\underset{\sim}{x},t) \equiv \begin{cases} 1, \text{ if } \underset{\sim}{x} \; \epsilon \text{ (domain occupied by remaining phases)} \\ 0, \text{ if } \underset{\sim}{x} \; \epsilon \text{ (domain occupied by foam)} \end{cases} \quad \text{(A-4)}$$

From Equations (A-2) and (A-3), it is seen that the foam bubbles are classified as either flowing or stationary, irrespective of their size. The characteristic functions are made time dependent to allow for deformation of their respective domains due to mass transfer, trapping, or mobilization.

From Figure A-1 it can be inferred that

$$\nabla \chi_i(\underset{\sim}{x},t) = -\underset{\sim}{n}_i \ \delta(\underset{\sim}{x} - \underset{\sim}{x}_{ik}) \ ; \qquad k \neq i, \quad k=f,t,r \qquad (A-5)$$

where
n_i is the outward unit normal for region i,
x_{ik} is the portion of the mathematical surface bounding region i
            in contact with region k, and
δ is Dirac's delta "function", $\delta(\underset{\sim}{x}-\underset{\sim}{x}_o) \equiv \begin{cases} + \infty, \text{ if } \underset{\sim}{x}=\underset{\sim}{x}_o \\ 0, \text{ otherwise} \end{cases}$

Volume-averaging leading to meaningful results requires an averaging volume whose characteristic length is much greater than the pore diameter but much less than the characteristic macroscopic scale of the medium. Additionally, the shape, size, and orientation of the averaging volume should be independent of space and time. (24,25).
Volume-averaging is performed most conveniently in a local coordinate system, ζ, whose origin is translated without rotation to a given point, $\underset{\sim}{x}$, in the medium. The location of the averaging volume with respect to the ζ coordinates becomes independent of $\underset{\sim}{x}$. Figure A-2 shows an example of the averaging volume.
The **domain average** $<F_i>$, i=f,t, of the bubble number density $F(\underset{\sim}{x}, m, t)$ is defined as

$$<F_i>(\underset{\sim}{x},m,t) \equiv \frac{1}{V} \int_V F(\underset{\sim}{x}+\underset{\sim}{\zeta},m,t) \ \chi_i \ (\underset{\sim}{x}+\underset{\sim}{\zeta},t)d^3\zeta = \frac{1}{V} \int_{V_i(\underset{\sim}{x},t)} F \ d^3\zeta \ ; \ i=f,t$$

$$(A-6)$$

since χ_i is zero outside domain i. (This average is called by Whitaker (25) the phase average. Since the flowing and stationary foam are not two different phases, the term domain average is used here.) The domain averaging leads to quantities that are not directly measurable.

The averaging volume V is the sum of volumes occupied by all the fluids and the solid

$$V \equiv V_f(\underset{\sim}{x},t) + V_t(\underset{\sim}{x},t) + V_r(\underset{\sim}{x},t) \qquad (A-7)$$

The **intrinsic domain average** $<F>^i$, i=f,t, of the bubble number density is defined as

$$<F>^i(x,m,t) \equiv \frac{1}{V_i(\underset{\sim}{x},t)} \int_{V_i(\underset{\sim}{x},t)} F(\underset{\sim}{x} + \zeta,m,t) \ d^3\zeta \ ; \ i=f,t \qquad (A-8)$$

Intrinsic averaging corresponds to calculating density, concentration, or other parameter directly obtainable from experiments. Comparison of (A-7) and (A-8) shows that

$$<F_i> = \varepsilon_i(\underset{\sim}{x},t) \ <F>^i \ ; \qquad i=f,t \qquad (A-9)$$

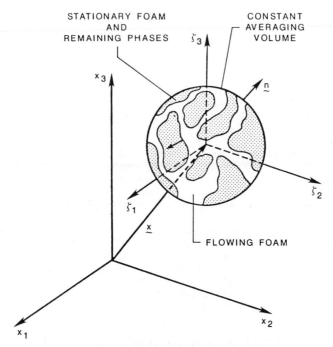

Figure A-2. Volume averaging is performed in a local coordinate
system ζ, whose origin is translated without rotation to a given
point x in the porous medium. The average volume (here a
sphere) is constant.

where

$$\varepsilon_i(\underset{\sim}{x},t) \equiv \frac{V_i(\underset{\sim}{x},t)}{V} = \frac{1}{V} \int_V \chi_i(\underset{\sim}{x} + \underset{\sim}{\zeta},m,t)d^3\zeta \quad ; \quad i=f,t \quad (A-10)$$

is the fraction of the bulk volume occupied by region i, e.g.

$$\varepsilon_f(\underset{\sim}{x},t) = \phi S_g X_f \qquad\qquad\qquad (A-11)$$

$$\varepsilon_t(\underset{\sim}{x},t) = \phi S_g X_t \qquad\qquad\qquad (A-12)$$

$$X_f + X_t = 1 \qquad\qquad\qquad\qquad (A-13)$$

     To volume-average Equation (A-1), the time derivative $\partial F/\partial t$
and the divergence $\nabla \cdot (\underset{\sim}{v}F)$ must be related to their analogs
operating on the averages $<F_i>$ and $<(\underset{\sim}{v}F)_i>$, i=f,t. The volume
average of the time derivative $\partial F/\partial t$ is

$$<\frac{\partial F_i}{\partial t}> \equiv \frac{1}{V} \int_V \frac{\partial F}{\partial t} \chi_i(\underset{\sim}{x} + \underset{\sim}{\zeta},t)d^3\zeta \quad ; \quad i=f,t \qquad (A-14)$$

By the chain rule,

$$\left\langle \frac{\partial F_i}{\partial t} \right\rangle = \frac{1}{V} \int_V \frac{\partial}{\partial t}(F\chi_i)d^3\zeta - \frac{1}{V} \int_V F\frac{\partial \chi_i}{\partial t}d^3\zeta \quad ; \quad i=f,t \quad (A-15)$$

If the ith domain is deforming, χ_i is a function of time and its convective derivative is

$$\frac{D\chi_i}{Dt} \equiv \frac{\partial \chi_i}{\partial t} + \underset{\sim}{w} \cdot \nabla\chi_i = 0 \quad (A-16)$$

where $\underset{\sim}{w}$ (in general $\underset{\sim}{w} \neq \underset{\sim}{v}$) is velocity of the mathematical surface bounding region i. Proof of Equation (A-16) follows from observation that the surface bounding region i is "material" (see Lamb, Reference 28, page 7), i.e., the fluid velocity relative to the surface must be either tangential or zero. Otherwise, one would have a finite flow across the surface contrary to its definition via the characteristic function (A-2)-(A-4) (this surface is not material in the strict sense because it does not bound a single phase).

Equations (A-16) and (A-5) can be combined together to yield

$$\frac{\partial \chi_i}{\partial t} = \underset{\sim}{w} \cdot \underset{\sim}{n}_i \, \delta(\underset{\sim}{x} - \underset{\sim}{x}_{ik}) \quad (A-17)$$

Substitution of Equation (A-17) into (A-15) gives

$$\left\langle \frac{\partial F_i}{\partial t} \right\rangle = \frac{\partial}{\partial t}\left(\frac{1}{V}\int_V F\chi_i d^3\zeta\right) - \frac{1}{V}\int_{A_i(\underset{\sim}{x},t)} F \underset{\sim}{w} \cdot \underset{\sim}{n}_i \, dA_\zeta$$

$$i=f,t \quad (A-18)$$

$$= \frac{\partial}{\partial t}\langle F_i \rangle - \frac{1}{V}\int_{A_i} F \underset{\sim}{w} \cdot \underset{\sim}{n}_i dA_\zeta$$

where $A_i(\underset{\sim}{x},t)$ is the mathematical surface bounding region i. In general, this surface is multiply connected. Note that the assumption V=const has been used to interchange the differentiation with integration.

The average of the divergence $\nabla \cdot (\underset{\sim}{v}F)$ can be related to the divergence of an average as follows

$$\langle \nabla \cdot (\underset{\sim}{v}F)_i \rangle \equiv \frac{1}{V}\int_V (\underset{\sim}{v}F)\chi_i d^3\zeta = \frac{1}{V}\int_V \nabla \cdot (\underset{\sim}{v}F\chi_i)d^3\zeta - \frac{1}{V}\int_V F\underset{\sim}{v} \cdot \nabla\chi_i d^3\zeta$$

$$i=f,t \quad (A-19)$$

Again, the volume integration is independent of ζ, and

$$\nabla\chi_i(\underset{\sim}{x} + \underset{\sim}{\zeta},t) = -\underset{\sim}{n}_i \, \delta(\underset{\sim}{x} + \underset{\sim}{\zeta} - \underset{\sim}{x}_{ik},t); \quad k\neq i, \quad k=f,t,r \quad (A-20)$$

With this in mind, Equation (A-19) can be rearranged to the form

$$\langle \nabla \cdot (\underset{\sim}{v}F)_i \rangle = \nabla \cdot \langle (\underset{\sim}{v}F)_i \rangle + \frac{1}{V}\int_{A_i(\underset{\sim}{x},t)} F \underset{\sim}{v} \cdot \underset{\sim}{n}_i dA_\zeta \quad ; \quad i=f,t \quad (A-21)$$

Certainly, the volume integration is independent of bubble mass, m

$$<\frac{\partial}{\partial m} (\dot{m} \; F)_i> = \frac{\partial}{\partial m} <(\dot{m} \; F)_i> \qquad (A-22)$$

Equations (A-18), (A-21), and (A-22) are now applied to the population balance (A-1) and the result is

$$\frac{\partial}{\partial t} <F_f> + \nabla \cdot <(\underset{\sim}{v}F)_f> + \frac{1}{V} \int_{A_f} F(\underset{\sim}{v}-\underset{\sim}{w}) \cdot \underset{\sim}{n}_f dA_\zeta + \frac{\partial}{\partial m} <(\dot{m}F)_f> = <H_f> \quad (A-23)$$

$$\frac{\partial}{\partial t} <F_t> \qquad\qquad + \frac{1}{V} \int_{A_t} F(\underset{\sim}{v}-\underset{\sim}{w}) \cdot \underset{\sim}{n}_t dA_\zeta + \frac{\partial}{\partial m} <(\dot{m}F)_t> = <H_t> \quad (A-24)$$

It only remains to express the average of the product $<(\underset{\sim}{v}F)_f>$ as the product of the averages $<\underset{\sim}{v}><F>_f$. This can be accomplished with use of the **fluctuation velocity**

$$\underset{\sim}{v}' \equiv <\underset{\sim}{v}> - \underset{\sim}{v} \qquad (A-25)$$

and the **fluctuation bubble density**

$$F' \equiv <F_f> - F \qquad (A-26)$$

The average bubble velocity is related to Darcy's velocity of the gas phase through relation (A-11)

$$<\underset{\sim}{v}> = \underset{\sim}{u}_g/\varepsilon_f = \underset{\sim}{u}_g/(\phi S_g X_f) \qquad (A-27)$$

By definition,

$$<\underset{\sim}{v}'> = \underset{\sim}{0} \qquad (A-28)$$

$$<F'> = 0 \qquad (A-29)$$

and the average $<(\underset{\sim}{v}F)_f>$ reduces to

$$<(\underset{\sim}{v}F)_f> = <\underset{\sim}{v}> <F_f> + <(\underset{\sim}{v}'F')_f> \qquad (A-30)$$

where $<(\underset{\sim}{v}'F')_f>$ is the **dispersion flux** of the bubble density distribution function F.

To close the present derivation of the continuum population balance equations, one needs to simplify the last two terms on the left-hand side of Equations (A-23) and (A-24). These terms describe various mechanisms of mass and/or bubble transfer among the regions defined by the characteristic functions (A-2)-(A-4).

The first term is

$$\frac{1}{V} \int_{A_i(\underset{\sim}{x},t)} F(\underset{\sim}{x} + \underset{\sim}{\zeta},m,t)(\underset{\sim}{v} - \underset{\sim}{w}) \cdot \underset{\sim}{n}_i dA_\zeta \qquad ; \quad i=f,t \qquad (A-31)$$

If the mathematical surface $A_i(\underset{\sim}{x},t)$ was an interface, i.e., if it bounded a region occupied by a single phase then expression (A-31) would be the rate of mass transfer (evaporation or condensation) at this interface. However, according to definitions (A-2) and (A-3), the respective surfaces A_f and A_t bound flowing and stationary foam which are not two separate phases. Therefore, the term "mass transfer" should be understood here as the rate of trapping (i=f) or mobilizing (i=t) the foam bubbles along the common portion of these surfaces. Along the portions of both surfaces in contact with the remaining phases, actual condensation or evaporation can occur.

The second term

$$\frac{\partial}{\partial m} <(\dot{m}F)_i> \quad ; \quad i=f,t \qquad (A-32)$$

describes the diffusive mass transfer among the bubbles of the same type, i.e., flowing or stationary. Indeed, the function $\dot{m}(x,m,t) F(x,m,t)$ represents the rate of growth of all bubbles with mass m under the influence of the environment. The derivative of this function with respect to bubble mass gives the net rate of bubble size redistribution. If only condensation or evaporation occurs, this net rate is zero because all the bubbles at a given position shrink or expand by the same fraction. The only phenomenon causing some bubbles to grow at the expense of others is diffusion.

The sum of expressions (A-31) and (A-32) can be written as

$$\frac{1}{V} \int_{A_i} F(\underset{\sim}{v}-\underset{\sim}{w}) \cdot \underset{\sim i}{n} dA_\zeta + \frac{\partial}{\partial m} <(\dot{m}F)_i> = \mp \tilde{\gamma}(\underset{\sim}{x},m,t) + \tilde{\delta}(\underset{\sim}{x},m,t) \qquad (A-33)$$

$$+ \frac{\partial}{\partial m} <(\dot{m}F)_i> \quad ; \quad i=f,t$$

where $+\tilde{\gamma}(\underset{\sim}{x},m,t)$ is the rate of bubble trapping per unit bulk volume (i=f), $-\tilde{\gamma}(\underset{\sim}{x},m,t)$ is the rate of mobilization (i=t), and $\tilde{\delta}(\underset{\sim}{x},m,t)$ is the rate of change of the bubble density distribution due to condensation ($\tilde{\delta} > 0$) or evaporation ($\tilde{\delta} < 0$).

With the help of Equations (A-9), (A-10), (A-27), (A-30), and (A-33), the volume averaged population balances (A-23) - (A-24) become

FLOWING FOAM

$$\frac{\partial}{\partial t} (\varepsilon_f <F>^f) + \nabla \cdot (\underset{\sim g}{u} <F>^f) + \nabla \cdot <(\underset{\sim}{v}'F')_f> + \varepsilon_f \frac{\partial}{\partial m} <\dot{m}F>^f \qquad (A-34)$$

$$+ \tilde{\gamma} + \tilde{\delta} - \varepsilon_f <H>^f = 0$$

STATIONARY FOAM

$$\frac{\partial}{\partial t} (\varepsilon_t <F>^t) \qquad\qquad + \varepsilon_t \frac{\partial}{\partial m} <\dot{m}F>^t - \tilde{\gamma} + \tilde{\delta} - \varepsilon_t <H>^t = 0 \qquad (A-35)$$

Where, ε_f, and ε_t are given by Equations (A-11)-(A-13).

This completes the derivation of the continuum form of the population balance of bubble number density suitable for modeling the flow of foam in porous media.

Acknowledgments

I thank G. J. Hirasaki for numerous discussions about this work. I
am also grateful to Shell Development Company for allowing me to
publish this paper.

Literature Cited

1. Kouskraa, V. A., "The Status and Potential of Enhanced Oil
    Recovery", SPE 14951,presented at SPE/DOE Fifth Symposium on
    EOR, Tulsa, Okla., April 20-23, 1986.
2. Bernard, G. G. and Holm, L. W., "Effect of Foam on Permeability
    of Porous Media to Gas", Soc. Petr. Eng. J. September 1964, pp
    267-274.
3. Bernard, G. G., Holm, L. W. and Jacobs, W. L., "Effect of Foam
    on Trapped Gas Saturation and on Permeability of Porous Media to
    Water", Soc. Petr. Eng. J. December 1965, pp 295-300.
4. Falls, A. H., Musters, J. J. and Ratulowski, J., "The Apparent
    Viscosity of Foams in Homogeneous Beadpacks", SPE 16048,
    submitted to Soc. Petr. Res. Eng., August 1986.
5. Falls, A. H., Gauglitz, P. A., Hirasaki, G. J., Miller, D. D.,
    Patzek, T. W. and Ratulowski, J., "Development of a Mechanistic
    Foam Simulator: The Population Balance and Generation by Snap-
    Off", SPE 14961 presented at the SPE/DOE Enhanced Oil Recovery
    Symposium, Tulsa, April 20-23, 1986.
6. Hirasaki, G. J. and Lawson, J. B, "Mechanisms of Foam Flow
    Through Porous Media -- Apparent Viscosity in Smooth
    Capillaries", Soc. Petr. Eng. J. April 1985 pp 176-190.
7. Holm, L. W., "The Mechanism of Gas and Liquid Flow Through
    Porous Media in the Presence of Foam", Soc. Petr. Eng. J.
    December 1968, pp 359-369.
8. Khatib, Z. I., Hirasaki, G. J., and Falls, A. H., "Effects of
    Capillary Pressure on Coalescence and Phase Mobilities in Foams
    Flowing Through Porous Media", paper SPE 15442 presented at the
    1986 SPE Fall Meeting, New Orleans, October 5-8, 1986.
9. Marsden, S. S. and Khan, S. A., "The Flow of Foam Through Short
    Porous Media and Apparent Viscosity Measurements", Soc. Petr.
    Eng. J., March 1966, pp 17-25.
10. Mast, R. F., "Microscopic Behavior of Foam in Porous Media", SPE
    3997 presented at 1972 SPE Fall Meeting, San Antonio, October 8-
    11, 1972.
11. Minissieux, L., "Oil Displacement by Foams in Relation to Their
    Physical Properties in Porous Media", J. Petr. Tech., January
    1974, pp 100-108.
12. Raza, S. H. and Marsden, S. S., "The Streaming Potential and
    the Rheology of Foam", Soc. Petr. Eng. J., December 1967,
    pp 359-368.
13. Dilgren, R. E., Deemer, A. R. and Owens, K. B., "Laboratory
    Development and Field Testing of Steam Foams for Mobility
    Control in Heavy Oil Reservoirs", SPE 10774, presented at the
    1980 California Regional Meeting of SPE AIME, San Francisco,
    Calif., March 24-26, 1980.
14. Patzek, T. W. and Koinis, M. T., "Shell's Steam Foam Pilots in
    Kern River", SPE 17380, to be presented at the 1988 SPE/DOE
    Sixth Symposium on Enhanced Oil Recovery, Tulsa, April 20-23.

15. Mohammadi, S. S., "Steam-Foam Project in Guadelupe Field,
    California", SPE 15054, SPE California Regional Meeting,
    Oakland, Calif., April 2-4, 1986.
16. Mohammadi, S. S., Van Slyke, D. C. and Ganong, B. L., "Steam
    Foam Pilot Project in Dome-Tumbador, Midway Sunset Field", SPE
    16736, 62nd Annual Technical Conference and Exhibition of SPE,
    Dallas, Texas, September 27-30, 1987.
17. Patler, W., "On the Integral Equation of Renewal Theory", Ann.
    Math. Stat., 1941, 12, pp 243-267.
18. Fredrickson, A. G and Tsuchiya, H. M., "Continuous Propagation
    of Microorganisms", Am. Inst. Chem. Eng. J., 1963, 9, No. 4,
    pp 459-468.
19. Himmelblau, D. M. and Bischoff, K. B., "Process Analysis and
    Simulation - Deterministic Systems", John Wiley & Sons, Inc.,
    New York, 1968, pp 191-192.
20. Katz, S. and Hulbert, H. M. "Some Problems in Particle
    Technology - A Statistical Mechanical Formulation", Chem. Eng.
    Sci., 1964, 19, pp 555-574.
21. Kendall, D. G.: "On the Generalized 'Birth-and-Death'
    Processes", Ann. Math. Stat., 1948, 19, pp 1-15.
22. Payatakes, A. C., Ng, K. M., and Flumerfelt, R. W., "Oil
    Ganglion Dynamics During Immiscible Displacement: Model
    Formulation", Am. Inst. Chem. Eng. J., 1980, 26 No. 3,
    pp 430-443.
23. Randolph, A. D., "A Population Balance for Countable Entities",
    Can. J. Chem. Eng., December 1964, pp 280-281.
24. Gray, W. G. and Lee, P. C. Y., "On the Theorems for Local Volume
    Averaging of Multiphase Systems", Int. J. Multiphase Flow, 1977,
    3, pp 333-340.
25. Whitaker, S., "The Transport Equations for Multi-Phase Systems,"
    Chem. Eng. Sci., 1979, 28, pp 139-147.
26. Kittel, C., "Elementary Statistical Physics," John Wiley and
    Sons, Inc., New York, 1958.
27. Chapman, S. and Cowling, T. G., "The Mathematical Theory of Non-
    Uniform Gases", Appendix B, Cambridge, Univ. Press, 1953.
28. Lamb, H., Hydrodynamics," Dover Publications, New York, 1945.

RECEIVED January 20, 1988

DISPERSION FLOODS
IN THE LABORATORY AND FIELD

Chapter 17

Surfactant-Induced Mobility Control for Carbon Dioxide Studied with Computerized Tomography

S. L. Wellington and H. J. Vinegar

Bellaire Research Center, Shell Development Company, P.O. Box 481, Houston, TX 77001

Computerized Tomography (CT) was used to study mobility control with CO_2 foam during tertiary horizontal corefloods at reservoir pressures and temperatures. CO_2 foam provided effective mobility control under first-contact miscible conditions. However, mobility control was not observed when the pressure was substantially reduced so that the oil and CO_2 were immiscible. If the beneficial effects of foam can be extended to developed–miscibility conditions, CO_2 foam will be an outstanding EOR process.

There are many attractive features of EOR using CO_2. First, carbon dioxide is a proven solvent for reconnecting, mobilizing and recovering waterflood residual oil. It is not completely miscible with most crude oils on first contact, as an ideal solvent is, but many studies show that it can achieve miscible–like displacement efficiency through multiple contacts (partitioning and extraction) with the crude oil (1).

Second, CO_2 is available naturally in large quantities and as a byproduct of lignite gasification and many manufacturing processes. It can be extracted from effluent gas production and reinjected to further reduce the average cost. CO_2 pricing is also low because there are no other large volume uses competing for CO_2.

GRAVITY AND VISCOUS INSTABILITIES

CO_2 works very well in steeply dipping reservoirs where it can be injected at the top of the reservoir near the gas–oil contact. Gravity can then stabilize the sweep–out of the pattern as the fluids move down to the production wells.

Unfortunately in most reservoirs the solvent must be injected horizontally because there is too little dip or too little vertical permeability. The greater mobility and lower density of the CO_2 versus the crude oil and brine make the displacement both viscous and gravity unstable. Under these conditions the fluid advances in irregular patterns called viscous fingers. The fingers may grow with time and break through to producing wells, allowing the displacing fluid to bypass portions of the reservoir. This is shown schematically in Figure 1.

On average, 5–10 MCF of CO_2 are required to displace 1 barrel of oil. A combination of poor sweep efficiency and poor volumetric displacement can therefore make the process uneconomical even with low CO_2 unit cost.

In addition to the viscous and gravity instability problems, there is a high front end capital cost associated with CO_2 projects, and many oil fields have pronounced heterogeneities that further reduce the sweep efficiency. These are often sufficiently severe that tertiary CO$_2$ flooding is uneconomical.

METHODS OF MOBILITY CONTROL

WAG, water–injected–alternately–with–gas, attempts to improve sweep efficiency during CO$_2$ flooding by lowering the CO$_2$ relative permeability. The daily CO$_2$ requirement is also lowered since water is injected in some wells while CO$_2$ is injected into others.

Although WAG reduces viscous instabilities, it does not prevent gravity segregation in thick layers. Further, high WAG ratios (e.g., 1:1) cause the water saturation to remain high which can prevent mobilization of trapped oil. Injectivity also decreases during the start of each alternate fluid cycle, and serious corrosion problems occur since the pH drops to about 3.5 as the CO$_2$ reacts with water to form carbonic acid. A recent review of 15 CO$_2$ field projects indicates less oil was recovered by WAG than by injection of a single CO$_2$ slug chased by brine (2).

CO$_2$ viscosifiers have been sought. Unfortunately these high molecular weight polymers are not soluble in CO$_2$. Low molecular weight materials require too great an amount to be cost effective.

Recoverable domestic oil reserves are estimated to be increased by at least another 5–10 billion barrels if enhanced CO$_2$ mobility control was available.

FOAM MOBILITY CONTROL

Surfactant–induced mobility control is a promising enhancement to WAG for reducing mobility and gravity instabilities during CO$_2$ flooding. The basic principle is to segment the CO$_2$ between aqueous surfactant lamella in the rock pores (3). Fluid mobility is reduced by the immiscible fluid/fluid contact, the so–called Jamin Effect (4), that occurs during segmented flow. The Jamin Effect is roughly quantified by the displacement pressure necessary to force one fluid through a capillary by another immiscible fluid. The overall flow resistance is proportional to the total number of interfaces that are displaced.

Displacement of the discrete CO$_2$ segments results in a much larger apparent viscosity inside porous media than the viscosity of any of the constituents. Small amounts of surfactant give large CO$_2$ mobility reductions even at low saturation and fractional flow of brine. The mobility level can be adjusted through surfactant concentration, structure and gas/liquid ratio.

The surfactant has two important roles in CO$_2$ foam. First, it increases the apparent viscosity of CO$_2$ so that brine and oil are displaced in a stable manner. Second, the surfactant lowers the interfacial tension between CO$_2$ and brine which promotes brine displacement. Reducing the brine saturation below S_{wc} allows bulk–phase CO$_2$ to completely access the oil–filled pore network. A high–saturation brine bank also retards CO$_2$ mobility by relative permeability effects. The brine bank carries surfactant and allows oil reconnection and mobilization ahead of the bulk CO$_2$ phase because of the favorable partitioning of CO$_2$ from brine into oil.

The differences between the miscible CO$_2$ foam process and a stable tertiary miscible solvent process are shown in Figures 2 and 3. In the miscible CO$_2$ foam process, oil mobilization occurs as CO$_2$ partitions into and swells the trapped oil above S_{orw}, allowing it to be displaced by the mobile brine. The carbonated brine in turn is displaced by CO$_2$ foam. In comparison, miscible N$_2$ and LPG do not transfer to oil through solution in the water phase, as CO$_2$ does. Instead of a brine bank, the solvent and oil are separated by a miscible dispersion zone. The brine saturation is not reduced below S_{wc}.

In the CO$_2$ foam process, the surfactant that is left behind, in the connate brine and adsorbed on the rock, is not entirely lost. In field processes, brine is injected following the CO$_2$ injection. That chase brine causes surfactant desorption and displaces connate brine

VERTICAL CROSS-SECTION

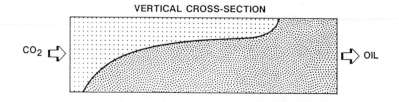

HORIZONTAL CROSS-SECTION

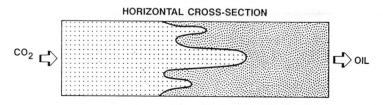

Figure 1. Viscous and gravity unstable displacements.

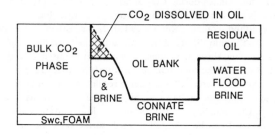

Figure 2. Idealized saturation profile for the CO_2 foam process.

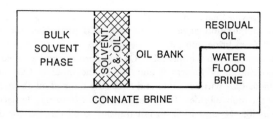

Figure 3. Idealized saturation profile for a stable solvent process.

containing the surfactant. The mobilized surfactant can form a CO_2 foam mobility buffer at the back side of the CO_2 bank. The surfactant also reduces the brine/CO_2 capillary pressure which reduces the amount of CO_2 trapped in the reservoir. These effects further improve the economics of the CO_2 foam process.

TECHNICAL DIFFICULTIES

A fundamental concern in CO_2 foam applications is how far foams can be transported at reservoir temperatures and salinities in the presence of crude oil. Oils that spread at gas/brine interfaces are known to have severe debilitating effects on foam stability. Another concern is that surfactants may retard oil droplet coalescence and therefore reduce tertiary oil reconnection and mobilization efficiency.

Additionally, the surfactants must be chemically stable for years at reservoir conditions, and adsorption losses on the rock must be modest. The chemical stability requirement can be achieved by choosing appropriate structures and by carefully designing the mixing process to exclude oxygen during injection. Once underground, the surfactants are usually safe from oxidation because most reservoirs are in a reducing condition unless there is high natural radioactive decay that causes free-radical initiation.

SURFACTANTS FOR CO₂ FOAM

Many surfactants have been suggested as candidates for CO_2 foam. However, at high salinity and temperature in the presence of oil, most surfactants foam poorly due to partitioning and emulsion formation and fail to control mobility during CO_2 injection. This behavior is analogous to that observed in chemical (microemulsion) oil recovery (5-7). As the salinity, hardness and temperature increase, surfactants form water/oil emulsions, precipitate surfactant-rich coacervate phases, and partition into the oleic phase. CO_2 decreases further the solubility of surfactant in the aqueous phase.

To be a promising candidate for CO_2 foam, the surfactant loss by adsorption, partitioning and emulsion formation must be low. In general, anionic surfactants have low adsorption on sandstones and high adsorption on carbonates, whereas the reverse is true for nonionics (8). Cationic surfactants are not considered because of their high adsorption on many surfaces.

Figure 4 shows a decision tree for choosing anionic versus nonionic surfactants. If the temperature and salinity are not too high, the choice between anionic and nonionic surfactant is based more on cost than on reservoir lithology. In general, however, anionic surfactants are more expensive than nonionics. Thus, an inexpensive nonionic surfactant might lower total surfactant cost, even though higher concentrations are required to overcome greater surface adsorption.

SURFACTANT SCREENING

The need to screen large numbers of surfactants to find oil-tolerant foam led to the following rapid surfactant screening test. Ten cc of surfactant solution was placed in a clean tared 25 cc graduated cylinder. The hydrocarbon phase: decane, decane/toluene (1:1 by volume), stock tank oil, or supercritical CO_2-extracted stock tank oil (3.0 cc), was then added. Samples were shaken after temperature equilibration, allowed to stand for 24 hours, and shaken again. Foam volume was then determined as a function of time. The results of screening tests, which included a systematic study of surfactant structure, have been described in a preceding chapter in this volume ("Surfactants for CO_2 'Foam' Flooding. Effects of Surfactant Chemical Structures on One Atmosphere Foaming Properties"). Refined paraffinic, mixtures of paraffinic and aromatic, and crude oils were tested.

Surfactants that did well in the initial screening tests were tested further in high-pressure sight cells at reservoir conditions. The best performers were then tested in microvisual cells, slim tubes, and finally in coreflood adsorption and oil displacement tests. One experimental surfactant structure (9–10),

$$CH_3 - (CH_2)_{11-14} \text{---} (O - CH_2 - CH_2)_{12} - OCH_2 - CH(OH) - CH_2 - SO_3 - Na$$

produced the most persistent foams even at high salinity and elevated temperatures in the presence of synthetic and crude oils. This surfactant is a particular example of a class of Alcohol Ethoxy Glycerol Sulfonates, abbreviated AEGS. AEGS 25–12 is an anionic sulfonate based on Neodol 25–12 ethoxy alcohol:

$$CH_3 - (CH_2)_{11-14} \text{---} (O - CH_2 - CH_2)_{12} - OH$$

and epichlorohydrin:

$$H_2C \overset{O}{\triangle} CH \text{---} CH_2Cl$$

   The AEGS structure is very salinity–tolerant because of the hydrophilic ethylene oxide middle group and the hydroxy glycerol sulfonate end group. Partitioning and oil emulsification with AEGS can be controlled by adjusting the number of ethylene oxide groups and by changing the length and structure of the hydrocarbon "tail." Being a sulfonate, AEGS is considerably more resistant to hydrolysis than sulfated surfactants that hydrolyze at high temperatures in low pH CO_2–saturated brines.
   Figure 5 shows the results of a typical surfactant transport study in a 2 ft long Berea sandstone core. The AEGS 25–12 surfactant, injected at 0.05 wt%, had a low loss on Berea sandstone of 0.008 meq/100 gm rock compared to ~0.05 meq/100 gm for typical petroleum sulfonates used in chemical flooding. Surfactant breakthrough occurred at 0.62 PV ($S_{orw} = 0.38$ PV). The surfactant concentration is consistent with about 10% transport with the brine front. Surfactant loss and transport were monitored using the hyamine titration technique.

CT SCANS OF THREE–PHASE FLOW EXPERIMENTS

Final laboratory testing of CO_2 foam was performed in Shell's CT facility (11–12). Tertiary miscible and immiscible CO_2 corefloods, with and without foam mobility control, were scanned during flow at reservoir conditions. The cores were horizontally mounted continuous cylinders of Berea sandstone. Table I lists pertinent core and fluid data.
   Two sets, i.e., four experiments, of core flow studies are compared. Sets No. 1 and No. 2 were tertiary miscible and immiscible CO_2 floods without mobility control. The same core from each set, after plain CO_2 injection, was restored to waterflood residual oil saturation and flooded with 0.05% AEGS 25–12 surfactant in brine. There was almost no difference between the oil saturation distributions in the cores between experiments, with the average S_{orw} values of 37 ± 1 saturation percent in both sets of experiments. CO_2 was injected continuously in all experiments at a nominal rate of 1 ft/day. No attempt was made to preform a foam, or to inject alternate slugs of surfactant solution and CO_2.
   The oil phase, Soltrol 130, a refined kerosene, was doped with iodated oils of similar molecular structure. The dopants are strong photoelectric absorbers and increase the accuracy of saturation determination by increasing the X-ray attenuation. The refined, nearly single–component, oil was also used to insure complete first–contact miscibility. Because this is a single–component oil, multiple contact developed miscibility is not observed below the miscibility pressure.

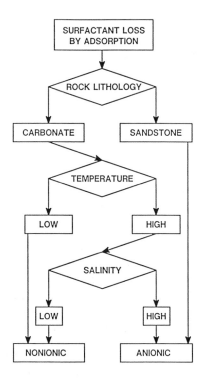

Figure 4. Surfactant choice decision tree.

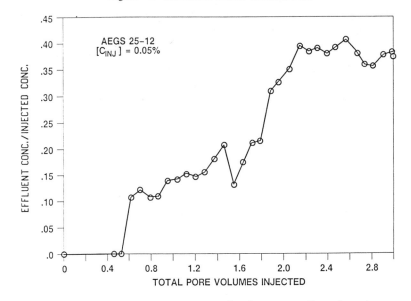

Figure 5. Normalized surfactant concentration from a core flow adsorption test.

Table I. Core and Fluid Data

	Core Flood Set No.	
	1	2
First Contact Miscible:	yes	no
Oil Composition (wt %):	60/40	80/20
	Soltrol 130/	Soltrol 130/
	Iodododecane	Iodododohexadecane
Backpressure (psi):	3000	1100
Oil Viscosity (cp):	0.875	0.815
Oil Density (gm/cc):	0.843	0.767
CO_2 Viscosity (gm/cc):	0.64	0.16
Core Length (in.):	23	23
Core Diameter (in.):	2	2
Brine Permeability (md):	280	540
Flood Rate (ft/day):	1	1
Temperature (deg C):	76	76
NaCl Brine (normality):	0.1	0.1

RESULTS

All cores were imaged dry, at 100% brine saturation, at S_{oi}, at S_{orw}, and at various times during CO_2 injection. Vertical CT reconstructions through the center of the cores, are shown in Figures 6, 9, 14, and 15 during CO_2 injection. Fluid flow is from left to right in all images. The aspect ratio is compressed in the figures: the cores were 2" OD x 23" long. The longitudinal reconstructions are not corrected for the 4–hour time period required to scan the core.

MISCIBLE CO_2/OIL, SET NO. 1 EXPERIMENT NO. 1. The low density and viscosity of CO_2 allows gravity to dominate this flood. The CO_2 entered the core as a finger at the middle of the inflow face (Figure 6a), and as more CO_2 was injected, the gas formed a single viscous and gravity–unstable finger (often called a gravity tongue) that buoyed up and overrode both the oil and water banks as shown in Figures 6b–d. The results of this flood reflect a low ratio of viscous to gravity forces. A value of approximately 8 for L. W. Lake's gravity number, N^*, predicts this behavior. N^* is the ratio of the time, t_L, for the fluid to traverse the reservoir length, divided by the time, t_V, to traverse the reservoir height.

$$N^* = t_L/t_V = \frac{K_V \Delta\rho \, g\lambda_r A(L/H)}{Q} \tag{1}$$

K_V is the vertical permeability, $\Delta\rho$ the fluid density difference (H_2O–CO_2), g the acceleration of gravity, λ_r the relative mobility of the least mobile phase (water), A the cross sectional area, Q the injection rate, L the length, and H the height.

Breakthrough of CO_2 occurred at 0.4 HCPV of CO_2 injected, and 22% of the oil left after waterflooding was recovered. The pressure differential across the core, shown in Figure 7, dropped suddenly after CO_2 breakthrough. Brine production halted completely. Oil recovery became extremely inefficient, as shown in Figure 8, and the gas–to–oil production ratio increased rapidly, exceeding 1 after 0.6 HCPV.

The oil saturation was reduced to near zero in the upper section of the core swept by the CO_2, but only 38% of the oil displaced from that region was produced. The balance of the mobilized oil moved downwards, increasing the oil saturation above waterflood residual in the lower unswept region of the core. Other experiments demonstrated that subsequent

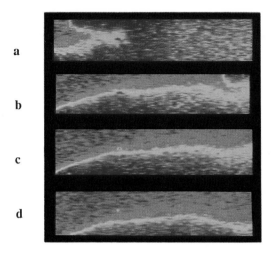

Figure 6a–d.   CT reconstructions from Set No. 1, Experiment No. 1.

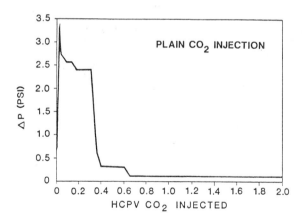

Figure 7.   Differential pressure during Set No. 1, Experiment No. 1.

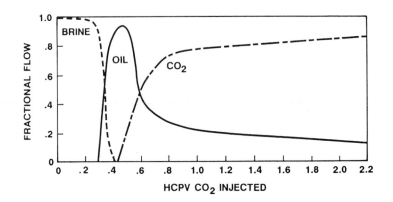

Figure 8.   Fractional flows from Set No. 1, Experiment No. 1.

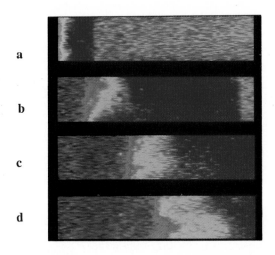

Figure 9a–d.   CT reconstructions from Set No. 1, Experiment No. 2.

brine injection displaced the oil saturation in excess of S_{orw}. Some of the mobilized oil buoyed into the upper oil–free section of the core and was retrapped.

After 2.6 HCPV, continued CO_2 injection allowed the flowing CO_2 stream to slowly extract all but a small wedge of oil at the bottom outflow end of the core.

SURFACTANT–INDUCED MOBILITY CONTROL, SET NO. 1 EXPERIMENT NO. 2. The injection of mobility–control surfactant prior to injection of CO_2, caused the bulk phase CO_2 to sweep through the core with a near vertical front in a piston–like manner. Oil and most of the brine were displaced ahead of the CO_2 foam. A series of vertical reconstructions through the center slice of the core are found in Figures 9a–d. CO_2 is colored blue, and oil is colored red. Bulk–phase CO_2 saturation graded from 80% at the core inlet to about 70 saturation percent at the leading edge of the advancing dark blue CO_2 front. More brine was displaced by foamed CO_2 than at S_{oi}. S_{wc} averaged 38%, whereas S_{wc} to foam is about 18 to 20%. The additional brine displacement was not due to evaporation since the CO_2 was fully saturated with water vapor at flood conditions before it was injected.

Comparing the fractional flow curves, Figures 8 and 10 show that oil and CO_2 production are delayed since the entire core cross section is swept and additional brine is displaced. Note the second "hump" in the brine production curve after 0.7 HCPV when foam is present. Without foam, brine production stops and oil production drops rapidly when CO_2 breaks through.

Saturation profiles calculated from the cross–sectional average CT data, Figures 11 and 12 which correspond to Figures 9b and c, reveal that most of the waterflood residual oil was displaced ahead of the bulk–phase CO_2. Mobilization occurs as oil swollen above S_{orw} is displaced by mobile brine.

The oleic phase saturation ahead of the bulk–phase CO_2 is composed of a mixture of oil and CO_2. In Figures 11 and 12, the amount of CO_2 in the oleic phase is represented by the cross–hatched area. The total oleic phase includes the cross–hatched area and the area above. The additional brine displaced by the CO_2 foam, i.e., the hump in the brine fractional flow curve in Figure 10, is saturated with CO_2 until it contacts and transfers its CO_2 to the oleic phase. Thus, displacement of oil by carbonated water continues until bulk–phase CO_2 arrives and directly displaces the remaining CO_2 diluted oleic phase. The solubilized CO_2 zone, i.e., CO_2 dissolved primarily in the oil, grows in length during the flood. The bulk–phase CO_2 foam bank is the least mobile phase. N^* changes from 8 to a stable 0.8, reflecting the apparent CO_2 viscosity increase of 25 times. (The pressure drop across the core was 0.1 psi without foam compared to 2.5 psi with foam, as shown in Figure 13.)

One important and as yet not fully understood feature of CO_2 foam concerns delineation of the length of the foam bank. Multiple pressure tap measurements indicated that the largest pressure drop occurred within 3 to 6 inches around the bulk CO_2 phase front. These and additional studies suggest that the foam bank can be relatively short compared to the length of the core. The pressure drop decreased to very low values as brine approaches its irreducible saturation. This is reasonable, since aqueous surfactant lamella cannot form or propagate when there is no mobile brine present.

IMMISCIBLE CO_2/OIL, SET NO. 2, EXPERIMENT NO. 3. This experiment should be compared with experiment No. 1. The CT scans showed a sharp, nearly vertical, CO_2 front without viscous fingering, Figures 14a–d. The CO_2 entered the core in the upper half of the inflow face and then swept down, forming a small layover angle. A CO_2–free oil bank developed during the flood but was overtaken by the faster moving CO_2. CO_2 and oil production were simultaneous at the outflow end of the core.

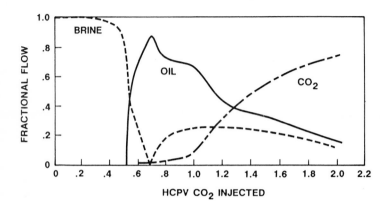

Figure 10. Fractional flows from Set No. 1, Experiment No. 2.

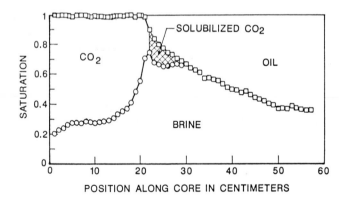

Figure 11. Oil, brine and CO_2 saturations corresponding to Figure 9b.

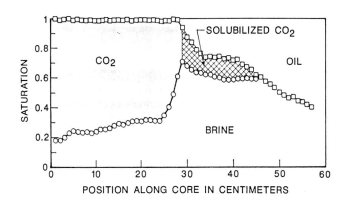

Figure 12.   Oil, brine and CO_2 saturations corresponding to Figure 9c.

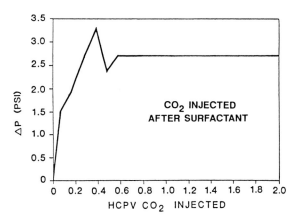

Figure 13.   Differential pressure during Set No. 1, Experiment No. 2.

The CO_2 had good cross–sectional sweep, even though the CO_2 density and viscosity were significantly lower than those of either the oil or the brine, and even though CO_2 entered the core at the upper edge of the inlet face. One reason for the good sweep in this coreflood is that capillary forces over the 2–inch (5 cm) high vertical cross section are strong enough to overcome the buoyancy forces. Another reason is that the high saturation of unmobilized oil effectively retards brine and CO_2 flow by relative permeability effects.

The good cross–sectional sweep observed during this coreflood will clearly not scale to reservoir conditions, unless the capillary pressure exceeds the gravitational forces over the height of the pay, i.e., the pay is very thin or is effectively divided into thin laminations by impermeable layers.

IMMISCIBLE CO_2/OIL WITH LOWERED CO_2/BRINE IFT, SET NO. 2, EXPT. NO. 4.

The core used in experiment No. 3 was restored to water flood residual oil saturation and flooded with 0.05 wt% AEGS 25-12 surfactant in brine, and then CO_2. Initially, the CO_2 began to sweep the entire core cross section and build an oil bank similar to the one observed in experiment No. 3. Then the CO_2 buoyed up and overrode the water and oil banks, Figures 15a–d. The result is both poor sweep and poor displacement of oil.

The large amount of unmobilized oil must have reduced or prevented lamella formation and propagation. Also, the surfactant reduced interfacial tension so that capillary forces no longer dominated the gravity forces as they did in experiment No. 3.

CONCLUSIONS

The CT images identified several mechanisms that help to explain how mobility–control surfactant causes CO_2 to be stable against both gravity and viscous forces during tertiary miscible core flow experiments.

The mobility–control surfactant increased the apparent viscosity of CO_2 sufficiently to prevent gravity override and viscous fingering. The bulk CO_2 phase passed through the core in a piston–like manner. Oil and most of the brine were displaced from the core ahead of the bulk–phase CO_2. Differential pressure measurement across the length of the core indicated an average gradient, 1.3 psi/ft, similar to that observed during the brine flood.

Connate brine was reduced below S_{wc}, 38 to 18 saturation percent ultimately, and caused a brine bank of 70+ saturation percent to immediately proceed and retard the mobility of the bulk–phase CO_2 by relative permeability effects. Oil mobilization ahead of the bulk CO_2 occurred as CO_2 partitioned into and swelled the trapped oil above S_{orw} and was displaced by mobile brine.

It also appears that interfacial tension lowering between the CO_2 and brine is beneficial. This allows brine displacement below S_{wc} which provides additional mobility control by relative permeability effects without requiring large pressure drops.

The beneficial surfactant effects observed during miscible CO_2 flooding did not extend to the case when the oil and CO_2 are immiscible. A viscous foam did not form or propagate, and the reduced capillary pressure due to surfactant IFT lowering allowed the CO_2 to override the brine and oil banks.

The important question is whether mobility control can be obtained in developed–miscibility CO_2 flooding. Further research is required to define CO_2 foam behavior under developed–miscibility conditions.

ACKNOWLEDGMENTS

We gratefully acknowledge fruitful discussions with Professor E. L. Claridge. Kermit Tschiedel and Larry Bielamowicz performed the CO_2 coreflood experiments.

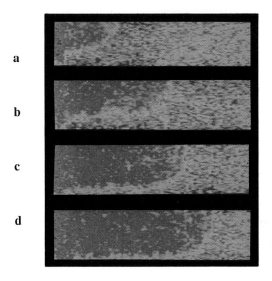

Figure 14a–d.   CT reconstructions from Set No. 2, Experiment No. 1.

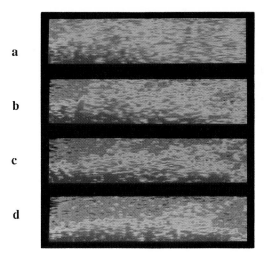

Figure 15a–d.   CT reconstructions from Set No. 2, Experiment No. 2.

LITERATURE CITED

1. Stalkup, F. I., Jr. Miscible Displacement; Society of Petroleum Engineers of AIME: New York, N.Y. 1983; Chapter 2.

2. Holm, L. W. J. Pet. Technol. 1987, 39, 1337–1342.

3. Heller, J. P.; Kuntamukkula, M. S. Ind. Eng. Chem. Res. 1987, 26, 318–325.

4. Jamin, J. M. Compt. Revd. 1860, pp. 50, 172–176, 311–314, 385–389.

5. Nelson, R. C.; Pope, G. A. Soc. Pet. Eng. J. 1978, 265, 325–338.

6. Healy, R. N.; Reed, R. L.; Stenmark, D. G. Soc. Pet. Eng. J. 1976, 261, 147–160.

7. Gash, B. W.; Griffith, T. D.; Chan, A. F. Second Joint Symposium, Enhanced Oil Recovery, Tulsa, OK 1981, SPE/DOE 9812, 547–562.

8. Lawson, J. B. Symposium on Improved Oil Recovery, Tulsa, OK 1978, SPE 7052, 159–170.

9. Wellington, S. L.; Reisberg, J.; Lutz, E. F.; Bright, D. B. U.S. Patent 4502538, (1985).

10. Borchardt, J. K.; Bright, D. B.; Dickson, M. K.; Wellington, S. L. 60th Annual Technical Conference & Exhibition of the Soc. of Pet. Eng. of AIME, Las Vegas.

11. Vinegar, H. J.; Wellington, S. L., Rev. Sci. Instrum. 1987, 58, (1), 96–107.

12. Wellington, S. L.; Vinegar, H. J., J. Pet. Technol. 1987, 36, 885–898.

RECEIVED March 3, 1988

Chapter 18

Carbon Dioxide Foam Flooding

Laboratory Model and Computer Simulation of the Process

E. L. Claridge, B. M. Lescure, and M. W. Wang[1]

Chemical Engineering Department, University of Houston, Houston, TX 77004

Injection of carbon dioxide as a foam or emulsion is under investigation as a means of reducing the mobility of the low-viscosity carbon dioxide and thus of increasing the oil recovery in oil field enhanced oil recovery floods. The prior method is alternate injection of water and carbon dioxide. In the work reported here, laboratory model results previously reported have been simulated using a finite difference computer program called Multiflood. The need for simulation of surfactant adsorption required that the components usually assigned to water phase or oil phase in using this program had to be reversed, and the role usually played by the mixing parameter in simulating effects of viscous fingering was replaced by partial solubility of carbon dioxide in the aqueous phase, while the surfactant was given a fictitiously high viscosity in its blend with water in order to match the viscosity of foam. With these adjustments, it was found possible to match the laboratory experiments with fair accuracy with a carbon dioxide solubility in the oil phase of 70-90 mol percent (the remainder having gas relative permeability and viscosity). It is believed that until computer programs specifically written to enable simulation of foam flooding become available, this simulator and method will be helpful in making preliminary estimations for field projects.

Enhanced oil recovery by injection of a slug of carbon dioxide, followed by water to complete the process, under conditions of temperature and pressure such that the carbon dioxide is apparently miscible with crude oil, is becoming a wide-spread process in the U.S.A., particularly in west Texas but also in certain other regions (1-9). Carbon dioxide is not unconditionally miscible with most crude oils; this is discussed below.

[1] Current address: Petroleum Engineering Department, University of Texas, Austin, TX 78712

Miscibility and Phase Behavior

In most cases, the temperature and pressure of application exceed the
critical values for carbon dioxide, but the temperature is lower than
the average critical temperature and above the average critical
pressure of the crude oil components. Hence, under conditions of use
the carbon dioxide is a fairly dense fluid which is neither a liquid
or a gas, but the crude oil is a liquid. There is also salty water
present in the oil reservoir in all cases, and water is used to drive
a slug of carbon dioxide through the reservoir from injection to
production wells which are typicAlly arranged in five-spot or nine-
spot patterns.

In general, the original reservoir temperature is not
significantly changed during the process, but the pressure may be
varied from the original level. In most cases, primary recovery by
pressure depletion as well as subsequent secondary recovery by water
flooding has already occurred. The carbon dioxide flooding process
is then a tertiary recovery process. By control of the input and
output during the waterflood, the pressure may be adjusted to the
pressure required for apparent miscibility between the carbon dioxide
and the remaining crude oil. This pressure is usually ascertained by
displacements of crude oil by carbon dioxide in long, slim tubes at a
series of pressures. The oil recovery at breakthrough of carbon
dioxide at the outlet, or at some designated amount of throughput
such as 1.2 pore volumes, varies from about 40-60 % at low pressures
to over 95 % at high pressures. The break in the curve at which the
oil recovery approaches the 95 % level is taken as the apparent
miscibility pressure (10).

This is termed the apparent miscibility pressure because there
is a considerable amount of evidence that asphaltenes, which are
colloidally dispersed in most crude oils, will aggregate under
appropriate dilution conditions and precipitate, so that the physico-
chemical definition of miscibility (single-phase for all proportions
of the fluids in question) is not realized. There are also
conditions, particularly found in oil reservoirs at temperatures
below about $135^{O}F$, at which two liquid phases, or two liquid phases
plus a gas phase, appear in addition to an asphaltene precipitate
(11-13). In general, however, this does not prevent the attainment
of 95 % oil recovery or more in slim-tube tests at the same pressures
at which these multiple phases appear. Hence, the process is deemed
"miscible," for all practical purposes.

Fluid Mechanics Effects

Either the asphaltene precipitation or the multiple phase behavior of
the process under "miscible" conditions appears to cause a certain
extent of reduction of the mobility of the carbon dioxide, as
compared to the mobility which would be calculated from the relative
permeability of the carbon dioxide divided by its viscosity (14-16).
However, it is usually considered that this is not sufficient to
resultinafavorable mobility ratio (a ratio of carbon dioxide
mobility to crude oil mobility less than one).

When the mobility ratio is greater than one, viscous instability
ensues during displacements in systems of sufficient width (>10 cm)
to permit the formation and propagation of viscous fingers. These

projections of the more mobile displacing fluid through the less mobile fluid being displaced can occur either with miscible or immiscible fluids (17,18), and result in much lower production of the fluid being displaced at any given throughput after breakthrough of the displacing fluid. It is therefore much to be desired to attain a mobility ratio of one or less, from the standpoint of oil recovery.

In order to reduce the carbon dioxide mobility still further, it is often planned to inject the carbon dioxide slug in small increments, alternated with similar small increments of water. This procedure, called WAG (water-alternating-gas) has been tested in comparison to straight slug injection (non-WAG) in laboratory model studies conducted at the University of Houston (19) with a glass bead pack in the form of a five-spot well pattern, with the beads in a relatively water-wet condition and in a relatively oil-wet condition. In the water-wet condition, it was found detrimental to oil recovery to use WAG, while an optimum WAG ratio (of water to CO_2) in the range of 0.5 to 2.0 was found for the oil-wet condition. These two wetting states are thought to correspond approximately to the wetting states of sandstones (mostly watwr-wet) and carbonate rocks such as limestones and dolomites (mostly oil-wet, or partially oil wet).

The density of the water involved (salt content 1-15%) is slightly over 1.0 g/cc (1000 kg/m³). That of the crude oil is about 0.7 g/cc, and the density of the carbon dioxide is usually slightly lower or slightly greater than that of the crude oil, at typical temperatures and pressures of application. Even if miscibility of carbon dioxide and crude oil is assumed, this does not mean that they are totally mixed as the carbon dioxide is injected in the reservoir. Separation by density is therefore possible even with miscible fluids when there is a density difference. In most cases, there is sufficient density difference between the carbon dioxide and the oil for this to occur if the injection rate is slow enough to allow gravity force to predominate over viscous force. The ratio of gravity force to viscous force is small near the injection wells (where the viscous force is large) but usually becomes greater than one at some distance from the injection wells, out in the middle of the well patterns. Of course, the density difference between injected water and either carbon dioxide or crude oil is sufficiently great that gravity segregation begins at a much shorter radius from the injection wells.

Such gravity segregation leads to early breakthrough of the injected fluid, compared to the case of no gravity segregation. Even more serious, when simultaneous flow of water and non-aqueous fluids over the same cross-sectional area is desired for control of the mobility of the non-aqueous fluid, e.g. carbon dioxide, the separation by gravity negates the mobility control over much of the reservoir volume.

Thus, both gravity segregation and viscous fingering hinder the attainment of good sweep efficiency of the well patterns. A third major factor adversely affecting oil recovery is the heterogeneity of the reservoir, as expressed in variations of porosity and permeability. In most cases, the heterogeneity is not random but is mostly present as layers of different permeability. In the present work, this is not dealt with; work is in progress on a layered laboratory model, however. A fourth factor affecting sweep efficiency is the use of well patterns. The fluids are injected at

what are practically point sources into rock formations which are
usually much greater in areal extent than they are thick, and fluids
are produced from point sinks in the areal system. With multiple
patterns, symmetry elements result which aid in confining fluids
within the patterns. However, the existence of relatively long and
short streamline flow paths in these symmetery elements results in
considerably less than 100% sweep efficiency when the injected fluid
appears at the producers along the shortest paths. This areal sweep
efficiency is about 71% for equal mobility injected and displaced
fluids for the five-spot well pattern, and it varies strongly with
the mobility ratio (20).

Recently, use of a surfactant in the injected water such that a
foam or emulsion is formed with carbon dioxide has been proposed
(20,21) and research is proceeding on finding appropriate surfactants
(22-24). The use of such a foam or emulsion offers the possibility
of providing mobility control combined with amelioration of the
density difference, a combination which should yield improved oil
recovery. Laboratory studies at the University of Houston (25) with
the same five-spot bead-pack model as used before show that this is
so, for both the relatively water-wet and relatively oil-wet
condition. We have now simulated, with a finite-difference reservoir
process computer program, the laboratory model results under non-WAG,
WAG, and foam displacement conditions for both secondary and tertiary
recovery processes. This paper presents the results of that work.

Laboratory Model Results

The model results (19,25) are presented below in Table I. The
design of the model and the fluids used are discussed briefly below.

The model is 15" square and 1" thick (38x38x2.5 cm), with an
injection well in one corner and a production well in the diagonally
opposite corner, thus forming a symmetry element of 1/4 of a five-
spot well pattern. It is filled with a log-normal distribution of
glass beads ranging in size from 10 mesh to 280 mesh (approx. 1 mm to
0.05 mm); the porosity is close to 0.3 and the permeability is in the
range of 4 to 8 darcys (4 to 8 E-12 square meters; different values
found in successive packings). An appropriate feed system and a
production system including an automatic phase separator and methods
of distinguishing "carbon dioxide" from oil were provided. For water-
wet beads, the initial bead coating to prevent caking was removed
with an aromatic wash followed by acetone, then de-ionized water,
then dilute hydrochloric acid followed again by de-ionized water.
For the oil-wet condition, DuPont Quilon in iso-propyl alcohol
solution was injected into the water-wet pack until all water was
displaced, allowed to stand for some hours, then displaced again with
water. This left the glass beads rather strongly oil-wet, but not as
completely as would be obtained by baking the Quilon coating on the
surface (26).

In order to avoid the necessity of operating at the high
pressures typical of carbon dioxide flooding (1500-3000 psig, or 10-
20 MPa), which would require heavy bracing of the model faces and
high-pressure inlet and production systems, substitute fluids were
used. In place of carbon dioxide, a non-aromatic kerosene-range
hydrocarbon was used, with a viscosity of 1.3 cp (or kPa-s). This is
about 20 times the viscosity of carbon dioxide under normal

conditions of use, so in order to preserve the same ratio of fluid
mobilities as in an oil reservoir, the water was thickened using
glycerine to 14.3 cp (kPa-s) and a mixture of white oils of 25.9 cp
was used to represent crude oil. The light oil representing carbon
dioxide was miscible with the heavier oil representing crude oil.
Thus phase behavior effects which might confound the observable
results were avoided.

The experiments were conducted at a flow rate such that most of
the dimensionless ratios of forces or of rates were made the same as
in a typical layer of a west Texas dolomite reservoir (about 10
millidarcys permeability, 105 $^{\circ}$F, 40 acre five-spot, 2000-3000 psig).
In particular, the gravity-viscous force ratio and the ratio of
transverse dispersion to longitudinal convection rate were made as
close as possible to the prototype oil reservoir. These ratios
control respectively the extent of gravity segregation and the extent
of viscous fingering (for any given unfavorable mobility ratio).

Table I.   Laboratory Model CO2 Flooding Results

Percent EOR Recovery Basis Original Oil In Place (OOIP)											
Wetting State:		Water-Wet					Oil-Wet				
Process:	Secondary		Tertiary			Secondary			Tertiary		
Slug Size* 0.1	0.2	0.4	0.1	0.2	0.4	0.1	0.2	0.4	0.1	0.2	0.4
1:1 WAG 2.5	4	5	2	3	3	1	2	3	6	10	13
1:1 Foam 5	10	12	4	8	10	10	16	19	10	15	18

*Equal size water slug injected in 0.05 HCPV increments in WAG or
continuously in foam operation.

In the foam displacements, IGEPAL 710 was used, at a
concentration of 0.05 % basis aqueous phase. This gave a foam or
emulsion containing equal proportions of each phase (we were advised
by an industrial supporter that this is the ratio observed with
carbon dioxide - brine foam systems). The emulsion was stable for a
week or more so long as no heavy oil (representing crude oil) was
present, but was unstable when as much as 10% of the heavy oil was
added, or when the heavy oil alone was used. These are desired
emulsion stability conditions. This particular surfactant was well
adapted for our substitute fluid system, but would not be appropriate
for reservoir systems where the temperature is higher and where the
water salinity is generally much higher than in our laboratory
system. The concentration of surfactant used is insufficient to
cause oil recovery by reduction of interfacial tension. This was
checked by performing waterfloods with and without surfactant in the
water; identical oil recovery curves were obtained. Thus, any sweep
efficiency improvement over that obtained using WAG had to be due to
the emulsifying or foaming effect of the surfactant.

The surfactant adsorbs on water-wet glass beads. Experiments
reported in Ref. 27 show that about 0.5 HCPV of surfactant solution
is required to satisfy the adsorption capacity of the bead pack.
Hence, when a slug size of 0.2 HCPV of carbon dioxide and an equal
slug of surfactant solution are injected together, the surfactant is
all adsorbed when about 40% of the volume of the bead pack has been
contacted. This is apparently sufficient to obtain the benefits of

the surfactant. No adsorption occurs on the oil-wet glass beads. Apparently the Quilon is so strongly adsorbed and the coverage is great enough that no Igepal 710 could be adsorbed.

Note that the oil-wet results are considerably better than the water-wet EOR recoveries, except for the secondary WAG process in the oil-wet model. The foam results were better for the oil-wet pack for both the secondary and tertiary cases than they were in the water-wet pack. We attribute the poorer performance of the water-wet pack to a higher proportion of oil which is blocked from contact with the CO2 at the water saturations which occur in WAG and foam flooding. This apparently cuts the EOR recovery to about half of that observed in the oil-wet pack for equivalent water saturations.

Computer Simulation

The computer simulation program which was available for miscible flood simulation is the Todd, Dietrich & Chase Multiflood Simulator (28). This simulator provides for seven components, of which the third is expected to be carbon dioxide and the seventh water. The third component is allowed to dissolve in the water in accordance with the partial pressure of the third component in the non-aqueous phase or phases. It is typically expected that the first two components will be gas components, while the fourth, fifth, and sixth will be oil components. There is provision for limited solubility of the sixth component in the non-aqueous liquid phase, so that under specified conditions of mol fraction of other components (such as carbon dioxide) the solubility of the sixth component is reduced and some of that component may be precipitated or adsorbed in the pore space. It is possible to make the solubility of the sixth component a function of the amount of precipitated or adsorbed component six within each grid block of the mathematical model of the reservoir. This implies, conversely, a dependence of the amount adsorbed or precipitated on the concentration (mol fraction) of the sixth component in the liquid non-aqueous phase, hence it is possible to use an adsorption isotherm to determine the degree of adsorption.

In this simulator, for incorporation of the effects of viscous fingering of carbon dioxide or of a hydrocarbon miscible solvent through the oil phase, the carbon dioxide or hydrocarbon solvent is given a fictitiously high equilibrium K value, so that it appears to be almost entirely in a fictitious or pseudo gas phase. This is solely for the purpose of invoking the mixing parameter method (29) for calculating a viscosity for the solvent which is in between the viscosity of pure solvent and of the mixture of solvent and oil in the proportions present in any given grid block, and similarly a viscosity for the oil which is intermediate between the pure oil viscosity and the mixture viscosity. The total relative permeability for the non-aqueous liquid phase is shared between the solvent and the oil in proportion to their volume fractions in the non-aqueous liquid phase (which is correct if there does not in fact exist a gas phase but only a liquid phase containing the miscible solvent and oil). The mixing parameter method thus causes the miscible solvent to move through the system at a rate relative to the oil which is characteristic of the behavior of viscous fingers of solvent, even when the number of blocks in the grid is too few to allow fingers to be seen in saturation maps. The sweep efficiency is affected in the same way as it is by viscous fingering.

For the purpose of simulating CO2 foam flooding, it is necessary to use the capability of adsorbing component 6 to represent the adsorption of surfactant from the water phase. This requires that the normally expected phases and phase components be reversed. That is, component 7 must now be oil (and the oil phase must contain both oil and carbon dioxide). Component 3 (carbon dioxide) can dissolve in component 7, now the oil phase, to an extent which can be specified as a function of system pressure. The aqueous phase can be made up of components 1,2,3,4,5, and 6. This is more than is needed for the present purpose. It was decided to use component 5 as the water component. Component 3 (carbon dioxide) can be kept as a separate phase from the water phase by introducing a large K-value for the third component, and a zero K-value for the other five of the six components mentioned, including components 5 and 6. Components 1,2, and 4 are not used. Hence, the system consists of components 3, 5, 6, and 7, representing carbon dioxide, water, surfactant, and crude oil, respectively, or their counterparts in the laboratory model fluid system: light oil, glycerine-thickened water, surfactant, and heavy oil, respectively.

With this reversal of phases, it is no longer possible to make effective use of the mixing parameter method present in the simulator. The value of the mixing parameter was therefore set to a very low value (0.01), but not to zero, because it was desired to use the table of CO2 solubility in the oil (seventh component) which would not be accepted if the mixing parameter were set to zero.

It was found that the solubility of CO2 in component 7 (oil) could be used in much the same way as the mixing parameter to vary the relative rate of movement of CO2 and oil, and thus control the extent of enhanced oil recovery in order to match the simulated oil recovery to the laboratory model results. Desirably, the solubility should fall in a narrow range in such matching, so that the value of solubility so obtained could be used for predictive purposes, just as a narrow range of mixing parameter is found to hold in matching laboratory and field miscible floods.

The pressure varies only over about a ten psi range between the injector and producer, even with the higher viscosity foam present as compared to the viscosity of the other fluids. The higher viscosity of the foam was obtained by assigning a fictitiously high viscosity to the surfactant, such that the blend of water and surfactant which was injected had a viscosity of 62 cp (kPa-s), which is the viscosity measured for foam in the laboratory experiments (with a Brookfield cone & plate viscometer).

Secondary as well as tertiary recovery (without or with prior water flooding) processes were simulated in both lab model and computer simulation experiments. In every case, injection of the carbon dioxide slug was followed by a final waterflood until an assumed economically limiting oil cut of 2 % was reached. The extra oil for secondary CO2 floods was determined by subtracting from the over-all oil yield the recovery which could have been obtained by a prior water flood. The secondary and tertiary extra oil yields are therefore comparable. A comparison of the extra oil recovery by computer simulation with that obtained in the laboratory model is shown in Table II, on the next page.

The water floods for both water-wet and oil-wet lab models were fitted by adjusting the relative permeability curves, starting with the relative permeability curves obtained with a 1-inch diameter, 12-inch long bead pack in either the water-wet or oil-wet state. Then the CO_2 floods were simulated. For simple WAG operation, the computer simulator components were used with the normal assignments of fluids, while for foam floods the assignments of oil and water components were reversed as mentioned above. In the latter case, the water floods were again matched.

Table II. Comparison of Computer Simulated and Lab Model Enhanced Oil Recoveries (Percent EOR Recovery Basis Original Oil In Place)

Wetting State:	Water-Wet						Oil-Wet					
Process:	Secondary		Tertiary				Secondary			Tertiary		
Slug Size	0.1	0.2	0.4	0.1	0.2	0.4	0.1	0.2	0.4	0.1	0.2	0.4
Lab.1:1 WAG	2.5	4	5	2	3	3	1	2	3	6	10	13
Sim.1:1 WAG	1.8	2.3	2.8	3.1	3.8	4.7	1.3	2.6	3.1	8.6	10.2	11
Lab.1:1 Foam	5	10	12	4	8	10	10	16	19	10	15	18
Sim.1:1 Foam	7.1	9.2	10.9	7.2	8.4	9.2	8.7	12.7	15.9	9.9	14.2	18

Figures 1 and 4 show the water flood matches to the water-wet and oil-wet lab model curves, respectively. The carbon dioxide flooding runs in the lab model were then matched by computer simulation. In the simulations, as in the lab model, the carbon dioxide slug was followed by waterflooding to an assumed economically limiting water cut of 98%, and the enhanced oil recovery was calculated as the difference between the ultimate total recovery at this point and that of a water flood starting from initial oil saturation and continued until a 98% water cut was reached. Secondary carbon dioxide floods started from the same initial oil saturation, while tertiary carbon dioxide floods started with the condition at the 98% water cut point in the simple water flood. Since the foam or emulsion tests involved a 1:1 ratio of water and carbon dioxide, comparisons are shown only for the case of 1:1 WAG operation vs foam.

Fitting of the WAG data was not entirely satiafactory, as can be seen from Table II and from Figures 2,3,5, and 6. The trapped oil saturation vs water saturation is important in the water-wet case, so we did not use the inverted phases for the water-wet WAG case, but instead varied the mixing parameter within the range of 0.6 to 0.8 and the value of a in the equation:

$$Sot = Sorw/[1 + a(kro/krw)]$$

- a version of the Raimondi-Torcaso equation (30) due to Todd (31), over the range from 5 to 0.5. The best compromise was found with a = 0.5 and with the mixing parameter = 0.8 for secondary and 0.6 for tertiary. In general, the computed EOR was too low for secondary and too high for tertiary, as compared to lab model results. In the oil-wet case, the inverted phase formulation was used and the solubility of carbon dioxide in the oil phase was varied. The fit shown for secondary WAG was obtained for 0.35 mol fraction carbon dioxide solubility, while for tertiary WAG a solubility of 0.75 mol fraction gave the fit shown. This large a variation in solubility is not

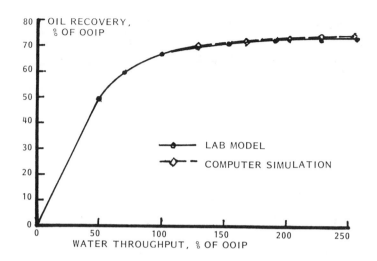

Figure 1. Water flood of water-wet system.

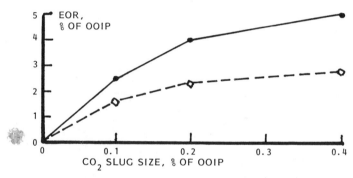

Figure 2. Secondary 1:1 WAG CO_2 flood of water-wet system.

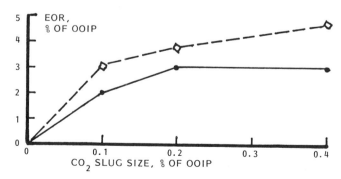

Figure 3. Tertiary 1:1 WAG CO_2 flood of water-wet system.

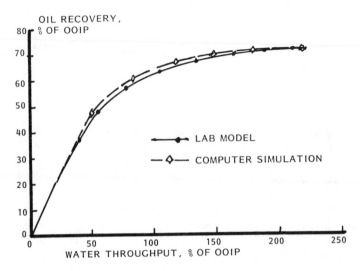

Figure 4. Water flood of oil-wet system.

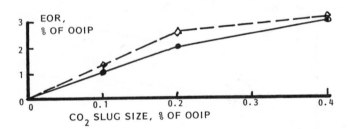

Figure 5. Secondary 1:1 WAG CO_2 flood of oil-wet system.

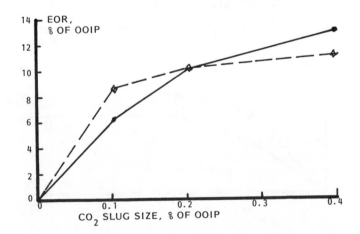

Figure 6. Tertiary 1:1 WAG CO_2 flood of oil-wet system.

reasonable. If the normal phase assignment is used, either a large variation in a or in the mixing parameter, or both, is required. Possibly the lab model results are in error. Further study of this question is planned.

In the case of the water-wet foam floods, the carbon dioxide solubility which gave the best match to the secondary floods was 0.8 mol fraction, while a value of 0.7 gave the best match for the tertiary foam floods, as shown in Figures 7 and 8. The best values for the oil-wet case were 0.9 mol fraction for both the secondary and tertiary cases, as shown in Figures 9 and 10.

Thus, in most cases a fit could be obtained with the inverted phases with carbon dioxide solubility in a narrower range than the range required of the mixing parameter if the normal phases were used and the mixing parameter was varied. On the other hand, the use of the inverted phases with carbon dioxide solubility as the only parameter available does not make use of the other features of the Multiflood simulator such as the trapped oil curve and the simulation of the effects of fingering in the way shown successfully before.

It must be concluded that it would be preferable to construct a new simulator, or a suitable adaptation of the Multiflood simulator or other miscible flooding simulator, to take into account the new physico-chemical aspects of foam or emulsion flow while retaining the other valuable capabilities of these simulators, rather than to use the inverted-phase form of the Multiflood simulator. This has been suggested to Todd, Dietrich & Chase.

In the meantime, the significant advantage indicated by our laboratory work for injection of carbon dioxide as a foam or emulsion rather than alternately with water (WAG) will, we hope, serve as a stimulus to the oil companies carrying out carbon dioxide floods to give consideration to this version of the process. Many of these companies have the Todd, Dietrich and Chase Multiflood simulator, or one which is similar. The option to precipitate or adsorb a component is necessary. Then the surfactant component can be adsorbed and can also provide a higher viscosity when it is not adsorbed (simulating foam).

In the meantime, using the guidance provided here, simulator predictions of field behavior under carbon dioxide foam flooding can be made. If these are sufficiently interesting to the oil company involved in any given case, then pilot testing would be encouraged.

It should be pointed out that foam could be employed in the case of hydrocarbon miscible flooding as well as in the case of carbon dioxide flooding. Based on the indications seen in our lab work, liquid hydrocarbons will probably make emulsions which are about 50% quality, or about 1:1 water and hydrocarbon. Hence this mixture could be injected as the slug material. However, it is not known whether the residue gas (90%+ methane), which is of low density and gas-like at the appropriate reservoiroditions for miscible flooding, will form a foam of this low a quality. If the quality is in the range of 95% or higher as is the case for atmospheric pressure gases and water plus surfactant, the viscosity is likely to be much too high for use in medium to low permeability reservoirs. However, water alone could be used to drive the slug through the reservoir. This would tend to leave a trapped saturation of NGL mixture, and this must be taken into account in the economic evaluation. It is still possible that the sweep efficiency improvement may make this

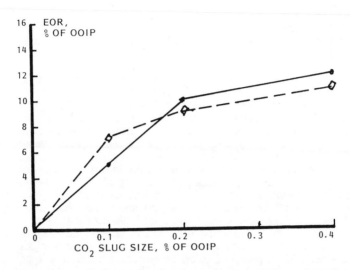

Figure 7. Secondary 1:1 foam CO_2 flood of water-wet system.

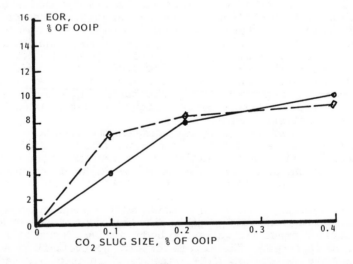

Figure 8. Tertiary 1:1 foam CO_2 flood of water-wet system.

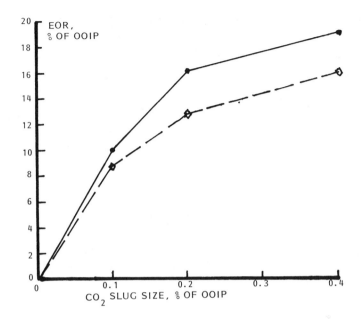

Figure 9. Secondary 1:1 foam CO_2 flood of oil-wet system.

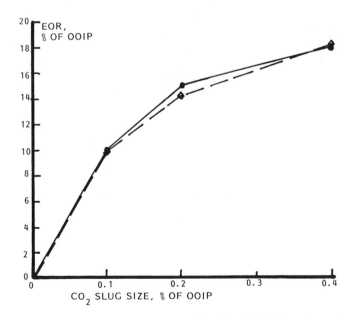

Figure 10. Tertiary 1:1 foam CO_2 flood of oil-wet system.

feasible. This is an important avenue for further research, particularly in connection with Canadian light oil reservoirs where carbon dioxide is not available in adequate quantities but NGL hydrocarbons are.

Conclusions

It is concluded that carbon dioxide foam or emulsion flooding has great promise. The oil recovery is significantly higher than with the use of WAG at the same water – carbon dioxide ratio.

We have been able to match laboratory model results with a computer simulation program (Todd, Dietrich & Chase's Multiflood) with values of the matching parameters which do not vary widely in most cases. For carbon dioxide foam simulation, it was necessary to invert the normal assignment of oil and water components.

It seems likely that predictions could be made for field cases using the guidance given here, which could serve for preliminary evaluation purposes to determine whether a pilot test would be justified.

The adaptation of the Multiflood simulator used in this work is not entirely satisfactory. A new simulator, or adaptation of an existing simulator to include the surfactant and foam behavior, is desirable.

Acknowledgments

This work was funded by the industrial consortium of companies which support the EOR Laboratory, a division of the Energy Laboratory of the University of Houston. The consortium has included: Amoco, Arco, Chevron, CNOOC, Conoco, Esso Canada, Exxon, Getty, Gulf Canada, Mobil, Schlumberger, Shell, Sun, Texaco, Todd Dietrich & Chase, and Unocal. Additional funding was provided by the Energy Laboratory. Further, personal contributions by E.L. Claridge, a pensioner of the Shell Companies, were matched two for one by the Shell Foundation.

Nomenclature

a = a parameter in the modified Raimondi-Torcaso equation,
kro = relative permeability of porous medium to oil phase,
krw =   "   "   "   "   "   " water phase,
OOIP = original oil in place in a given system, volume units.
Sot = trapped oil saturation, dimensionless,
Sorw = residual oil saturation after exhaustive water flood,
       dimensionless

References

1. Leonard, Jim: "Increased Rate of EOR Brightens Outlook," Oil & Gas J., April 14, 1986, pp. 71-102.
2. Soc. Pet. Eng. Reprint Series Book No. 18: Miscible Processes.
3. Stalkup, F.I.: Soc. Pet. Eng. Monograph No. 8, Miscible Displacement.
4. Claridge, E.L.: "Current Status, Economics, and Future of CO2 Enhanced Oil Recovery," Paper presented at the 2nd World Oil & Gas Show & Conference, Dallas, Texas, June 4-7, 1984.

5. Claridge, E.L.: "Prospects for Enhanced Oil Recovery in Texas," IOCC Bulletin 23, No. 2 (Dec. 1981).
6. Claridge, E.L.: "Prospects for Enhanced Oil Recovery in the United States," Proceedings IECEC'82, The 17th Intersociety Energy Conversion Engineering Conference, Los Angeles CA August 1982.
7. Goodrich, John: "Target Reservoirs for CO2 Miscible Flooding," Dept. of Energy Report DOE/MC/08341-12, Vols. I-IV.
8. National Petroleum Council Report: "Enhanced Oil Recovery," National Petroleum Council, 1625 K Street N.W., Washington D.C. 20006, June 21, 1984.
9. National Petroleum Council, "Factors Affecting U.S. Oil & Gas Outlook," February 1987.
10. Reference 2, pp. 65-89.
11. Shelton, J.L., and Yarborough, L.: "Multiple Phase Behavior in Porous Media During CO2 or Rich Gas Flooding," J. Pet. Tech. (Sept. 1977), 1171-1178.;
12. Stalkup, F.I.: "Carbon Dioxide Miscible Flooding: Past, Present, and Outlook for the Future," J. Pet. Tech. (Aug. 1978), 1102-1112.
13. Gardner, J.W., Orr, F.M., and Patel, P.D.: "The Effect of Phase Behavior on CO2-Flood Displacement Efficiency," J. Pet. Tech. (Nov. 1981), 2067-2081.
14. Schneider, F.N. and Owens, W.W.: "Relative Permeability Studies of Gas-Water Flow Following Solvent Injection in Carbonate Rocks," Soc. Pet. Eng. J. (Feb. 1976), 23-30.
15. Pontious, S.B. and Tham, M.J.: "North Cross (Devonian) Unit CO2 Flood - Review of Flood Performance and Numerical Simulation Model," J. Pet. Tech. (Dec. 1978), 1706-1716.
16. Reference 2, pp. 648-655.
17. Chuoke, R.L., van Meurs, P., and van der Poel, C.: "The Instability of Slow, Immiscible, Liquid-Liquid Displacements in Permeable Media," Petr. Trans. AIME 216 (1959), 188.
18. Perrine, R.L.: "A Unified Theory for Stable and Unstable Miscible Displacement," Pet. Trans. AIME 228 (1963), 205.
19. Jackson, D.D., Andrews, G.L., and Claridge, E.L.: "Optimum WAG Ratio vs Rock Wettability in CO2 Flooding," SPE Paper No. 14303, presented at the 60th Annual Technical Conference of the Society of Petroleum Engineers, Las Vegas, Sept. 1985 (being revised for publication).
20. Claridge, E.L.: "Prediction of Recovery in Unstable Miscible Flooding," Soc. Pet. Eng. J. (April 1972), 143.
21. Bernard, G.G., Holm, L.W., and Harvey, C.P.: "Use of Surfactant to Reduce CO2 Mobility in Oil Displacement," Soc. Pet. Eng. J. (Aug. 1980), 281-292.
22. Casteel, J.F. and Djabbargh, N.F.: "Sweep Improvement in Carbon Dioxide Flooding Using Foaming Agents," SPE Paper No. 14392, presented at the 60th Annual Technical Conference of the Society of Petroleum Engineers, Las Vegas, Sept. 1985.
23. Heller, J.P., Lien, C.L., and Kuntamukkula, M.S.: Foamlike Dispersions for Mobility Control in CO2 Floods," Soc. Pet. Eng. J. (Aug. 1985),603-13.
24. Borchardt, J.K., Bright, D.B., Dickson, M.K., and Wellington, S.L.: "Surfactants for CO2 Foam Flooding," SPE Paper 14394, presented at the 61st Annual Technical Conference of the Society of Petroleum Engineers, Las Vegas, September 1985.

25. Lescure, B.M. and Claridge, E.L.: "CO2 Foam Flooding Performance vs. Rock Wettability," SPE Paper No. 15445, presented at the 61st Annual Technical Conference of the Society of Petroleum Engineers, New Orleans, Oct. 1986.
26. Tiffin, D.L., and Yellig, W.F.: "Effects of Mobile Water on Multiple-Contact Miscible Gas Displacements," Soc.Pet.Eng.J. (June 1983), 447-455.
27. Lescure, B.M.: "CO2 Foam Flooding," M.S. Thesis, Ch.E. Dept., Univ. of Houston, May 1987.
28. Chase, C.A. and Todd, M.R.: "Numerical Simulation of CO2 Flood Performance," Soc. Pet. Eng. J. (Dec. 1984), 597-605.
29. Todd, M.R. and Longstaff, W.J.: "The Development, Testing, and Application of a Numerical Simulator for Predicting Miscible Flood Performance," J. Pet. Tech. (July 1972), 874.
30. Raimondi, P. and Torcaso, M.A.: "Distribution of the Oil Phase Obtained Upon Imbibition of Water," Soc. Pet. Eng. J. (March 1964), 49-55
31. Todd, M.R. et al: "CO_2 Flood Performance Evaluation for the Cornell Unit, Wasson San Andres Field," J.Pet.Tech. (Oct. 1982), 2271-2282.

RECEIVED February 9, 1988

Chapter 19

Carbon Dioxide–Foam Mobility Measurements at High Pressure

Hae Ok Lee and John P. Heller

New Mexico Petroleum Recovery Research Center, New Mexico Institute of Mining and Technology, Socorro, NM 87801

Mobility control of high pressure CO_2 floods by use of foam is a promising technique of enhanced oil recovery. Critical information for the use of CO_2-foam as a thickened displacement fluid is the ratio of combined CO_2 and surfactant solution flow rate to pressure drop in the swept region. In this work, foam mobility measurements have been carried out on a laboratory scale. A high pressure apparatus has been designed and carefully calibrated to measure pressure drop across a core sample. The measured steady-state data are used to evaluate the mobility of foam. These measurements are made during simultaneous flow of the dense CO_2 and surfactant solution through core samples. Dependences of the mobility of CO_2-foam on aqueous surfactant concentration and CO_2 volume fraction have been investigated.

The addition of surfactant to the flowing water during the flood reduces CO_2 mobility and should improve both areal and vertical sweep efficiencies by stabilizing viscous fingering and flow through the more permeable zones (1-3). Numerous laboratory studies have demonstrated that, if contact is made with the oil, dense supercritical CO_2 can develop multicontact miscibility with many crudes (4,5). Most of the time, though, oil recoveries with CO_2 have been much higher in the laboratory than in the field because the field conditions are more severe for all oil recovery processes, permitting much more non-uniform flow.

  CO_2-foam presents an immediately available, and perhaps a more efficient way of attaining uniform displacement in the reservoir because a foam consists of at least 80% CO_2, and because the cost of needed surfactant promises to be minor. Furthermore, foam possesses properties that are favorable for oil recovery,

0097–6156/88/0373–0375$06.00/0

especially by CO_2 flooding. The apparent viscosity of foam in
porous rock is greater than the actual viscosities of its components, and increases with the rock permeability. Also, foam
lamellae increase trapped gas saturation. As gas saturation
increases, oil saturation decreases. Usually, a high-trapped gas
saturation reduces gas mobility.
    Foam flooding is a method that modifies the flow mechanism by
changing the structure of the displacing fluid at the pore level.
A critical literature review on foam rheology is given elsewhere
(6). The injection of foam-like dispersions or CO_2-foams is a
useful method in enhanced oil recovery (7). This method of
decreasing the mobility of a low-viscosity fluid in a porous rock
requires the use of a surfactant to stabilize a population of
bubble films or lamellae within the porespace of the rock (8). The
degree of thickening achieved apparently depends to some extent on
the properties of the rock itself. These properties probably
include both the distance scale of the pore space and the
wettability, and so can be expected to differ from reservoir to
reservoir, as well as to some extent within a given field (9,10).
    In the case of CO_2 floods, a partially compensating feature
has been reported (11). Unexpectedly low mobilities have been
observed during CO_2 injection in both field and laboratory
experiments, and some authors have concluded this effect is caused
by "mixed-wettability" of the rock. The decreased mobility also
might possibly be connected with the high solubility of CO_2 in
crude oil, which adds to the effectiveness of transverse
dispersion in tending to dissipate at least the closely spaced
fingers. Despite these mitigating features, most CO_2 floods do
show early breakthrough, indicating higher flow rate in a CO_2
finger or channel connecting the injection and production wells.
So long as this finger expands laterally, thereby entraining
enough additional oil to make continuance of the flood economic,
the produced CO_2 can be reinjected. The costs of gathering,
processing and recompression are an additional operating expense
that must necessarily hasten the day of abandonment. Because of
this the overall recovery efficiency is reduced - a loss of oil
that can be considered a result of the unfavorable mobility ratio.
    Researchers have investigated the nature of the foam flow by
examining the mechanisms of foam generation (12). An extensive
study (13), that is quite relevant to the mechanism of foam flow
in porous media, has shown that the apparent viscosity of foam in
a capillary tube decreases rapidly as the ratio of bubble radius-
to-tube radius is increased.
    Beside the investigation on foam flow mechanisms, researchers
have endeavored to find good foaming agents, especially applicable
to enhanced oil recovery (14). Furthermore, due to the harsh
reservoir conditions, studies on the compatibility of the
surfactant with oilfield brines at reservoir temperatures and
pressures have been made (15).
    Although there has been considerable effort to calculate the
mobility of foam in porous media from first principles, utilizing
usually measured rock properties (16), a different approach is
used here. In this research, major emphasis has been on
measurement of the mobility of CO_2-foam in rock core samples. The

work on this method of thickening has also included the testing of
several features of surfactant suitability. The work reported
describes the development of apparatus and experimental methods
and presents available results.

Experimental

Description of the Foam Mobility Measurement Apparatus. A
schematic of the CO_2 foam mobility measurement apparatus is given
in Figure 1. The Ruska pump pressurizes the liquid CO_2,
maintained at a constant temperature by circulating antifreeze
inside the jacketed pump. The CO_2 flows through the capillary
tube; the pressure drop across the tube is measured by a Validyne
differential pressure transducer. An Isco pump is used to
pressurize the brine surfactant solution, which also flows through
the foam generator and the core. As a matter of procedure, the
core is initially saturated fully with brine-surfactant solution.
The foams are generated inside the short core used as a foam
generator, where the mixing between CO_2 and surfactant solution
occurs. The mixed CO_2-foam flows through the core. The pressure
drop across the core is recorded by a second Validyne differential
pressure transducer. The arrangement of zeroing valves at each of
these transducers is important to point out. In addition to the
digital readout of the values of ΔPcap and ΔPcore, a two-pen
recorder is used to record the simultaneous measurements. The
needle valve is operated by the electronic flow regulator system.
     To maintain the macroscopic steady-state condition of the
flow of dense CO_2 through the capillary tube, several tests have
been made in the development of an optimum flow device. Two fine
tapered needle valves in series are used to regulate the flow rate
of the CO_2 at high pressure. A fine adjustment dial on one of the
valves permits repeatable manual settings. Steady-state is
accomplished manually by setting the dial to give a desired
pressure drop across the calculated capillary that carries dense
CO_2 from the pump. To accomplish this automatically, a controller
set point that corresponds to the desired pressure drop across the
capillary is chosen. After opening the manual needle valve to an
approximately desired value, the appropriate PID gains
(proportional, integral, differential) are optimized to open or
close the motorized needle valve electronically, according to the
set point. All calculations were performed with steady-state data
which were averaged over a period of time.

Interpretation of the Experimental Data. In this experiment, the
measured variables are ΔPcap and ΔPcore. Besides these variables,
other data, such as the Ruska pump flow rate, Isco pump flow rate,
inlet pressure of CO_2, and the temperature of the CO_2 at the inlet
are recorded. Knowing the ΔPcap, the flow rate of pure CO_2 into
the core is computed by using the calibration constant obtained
earlier. The Isco pump flow rate gives the flow rate of the
aqueous surfactant solution, and the total flow rate is simply the
sum of these two flow rates. From the total flow rate and the
ΔPcore, the mobility can be evaluated from:

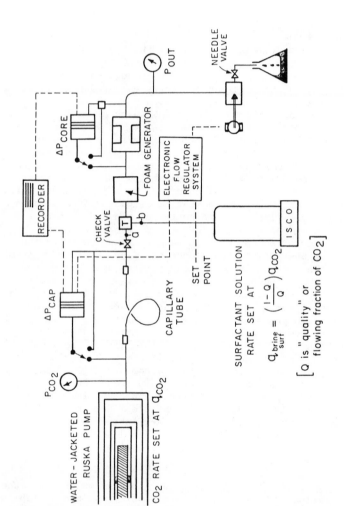

Figure 1.   Schematic of the CO_2-foam mobility measurement apparatus.

$$\lambda = \frac{Q/L}{\Delta Pcore/L} \tag{1}$$

where   Q is the total flow rate
         A is the cross-sectional area of the rock sample
         L is the length of the rock sample
         $\Delta Pcore$ is the pressure drop across the rock sample.

CO₂-Foam Mobility Measurements

Effect of Surfactant Concentrations. The effect of the surfactant
concentration on foam mobility has been studied extensively. The
surfactant under investigation for this effect was Varion CAS, a
zwitterionic surfactant from Sherex. The rock under study was
Berea sandstone which has a permeability of 308 ± 9 md measured by
using 1% brine solution. The permeability using N_2 gas at
atmospheric pressure was 1000 ± 6.2 md.
      The effect of concentration of the Varion CAS was studied.
The initial concentration (0.001 wt% active) was selected from the
measurement of surface and interfacial tensions. It lies in the
range of CMC value for these measurements, which were performed by
using the Wilhelmy plate method in a Rosano surface tensiometer.
The standard procedure described in the manual was carefully
followed. Surface tension was measured against the air, whereas
the interfacial tension was measured against isooctane, which was
used to simulate dense CO_2. It is well known fact that good
foamers are most effective well above the CMC range and in
subsequent tests the concentrations were increased well above that
range.
      Two significant conclusions can be drawn from this
experiment. Figure 2 shows the measurements of mobility of CO_2-
foam, at various levels of surfactant concentration in NaCl +
$CaClC_2$ brine. The mobility generally decreases with the increase
of the surfactant concentration. The second conclusion is that
there exists a critical concentration above which further increase
in concentration causes no significant reduction in the magnitude
of the foam mobility. Because it is difficult to read all of the
data at the higher concentrations on this curve, they are all
given in Table I. In addition, the points measured at different
velocities and the same concentration have been averaged, and
these average mobilities plotted in Figure 3 on a log-log scale,
against the concentrations. The general form of the effect of
concentration is easier to see in this plot. The line
extrapolated towards the left is based on one additional mobility
measurement made during simultaneous flow of dense CO_2 and brine
containing no surfactant at all. The remainder of the line drawn
through the five points, and extrapolated to higher
concentrations, might be considered more speculative.
      But such a lower limit of attainable mobility, as is shown on
the Figure, has also been observed with one other surfactant. If
it proves to be a general feature of all surfactants that are
effective in stabilizing CO_2 foams, it will be of great economic
interest, since it will fix the maximum concentrations of
particular surfactants that could be useful in the field.

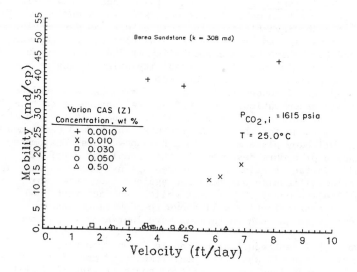

Figure 2.   Effect of surfactant concentrations.

Table I. CO_2-Foam Mobility Measurements

Effect of Surfactant Concentrations
Varion CAS (Z) in 1% Brine *

Rock type: Berea Sandstone
Permeability of the rock: 308 ± 9 md
Inlet pressure of the pumps: 1615 ± 5 psia
Temperature of the CO_2 in Ruska pump: 25.0 ± 0.2°C

Concentration (wt% active)	v (ft/day)	λ (md/cp)	λ_r (cp^{-1})	CO_2 (quality,** %)
0.0010	3.60	39.4	0.128	82.7
	4.86	37.8	0.123	80.8
	8.10	44.5	0.144	81.5
				81.7 ± 1.0
0.010	2.84	10.5	0.0340	78.1
	5.76	13.2	0.0428	78.4
	6.14	14.2	0.0461	81.7
	6.86	17.5	0.0568	82.2
				80.1 ± 2.1
0.030	1.73	0.953	0.00309	81.2
	2.97	1.64	0.00531	83.2
	3.60	1.28	0.00414	82.7
				82.4 ± 1.0
0.050	2.46	0.520	0.00169	79.7
	3.54	0.643	0.00209	82.4
	3.74	0.582	0.00189	83.3
	3.82	0.755	0.00245	83.7
	3.88	0.662	0.00215	83.9
	4.53	0.762	0.00247	83.5
	4.85	0.927	0.00301	83.3
	5.16	0.854	0.00277	83.1
				82.9 ± 1.4
0.50	2.37	0.743	0.00241	81.0
	4.14	0.549	0.00178	81.9
	4.80	0.648	0.00210	81.8
	6.36	0.753	0.00244	80.7
				81.4 ± 0.6

*Standard brine solution is made of 0.5% NaCl and 0.5% $CaCl_2$.
**Quality is the volume fraction of the CO_2.

Some observations were recorded during the experiment. Through the transparent, low pressure outlet tubing, no foam lamellae have been observed when the core was flushed with concentrations of 0.001 wt% and 0.01 wt% actives. Furthermore, during the simultaneous flow of dense CO_2 and these concentrations of surfactant solution, the bubbles exiting the outlet tubing did not have much elasticity, either. It seemed as if insufficient surfactant molecules were present to generate durable foam lamellae to pass continuously through the rock to have an appreciable effect on the flow resistance in the core. With 0.03 wt% active and up, well-formed foam lamellae were observed in the output tube as the core was washed with each new concentration before the actual experiment with dense CO_2. With simultaneous run of dense CO_2, elastic bubbles were seen through the outlet tubing. As the concentrations were increased from 0.001 wt% and up to 0.05 wt%, the measured pressure drops across the core increased significantly. However, from 0.05% and up, the pressure across the core was fairly constant. This is an indication that sufficient surfactant molecules are present and further increase in concentration seems to have further no effect on generating more foam lamellae. This phenomenon is analogous to the critical micelle concentration (CMC) where further increase in surfactant concentration does not lower the value of the surface tension.

Whenever the concentration was increased, sufficient pore volumes were used to wash the core thoroughly. Some of the data were repeated in order to assess the reliability of the apparatus. The order of magnitude of the mobility was the same for all measurements at each particular concentration, independent of total flow rate. This indeed demonstrates the sensitivity and viability of this foam mobility measuring apparatus.

Varion CAS foams well but the foam also dies out rather quickly. It must be a fast draining foam. In general, the critical concentration (above which no further mobility reduction takes place) is well above the CMC range of the surfactant solution.

The Effect of CO_2 Fraction. The effect of CO_2 fraction (sometimes termed "quality") on foam mobility measurements has also been studied. The surfactant under the investigation in this case was the anionic Enordet X2001 from Shell. Four different CO_2 fractions were tested at constant surfactant concentration of 0.05 wt% active. Again, the concentration of 0.05% was chosen from the measurements of surface and interfacial tensions. The rock sample used for the test was Berea sandstone with k = 302 ± 16 md. The variation of foam mobility with different values of CO_2 fraction is clearly seen in Figure 4, where the mobilities are plotted against total flow rate. The corresponding numerical data are presented in Table II. In Figure 4, the effect of CO_2 fraction is quite definite. As has been expected, the foam mobility decreases with decreasing CO_2 fraction: greater pressure drop is observed across the core with the increase of surfactant fraction. This verifies the fact, long known in the literature, that the presence of surfactant solution along with CO_2 lowers the mobility.

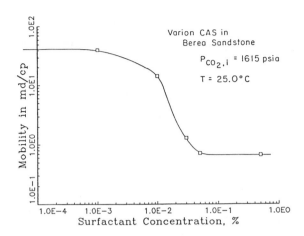

Figure 3.   Effect of surfactant concentration on foam mobility.

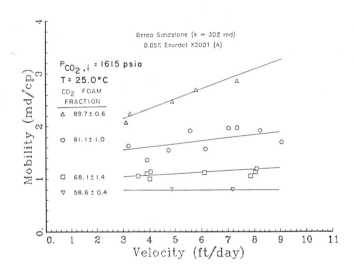

Figure 4.   Effect of CO_2 fraction.

Table II

CO_2-Foam Mobility Measurements

Effect of CO_2 Foam Fraction
0.05% Enordet X2001 (A) in 1% Brine*

Rock Type: Berea Sandstone
Permeability of the rock: 302 ± 16 md
Inlet pressure of the pumps: 1615 ± 5 psia
Temperature of the CO_2 in Ruska pump: 25.0 ± 0.2°C

CO_2 (Quality,** %)	v (ft/day)	λ (md/cp)	λ_r (10^{-3} x cp^{-1})
58.6 ± 0.4	4.86	0.785	2.60
	7.17	0.791	2.62
68.1 ± 1.4	3.57	1.05	3.48
	3.91	1.10	3.64
	4.02	0.998	3.31
	4.05	1.13	3.75
	6.12	1.11	3.67
	7.86	1.05	3.47
	8.04	1.13	3.73
	8.10	1.18	3.91
81.1 ± 1.0	3.20	1.62	5.37
	3.92	1.35	4.48
	4.75	1.54	5.10
	5.57	1.91	6.32
	6.17	1.56	5.16
	7.03	1.95	6.46
	7.36	1.96	6.49
	8.23	1.91	6.32
	9.04	1.69	5.60
89.7 ± 0.6	3.12	2.07	6.85
	3.27	2.24	7.42
	4.88	2.47	8.18
	5.79	2.68	8.87
	7.36	2.85	9.44

*Standard brine solution is made of 0.5% NaCl and 0.5% $CaCl_2$.
**Quality is the volume fraction of the CO_2.

It is also interesting to observe that the slope of the
fitted lines (that is, the dependence of mobility on overall flow
rate) decreases as the surfactant fraction increases. A possible
explanation is that the lamellae formed in the pore space between
the CO_2 and surfactant mixture become more durable as the aqueous
fraction is increased. From a macroscopic viewpoint, more uniform
displacement would be expected as a result of the decreased
mobility. Furthermore, greater scattering of data is observed for
the CO_2 fraction of 81.1 ± 1.0%. It is possible that this
particular mixture may be more thermodynamically unstable than
"foams" of different quality.

This variation of mobility with CO_2 fraction has also been
observed with other surfactants, such as Varion CAS and
Chembetaine BC-50. Although those data are not presented here,
similar behavior consistent with such a significant decrease in
mobility with increase in surfactant fraction was observed. The
results of an extensive study with Enordet X2001 have supported
the previous observations.

To verify the reproducibility of all these experimental
points, some of the measurements were purposely repeated. A
typical data set, such as those with a CO_2 fraction of 81.1 ± 1.0%
of 0.05% Enordet X2001 was used to do some standard error
analysis. The relative error on the mobility measurement, itself,
was approximately 4%. The error was judged to arise primarily
from the uncertainties in the total flow rate due to the
difficulty of maintaining "ideal" steady-state because CO_2 is a
compressible fluid. Also, the flow rate is measured by using a
Validyne differential pressure transducer which also produces some
uncertainty even though it has been calibrated against a dead-
weight tester, and its zero is frequently checked during the
experiment. In addition to these, inaccuracy occurs as a result
of the heterogeneities of the rock.

In our laboratory, we have screened a number of surfactants.
The standard surfactant screening test included foam height
measurement, thermal aging test, and pH evaluation. Varion CAS
and Enordet X2001 performed very well on the foam height
measurement. Both also showed very promising results on the
thermal aging test, even though their results are not presented
here.

Conclusions

From the research on CO_2-foam mobility measurements, a few
significant conclusions on foam flooding have been drawn. The
effect of concentration on foam mobility has been clearly seen
with the Varion CAS. With this surfactant an increase beyond
0.05% of its concentration seemed to secure no further decrease in
mobility. The same general pattern of variation was observed with
other surfactants. The effect of CO_2 fraction was measured in a
comprehensive study using Enordet X2001. Again, the same trend
was observed with other surfactants, such as Chembetaine BC-50 and
Varion CAS, although the data were not presented here. The
mobility is definitely lowered by increasing the surfactant
fraction in the foam. It seems clear that, in the presence of

surfactant solutions, dense CO_2 forms foam lamellae, in the
porespace, and that this is the responsible mechanism of mobility
reduction. At surfactant concentrations lower than the CMC,
lamellae are so rare as to have no influence on the flow.
Although not pointed out in the test, it may be observed from the
data that there is some effect on mobility of velocity or total
flow rate. Some shear thinning or pseudoplastic behavior has been
observed under certain conditions. This is, of course, the more
favorable of possible non-Newtonian behaviors for foam mobility
control, since it would mean that less thickening occurred in the
vicinity of the injection well, than further out in the formation.

Literature Cited
1.   Bernard, G.G.; Holm, L.W.; Harvey, C.P. Soc. Pet. Eng. J.
     1980, 20, 281-92.
2.   Wang, G.C., presented at the SPE/DOE Fourth Symposium on
     Enhanced Oil Recovery, Tulsa, OK, April 1984.
3.   Wellington, S.L.; Vinegar, R.J. Soc. Pet. Eng. J. 1987, 27,
     885-98.
4.   Hydrocarbon Displacement by Carbon Dioxide Dispersion,
     Technical Note, U.S. Department of Energy, March 1986.
5.   Holm, L.W.; Josendal, V.A. J. Pet. Tech. 1974, 26, 1427-37.
6.   Heller, J.P.; Kuntamukkula, M.S. Ind. Eng. Chem. Res. 1987,
     26, 318-25.
7.   Heller, J.P.; Lien, C.L.; Kuntamukkula, M.S. Soc. Pet. Eng.
     J. 1985, 25, 603-13.
8.   Radke, C.J.; Ransohoff, T.C., presented in part at the 61st
     SPE Annual Technical Conference and Exhibition, New Orleans,
     LA, Oct. 1986.
9.   Heller, J.P., presented at the SPE/DOE Fourth Symposium on
     Enhanced Oil Recovery, Tulsa, OK, April 1984.
10. Heller, J.P.; Boone, D.A.; Watts, R.J., presented at the 60th
     Annual Technical Conference and Exhibition of the Society of
     Petroleum Engineers, Las Vegas, NV, Sept. 1985.
11. Patel, P.D.; Christman, P.G.; Gardner, J.W., presented at the
     60th SPE Annual Technical Conference and Exhibition, Las
     Vegas, NV, Sept. 1985.
12. Falls, A.H. et al., presented at the DOE/SPE Fifth Symposium
     on Enhanced Oil Recovery, April 1986.
13. Hirasaki, G.J.; Lawson, J.B. Soc. Pet. Eng. J. 1985, 25, 176-
     90.
14. Borchardt, J.K. presented at the SPE International Symposium
     on Oilfield Chemistry, San Antonio, TX, Feb. 1987.
15. Maini, B.B., Ma, V. J. Can. Pet. Tech. 1986, 25, 65-9.
16. Khatib, Z.I., Hirasaki, G.J., and Falls, A.H., presented at
     the 61st SPE Annual Technical Conference and Exhibition, New
     Orleans, LA, Oct. 1986.

RECEIVED January 5, 1988

Chapter 20

Enhancement of Crude Oil Recovery in Carbon Dioxide Flooding

John T. Patton and Stanford T. Holbrook

Department of Chemical Engineering, New Mexico State University, Las Cruces, NM 88003

Tertiary oil was increased up to 41% over conventional CO_2 recovery by means of mobility control where a carefully selected surfactant structure was used to form an in situ foam. Linear flow oil displacement tests were performed for both miscible and immiscible floods. Mobility control was achieved without detracting from the CO_2-oil interaction that enhances recovery. Surfactant selection is critical in maximizing performance. Several tests were combined for surfactant screening, included were foam tests, dynamic flow tests through a porous bed pack and oil displacement tests. Ethoxylated aliphatic alcohols, their sulfate derivatives and ethylene oxide - propylene oxide copolymers were the best performers in oil reservoir brines. One sulfonate surfactant also proved to be effective especially in low salinity injection fluid.

As the world's oil reserves become depleted, the role of more sophisticated oil recovery methods increases in importance. Modern methods for enhanced oil recovery employ the injection of fluids into a oil bearing stratum, the petroleum reservoir. The injection of selected gases has proved to be effective, especially when some form of gas mobility control has been used. The usual method, demonstrated to be moderately effective in lowering gas mobility, consists of injecting gas alternated with slugs of water[1]. Various gases have been used including CO_2, steam, nitrogen, light hydrocarbons and flue gas.

For example, steam is being injected for enhanced oil recovery by numerous operators in Kern County, California[2]. Carbon dioxide flooding is being used in the SACROC field of West Texas and the Bati Kaman field, Turkey's largest oil reservoir[3,4]. The injection of light hydrocarbons is being initiated in Prudhoe Bay, Alaska, and a significant commercial venture is underway utilizing nitrogen injection in Alabama[5,6]. Numerous other field tests are underway or are in the planning stages at the present time.

0097–6156/88/0373–0387$06.00/0
© 1988 American Chemical Society

Since gases have much higher mobility than crude oil or water, the injected gas has a strong tendency to channel through the reservoir and bypass the reservoir oil. This bypassing effect decreases, significantly, the efficiency of the recovery process and, in some cases, even makes any gas injection unprofitable.

Previous studies have shown that foam can be effectively used to mitigate oil bypassing[7]. Foam can be injected into a reservoir, or, preferably, it can be generated in situ. The in situ generation is accomplished by injecting a bank of surfactant solution followed by the injection of the gas. The latter method avoids increased injection pressure that would result from the foam's low mobility in or near the injection well.

Work previously reported has shown very significant reductions in gas mobility when in situ generated foam was used in laboratory core tests. Using unconsolidated core models, investigators measured ten to several hundredfold decreases in gas mobility[8]. The effectiveness of foam for mobility control is vitally dependent on the choice of the surfactant used. Early tests using 1% ammonium lauryl sulfate were not totally successful in demonstrating foam effectiveness[9].

Extrapolation of the linear experimental laboratory results is less than accurate unless the effects of areal displacement and gravity are taken into account. A recent study by Chilton (1987)[10] embodies computer support for predicting two-dimensional and three-dimensional production curves for oil recovery with mobility-controlled carbon dioxide. Figure 1 shows the three-dimensional results and indicates a 26% increase in oil production when the mobility of the gas is decreased tenfold by appropriate means. In contrast, the equivalent linear model predicted only 5.4% increase in oil production with the same decrease in mobility.

Surfactant Requisites

There are several factors that can affect a given surfactant's performance in a reservoir environment. First, the effect of inorganic ions is significant. Most oil reservoirs have an aqueous phase of saline brine that may vary in concentration from 0.5% to upwards of 15% NaCl. Also, there are divalent ions, such as Ca^{++} and Mg^{++} present in significant concentrations. Most of the experimentation, which serves as the basis for this paper, was conducted utilizing a brine of 3% NaCl with 100 ppm (mg/l) of Ca^{++}. This composition is typical of many natural reservoir brines, and those surfactants that will perform well with this brine will also do well in the majority of reservoirs.

Adsorption or possible chemical reaction of the surfactant with reservoir minerals is also a limitation. In general, the cationic surfactants are of very limited value because of this factor. Nonionic surfactants are usually the best, particularly where significant clay deposits are present; however, anionic surfactants are in some cases equally effective.

Because of adsorption effects, it is not recommended that surfactant mixtures be used. These mixtures tend to be chromatographically fractionated as the solution passes through the reservoir. The justification for the use of a mixture, e.g., synergistic enhancement, would not be achieved once the additives were separated.

The effect of crude oil can also be deleterious. In some foam systems the presence of crude oil, even in small amounts, is sufficient to greatly inhibit foam formation and foam stability. A preferred surfactant is either unaffected or its effectiveness is enhanced by the presence of crude oil.

Temperature limitations are also important. In order for the surfactant to be effective it should be soluble at the temperature of the reservoir. Materials, such as some of the petroleum sulfonates, have very low solubilities in brines at temperatures below about 150°F (71° C). Consequently, they produce almost no foam below this temperature. Other surfactants, e.g., ethoxylated alcohols, become less soluble as the temperature is increased. These materials become limited in their effectiveness in higher temperature applications. An increase in temperature may also decrease the surface-active property of some surfactants.

Screening Tests

A sequence of tests was designed to rapidly identify those additives that have high potential for commercial application. Static blender foam tests eliminated additives with low potential. The severity of the tests was increased stepwise, resulting in the elimination of additional candidate additives. The final tests involved the displacement of tertiary oil employing those additives that still looked promising after the preliminary screening studies.

From approximately 130 carefully selected candidate materials, a relatively small number of surfactants emerged as being superior for mobility control applications.

The surfactants were first tested to determine their foamability and foam stability using a 0.5% solution in the standard 3% brine solution. The foam tests were conducted in accordance with the standard ASTM test, D3519 76, except that the blender chamber was modified. In this test a solution containing the additive is blended in a commercial Waring blender, and the height of the foam generated in the blender container was measured. It was soon found that the height-to-diameter ratio of the blender container was too low to allow precise evaluation of the additive under test. Accordingly, the apparatus was modified by constructing a special blender chamber three inches in diameter and twenty-eight inches high. The increased height-to-diameter ratio greatly eliminated the human error in reading the foam height and improved the reproducibility of the test. In these tests the initial foam height is a measure of foamability, and the rate of decrease of height with time is a measure of foam stability.

Surfactants with low foamability were found to be poor candidates for mobility control. In general, anionic surfactants appear superior in the static foam tests. Sulfate esters of ethyoxylated linear alcohols were slightly better than the other classes of surfactants. Most sulfonates do not appear compatible with even small amounts of calcium and, therefore, produce very little foam at the temperatures used, 75-120°F.

Dynamic Foam Tests

Apparatus and Procedure. It was necessary to design more definitive tests to further evaluate the better candidate surfactants. This was accomplished by means of a multi-phase dynamic-flow test that consists of a small packed bed through which surfactant solution can be passed followed by gas to produce in situ foam. The pressure drop through the column is measured as the fluid is drawn through the column at a constant volumetric flow rate. From the recorded data, relative mobilities of the liquid and gas phases may be calculated. The change in gas mobility due to the presence of the surfactant is very closely related to the effectiveness of that surfactant for mobility control in oil core studies. A schematic drawing of the apparatus is shown in Figure 2.

The flow column consists of a 1 1/4 inch (3.18 cm) inside-diameter plexiglass tube which contains the porous bed. The porous bed used consists of −150+200 Tyler-mesh glass beads (0.105 mm to 0.074 mm) supported on a thin layer of −48+65 Tyler-mesh glass beads (0.297 mm to 0.210 mm) supported by a 65-mesh screen. A bed approximately four inches in depth was used to test the effectiveness of the surfactants.

Solution is withdrawn from the bottom of the flow column with a Ruska proportioning pump. The pump is capable of pumping or withdrawing fluid at a rate of 10 to 1200 cc/hr. by the action of a 500 cc piston in a cylinder, powered by a 1/4 horsepower drive motor.

Pressure readings were taken with a Gould-Statham, model number PM6TC, 10 psi differential pressure transducer connected to a Gould-Statham, model SC1012, digital readout. The transducer readout is a self-contained portable unit which provides DC excitation for a bridge transducer, digitizes the transducer output, and also provides a high-level DC output to drive a recorder or analog meter. For this work, a Cole-Parmer, model number 385, x-y strip chart recorder was connected to the digital readout to record the pressure profile as an experiment was being run. The transducer used in this work had a sensitivity of 4 millivolts/volt, and the offset was set at +20% of full scale.

During the flow tests, transducer output vs. time data were automatically recorded into a computer (HP-1000). It was programmed to analyze the data and provided calculated relative mobilities as a function of flow volume. The resulting values were presented in both tabular and plot form.

The ability of various surfactants to lower the mobility of a gas in a porous medium was evaluated by comparing their mobility ratios, which is the ratio of the gas mobility to that of the liquid as determined during the course of the experiment.

Mobility was chosen as the basis for analysis as this avoids the necessity of defining the mechanism of flow restriction by foam in porous media. The mobility of a fluid is defined as the ratio of the permeability of the porous bed to the fluid divided by the fluid viscosity.

Results of the Dynamic Foam Tests. Typical plots of pressure drop vs. time for the dynamic foam tests are given in Figure 3. The initial sharp rise in

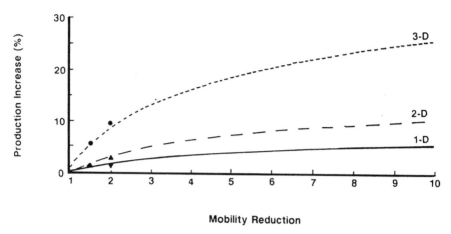

Figure 1.   Predicted increase in oil production with mobility control.

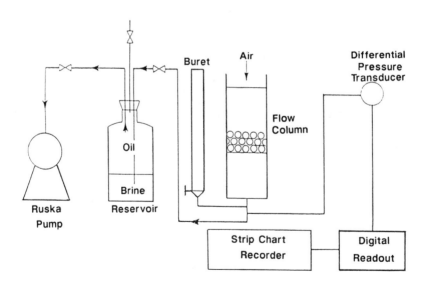

Figure 2.   Dynamic flow apparatus.

pressure differential occurs when the gas-liquid interface first enters the bed. The rise is due to both capillary pressure, which is a function of surface tension, and flow effects. The brine curve has the sharpest rise because of its greater surface tension. When the gas phase breaks through the bed, the pressure drop decreases in the brine case because capillary backpressure is removed. Conversely, when surfactant is present, the pressure continues to increase after gas breakthrough because of the low mobility of the in situ generated foam.

A plot showing the ratio of gas mobility to liquid mobility for air flow in the presence of brine and some surfactant solutions in brine is shown in Figure 4. The results indicate that these surfactants should be effective agents for reducing gas mobility, particularly in water-wetted sandstone reservoirs. A list of the surfactants used are given in Table I.

Table I - Mobility Control Surfactants

Surfactant	Company	Chemical Structure
Pluronic F-68	BASF-Wyandotte Corp.	Ethylene oxide-propylene oxide co-polymer
Alipal CD-128	GAF Corp.	Sulfated ethoxy-alcohol
Neodol 23-6.5	Shell Chemical Co.	Ethoxylated alcohol
Witcolate 1276	Witco Chemical	Alcohol ether sulfate
Pluradyne SF-27	BASF-Wyandotte Corp.	Ethoxylated alcohol
Stepanflo-50	Stepan Chemical Co.	Sulfate ester of ethoxylated alcohol
Exxon LD 776-52	Exxon Chemical Co.	Sulfonate

NOTE: Surfactant names are registered trademarks.

In spite of the fact that sulfonated-hydrocarbon type surfactants have poor tolerance to brine containing polyvalent cations, a few specially tailored sulfonate additives look promising. Of this group, one sample was found to possess interestingly superior performance. Although the foam produced in the static screening test with Exxon LD 776-52 would rate about average, its performance was significantly better than all other sulfonates. For this reason it was evaluated in the dynamic screening test. In spite of only modest foam produced in the static tests, LD 776-52 provided mobility control to gas flow comparable to that of some of the better additives previously identified.

On the basis of this surprising result and because it is known that sulfonates perform better in fresh water, tests were also made using tap water for the solution instead of brine. The resulting decrease in mobility ratio was even more pronounced as the results shown in Figure 5 indicate. Note that the tap-water solution was also tested at 170°F (77°C). These latter tests are indicative of mobility control in reservoirs where fresh water is available for injection.

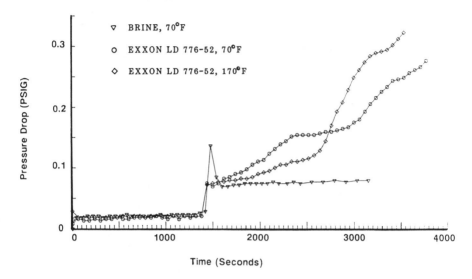

Figure 3.   Pressure drop for brine-air flow in packed bed.

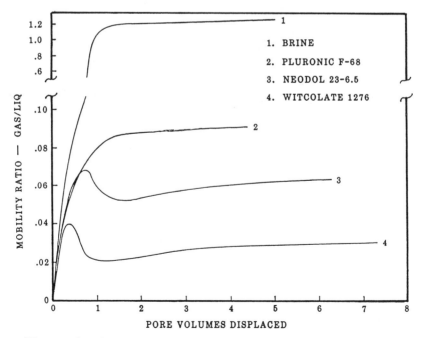

Figure 4.   Gas mobility reduction for surfactants in brine.

Linear Oil Displacement Tests

Apparatus and Procedure.   Tests for the actual displacement of tertiary oil
were made in horizontal flow models containing tight, unconsolidated sandpack cores. A schematic drawing of the flow apparatus used for the tests
reported here is shown in Figure 6.

The flow model is 54.3 in. (138 cm) long made from 1 1/2 in. NPS,
schedule 40, seamless, type-316L stainless-steel pipe. The ends are closed with
900 pound flanges welded to the pipe. The inside diameter of the column is
1.59 in. (4.03 cm), and the cross-sectional area of the model for flow is 1.975
square in. (12.73 square cm). The flange gaskets are teflon-stainless spiral
type gaskets. A sintered stainless-steel disk confines the fine, unconsolidated
sand packing. The column is filled with silica sand, elutriated to obtain a size
range of 20μm to 80μm. The pore volume of the model was determined to be
42.0 cubic in. (688 cubic cm) by fluid displacement tests, and the porosity was
calculated to be 39.2 percent. The model is mounted in a constant temperature
chamber in which air is circulated to maintain the temperature of the model.
The fluid entering the column is filtered through two different types of filters
in series (not shown). The first is a sand filter and the second, a millipore
filter.

A piston-type transfer cylinder is used for the injection of fluid into
the model through a type-304 stainless steel transfer cylinder, 800 cc, manufactured by Welker Engineering Company. The cylinder is located inside the
temperature chamber to equilibrate the contents to the test temperature prior
to injection. Fluid is injected into the model by pumping hydraulic oil into
the transfer cylinder. Mixing of the hydraulic oil with the injection fluid is
eliminated by a sliding piston built into the cylinder.

For the injection, a modified Ruska pump was used. The body of
the pump was manufactured by the Ruska Instrument Company of Houston,
Texas. It is equipped with an electronically controlled drive train developed at
New Mexico State University. It can be controlled to run in either a constant
rate, a constant pressure or a constant differential pressure mode.

A sample system is provided to enable the model to operate unat-
tended. The system consists of a rotary valve with sample containers and a
gas collection system connected to a wet test gas-flow meter. The rotary valve
is a six-position piston-operated valve which is available from Pierce Chemical
Company or Laboratory Data Control Division of Milton Roy Company (catalog number 4142530). The piston is operated with 85 psi nitrogen controlled
by a multiport solenoid valve. The solenoid valve is operated by an electrical timing clock. The clock can be set for intervals ranging from 1/4 hour
to several days. Typically, samples are taken every 2 to 12 hours depending
on the flow rates. The gas flow is measured by passing it from the sample
containers through a manifold and then through a laboratory type precision
wet-test meter.

Test Procedure.   The test procedure followed in this study consists of saturating the model with brine and then oil to produce a condition similar to that

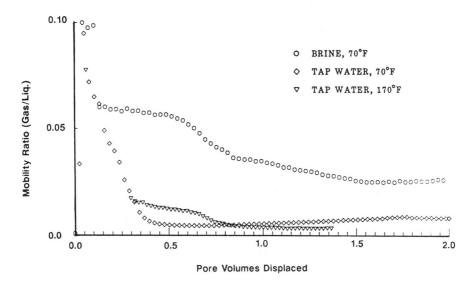

Figure   5.   Mobility ratios for EXXON LD 776-52.

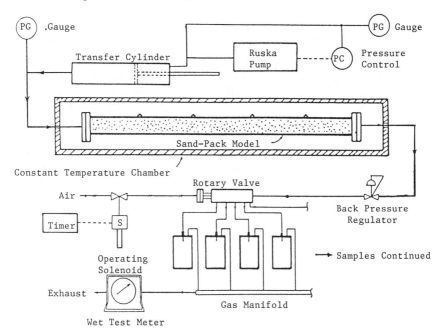

Figure   6.   Flow apparatus for displacement tests.

encountered in a virgin oil reservoir. The model is then flooded with brine or water to recover primary production plus the oil produced by a waterflood. In mobility tests, brine containing a mobility control additive is then injected; the delay is to insure no additional oil could be attributed to the presence of the additive in the brine phase during a conventional waterflood. A carbon dioxide flood is next initiated and continued to carbon dioxide breakthrough. Brine is then injected to recover the tertiary oil by a second waterflood. In some cases this step is not needed.

The phase behavior of supercritical carbon dioxide with most crude oils is such that miscibility of the two fluids occurs above a certain pressure. Above this pressure, if crude oil is contacted with carbon dioxide, it is dissolved and swept forward, and complete recovery of the oil can be accomplished in the regions where good fluid contact is achieved. As the API gravity of the crude decreases, the pressure needed to achieve miscibility increases. Most crude oils are not totally miscible with supercritical carbon dioxide. They require a certain concentration of light hydrocarbons to achieve miscibility. This is accomplished by multiple contacting of the carbon dioxide with the residual oil phase that gradually extracts some of the more volatile organic constituents. This process produces an oil-rich gas which slowly passes through the bed. Oil subsequently contacted by the mixture is miscible, and almost complete displacement of oil can then be accomplished. In contrast, if carbon dioxide is injected at a pressure below the miscibility pressure, the carbon dioxide will dissolve into the oil resulting in a thinning and swelling of the oil, but no solvent effect will occur. Both the miscible and immiscible processes are considered commercial and several large field applications have been reported[11].

Two different oils were used. A heavy crude chosen was from the Los Angeles Basin, Wilmington Reservoir, and had a gravity of 14° API. A light crude was represented by n-octane which has a gravity of 71° API. These were chosen because they interact differently to high-pressure carbon dioxide. At 100°F, (38°C), n-octane is miscible with carbon dioxide at pressures above 1080 psi. In contrast the heavy crude is not miscible with carbon dioxide at pressures below 2500 psi. Thus, two significantly different cases are represented by the two crudes. The n-octane was tested at 1300 psi to investigate if surfactants affect the miscible displacement process, and the Wilmington crude was tested at 1000 psi to provide similar information for the immiscible case.

In these displacement studies it was necessary to limit the presssure drop in the test core to a value comparable to that encountered in reservoir, about 1 psi/ft. Excessive pressure gradients can artificially cause oil to flow as a consequence of providing sufficient force to overcome the lowered interfacial tension produced by the surfactants. The swelling of the oil accompanied by a significant decrease in viscosity makes it possible to follow carbon dioxide with water injection to effectively recover additional oil in the immiscible oil displacement process. Tests were performed in this manner to evaluate the effect of mobility control agents on this process.

Results of Immiscible Displacement Tests.   The results of the tests show that immiscible carbon dioxide flooding followed by waterflooding is effective in increasing the oil recovered from a core. The oil recovered by a conventional waterflood was equal to about 30.4% of a pore volume, PV. Immiscible carbon dioxide flooding increased the recovery to a total of 50.5% PV. The addition of a mobility control agent increased the recovery further to 58.3% PV; this amounts to 39% additional tertiary oil due to the effectiveness of the mobility control in the carbon dioxide immiscible process.

Plots of the cumulative oil produced versus the cumulative volume injected are shown for core floods without and with mobility control; see Figures 7 and 8. Note that in the immiscible flood without mobility control, Figure 7, little additional oil was produced during the carbon dioxide injection. Almost all of the tertiarty oil occurred during the final waterflood. In contrast, the effect of mobility control is reflected in greater oil production during carbon dioxide injection as shown in Figure 8, where 0.5% Pluronic F68 in brine was used.

The average oil produced in the two initial waterfloods was 305 cc. With an initial oil volume of 870 cc, the recovery in the initial waterflood phase is 35% of the original oil in place. The total oil produced in the case of the conventional carbon dioxide flood was 510 cc which is 58.1% of the initial oil. When mobility control was used the total oil recovered increased to 585 cc which represents 67.4% of the original oil. A summary of the results are shown in Table II.

Table II - CO_2 Flood Results

Process	Oil Recovery, PV	% Original Oil
Waterflood Only	0.304	34.8
Water, CO_2, Waterflood	0.505	57.9
Water, Mobility Controlled CO_2, Waterflood	0.583	67.4

NOTE: Original oil saturation was 0.86 PV (Pore Volume).

The tests cannot be extrapolated directly to field reservoir performance because the spatial geometry is different. The core tests have a linear, 1-dimensional flow geometry while the actual reservoir has radial, 3-dimensional flow. In 3-dimensional flow the displacement efficiency is typically less than that measured in linear displacement studies. Chilton (1987)[10] showed in his computer simulation studies that, as compared with the linear flow case, the predicted oil produced was 10% less for the two-dimensional model and 27% less for the three-dimensional model. However, when mobility control was used with a tenfold decrease in carbon dioxide mobility, the calculated improvement in displacement efficiency was much less for the linear case than the three-dimensional case. This result indicates that the increase in displacement efficiency under field conditions should be greater than that recorded in these linear laboratory tests.

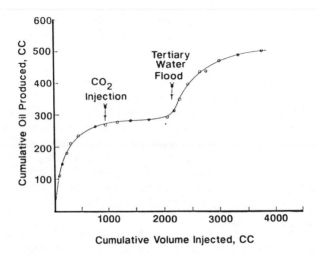

Figure   7.   Conventional immiscible CO_2 flood.

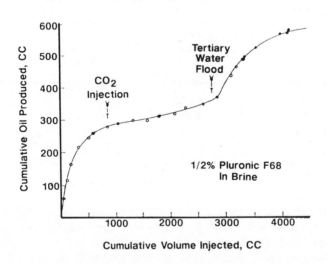

Figure   8.   Immiscible CO_2 flood with mobility control.

Miscible Displacement Tests

The purpose of the miscible displacement tests was, primarily, to evaluate the effect of the surfactant on the interaction between the high pressure CO_2 and the crude oil. The miscible displacement process is so efficient in linear laboratory tests that any improvement would be minor. However, any interference with the mechanism of miscibility should be detrimental to the recovery efficiency and easily observed.

Equipment and Procedure.   The tests in this study were run in the same model as the immiscible tests, for comparison of the two cases.

In these tests the same procedure was followed as that used in the immiscible case. In order to assure complete miscibility of the oil and CO_2 phases n-octane was used to simulate a typical light crude oil.

The analytical procedure was modified from that of the previous study. Since the octane is colorless, a dye was added. The dye chosen was Red LS672, a red dye used to color gasoline for market. It was found to be stable and have no measurable solubility in the water phase. The colored octane was then analyzed colorimetrically in a similar manner to that used for the heavy crude oil. The dye was obtained from Tricon Colors, Inc., Elmwood Park, N. J.

Three different types of surfactants, Pluradyne SF-27, Stepanflo-50 and Exxon LD 776-52, were evaluated. All are commercially available at present. Specific information on these surfactants is given in Table III. In addition to the tests with brine as shown in the table, the Exxon LD 776-52 was also tested with tap water instead of brine. The additional test with LD 776-52 in tap water was chosen as a consequence of the results of the dynamic flow test that indicated a significant influence of salinity on its performance.

Table III - Surfactants used in Miscible Displacement Tests

Surfactant	Company	Chemical Structure	Solution
Pluradyne SF-27	BASF Wyandotte Corp.	Ethoxylated alcohol	0.5% in brine
Stepanflo-50	Stepan Chemical Co.	Sulfate ester of ethoxylated alcohol	0.5% in brine
Exxon LD 776-52	Exxon Chemical Co.	Sulfonate	0.5% in both brine and tap water

Results of the Miscible Displacement Tests.   The results of the tests are shown in Figures 9-12. Referring first to Figure 9, the CO_2 flood of a waterflooded light oil reservoir, it can be seen that the waterflood, represented by the data prior to CO_2 injection, recovered 0.5 PV of oil. This amount represented 76% of the oil initially in place and produced a residual oil saturation

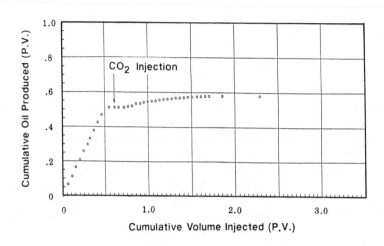

Figure   9.   Miscible carbon dioxide flood.

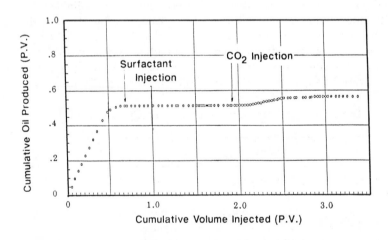

Figure 10.   Stepanflo-50 mobility controlled carbon dioxide flood.

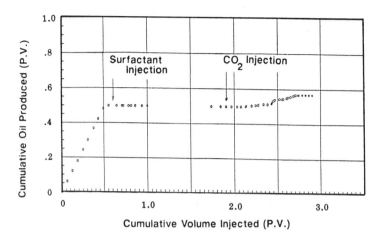

Figure 11.   Pluradyne SF-27 mobility controlled carbon dioxide flood.

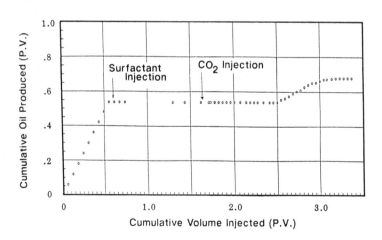

Figure 12.   Mobility controlled carbon dioxide flood using EXXON LD 776-52 in tap water.

at the end of a waterflood of 0.16 PV. Both the recovery and the residual saturation are consistent with values expected in a waterflood of a light crude oil.

Injection of 0.6 PV of carbon dioxide produced only 6% of a pore volume of additional oil. Linear laboratory miscible floods normally produce over 10% additional oil. The lower value of oil recovery reflects the inefficient macroscopic oil displacement induced by an unfavorable mobility ratio that exists between carbon dioxide and the oil. This phenomenon has not been reported for experiments conducted in long, slim tube models, where their smaller diameter, 0.2 in. vs. 1.6 in., masks the effect.

The results presented in Figure 10 were obtained using Stepanflo-50 as a mobility control additive. It is a sulfated ethoxylated derivative of a linear alcohol. It is one of the most effective additives identified in this program. As was the case in the conventional CO_2 flood, the preliminary waterflood produced 0.5 pore volumes of oil. Prior to injecting carbon dioxide, the interstitial water was replaced by injecting a brine solution containing 0.5% Stepanflo-50 at a low presure drop, 2 psi/ft. This surfactant does not reduce interfacial tension between n-octane and brine enough to effect displacement of residual oil at this gradient. This is verified by the long period of surfactant injection without any oil production.

A comparison of the Figures 9 and 10 shows almost identical results in oil recovery for the two cases.

It is apparent that the addition of the mobility control agent did not affect the miscibility process, as evidenced by the comparable oil recovery to that of the no mobility control case. At the same time the mobility of carbon dioxide was decreased more than twenty-fold. This important finding demonstrates that the mobility control method produces effective mobility control without adversely affecting the miscible recovery mechanism.

The results obtined for mobility control additive, Pluradyne SF-27, are shown in Figure 11. Pluradyne SF-27 is a blend of ethoxylated alcohols having chain lengths varying between C_{11} and C_{15}. It is superior to all other ethoxylated alcohol additives tested to date. However, additives having a similar chemical structure are also effective at reducing the mobility of carbon dioxide. The results of this test are very similar to those of the Stepanflo-50 test and support the conclusion that mobility control does not adversely affect the miscible interaction between CO_2 and oil.

A significant improvement in tertiary oil recovery was obtained using Exxon's LD 776-52. This additive is a synthetic sulfonate whose performance in brine is only average. Screening tests conducted in tap water containing some calcium ion indicated that its potential for mobility reduction of gas flow in oil reservoirs flooded with freshwater was significantly higher than any sulfonate type additive previously investigated. As shown in Figure 12, the waterflood produced about the same amount of primary plus secondary oil as observed in the other three experiments. However, the mobility controlled CO_2 flood increased oil recovery substantially. Of the approximately 0.16 PV of oil which is considered residual oil, 95% was recovered when Exxon LD

776-52 was used with CO_2 as compared with an average of 43% recovery for the previous three tests. Table IV shows a summary of the results of the flood tests, where S_{oi} is the initial oil saturation, S_{owf} is the oil saturation after waterflooding and S_{oCO2} is the oil remaining after the CO_2 miscible flood.

Table IV - Mobility Control Flood Results

	Conventional CO_2 Flood	Exxon LD 776-52	Pluradyne SF-27	Stepanflo-50
S_{oi}; cc	464.	465.	477.	445.
S_{oi}	0.67	0.68	0.69	0.65
WATERFLOOD				
Oil Produced:				
cc	352.	364.	342.	350.
PV	0.51	0.53	0.50	0.51
S_{owf}	0.16	0.15	0.19	0.16
CO_2 FLOOD				
Oil Produced:				
cc	46.	98.	48.	34.
PV	0.067	0.142	0.070	0.049
S_{oCO_2}	0.09	0.01	0.12	0.11

In none of the tests was there any indication that surfactants exert any adverse effect on the interaction of the CO_2 and crude oil phases to achieve miscibility and produce a high level of recovery enhancement.

Conclusions

The results of this study show that the static foam tests are helpful for screening candidate additives for effectiveness as mobility control agents. It was found to be important to consider chemical structure in the interpretation of foam test results. In general, surfactants which had greater foamability and foam stability were better mobility control agents; but, when surfactants having different basic chemical structures were compared, some moderate foamers were as good or better than excellent foamers of a different structure type. No poor foamers were found to be good mobility control additives.

Dynamic foam tests and the displacement tests are needed to complete the screening of the candidate surfactants. Good correspondence was obtained between the two tests.

The displacement tests show that a significant increase in tertiary oil production can be achieved when one of the better additives is utilized in 1-dimensional laboratory immiscible carbon dioxide floods. In the miscible displacement case no adverse effect on the miscibility process was found. In addition, the demonstrated CO_2 mobility reduction indicates improved areal and vertical conformance will be achieved. This greater reservoir coverage would yield increased recovery providing the pressure required for miscible displacment the oil can be maintained.

An overall comparison of the results of tests using brine indicates that ethoxylated alcohols and ethoxylated alcohol sulfates were superior for mobility control in enhanced oil recovery. It was found that the optimum hydrophobe for ethoxylated alcohol is a mixture with an average chain length of about 13 carbon atoms. The degree of ethoxylation is also important. It was found that 6 ethoxylate groups is close to the optimum. The formation of the sulfate ester tended to lessen the effect of chain length and reduce the optimum number of ethoxylate groups to 5.

The effectiveness of the sulfonate, LD 776-52, in low salinity injection water indicates the advantage of a carefully chosen structure for mobility control. The result suggests the possibility of improving the additive performance by tailoring molecular structure.

Literature Cited

1. Shelton, J.L. and Schneider, F.N.: "The Effects of Water Injection on Miscible Flooding Methods Using Hydrocarbons and Carbon Dioxide," SPEJ. June 1975, 217-26
2. Bursell, C.G. and Pittman, G.M.: "Performance of Steam Displacement in the Kern River Field," J. Pet. Tech. Aug. 1975, 997-1004.
3. Smith, R.L.: "Sacroc Initiates Landmark CO2 Injection Project," Pet. Eng., Dec. 1971, 4347.
4. Kantar, K. and Conner, T.E.: "Turkey's Largest Oil Field Poised for CO2 Immiscible EOR Project," Oil and Gas Journal, Nov. 26, 1984.
5. "Prudhoe Bay Producers Expand to Enhanced Oil Recovery, " Oil & Gas Journal, Jan. 16, 1984, 47.
6. Langston, E.P. and Shirer, J.A.: "Performance of Jay/LEC Fields Unit Under Mature Waterflood and Early Tertiary Operations," SPE 11986, 58th Annual SPE Meeting, San Francisco, CA, Oct. 5-8, 1983.
7. Bernard, G.C. and Holm, L.W.: "Method for Recovering Oil from Subterranean Formations," U.S. Patent No. 3,342,256, 1967.
8. Dellinger, S.E., Patton, J.T. and Holbrook, S.T.: "Carbon Dioxide Mobility Control," SPE Journal, v. 24, Apr., 1984, 191-196.
9. Holm, L.W.: "Foam Injection Test in the Siggins Field, Illinois," J. Pet. Tech., v. 22, 1970, 1449-1506.
10. Chilton, W.D.: "Computer Simulation of Mobility Control in CO_2 Enhanced Oil Recovery," Master's Thesis, New Mexico State University, Las Cruces, NM, 1986.
11. Patton, J.T., and Saner, W.B.: "CO_2 Recovery of Heavy Oil: The Wilmington Field Test," Society of Petroleum Engineers of AIME, 58th Annual Technical Conference and Exhibition, San Francisco, California, 10-5-83.

RECEIVED January 29, 1988

Chapter 21

Use of Crude Oil Emulsions To Improve Profiles

T. R. French

IIT Research Institute, National Institute for Petroleum and Energy Research, P.O. Box 2128, Bartlesville, OK 74005

Petroleum production from subterranean reservoirs can be increased by injecting water as liquid or steam. Various chemicals have been added to the water or steam to increase volumetric sweep efficiency. One alternative is the use of emulsions which serve as diverting agents to correct the override and channeling problems that occur during fluid injection. Laboratory results show that it may be possible to control channeling and steam override with an emulsion blocking technique. The emulsion can be formed with the aid of a surfactant mixture or by use of natural surfactants that exist in some crude oils. Core-flood experiments to demonstrate the effectiveness of flow diversion by emulsions were performed at temperatures ranging from ambient to 194°C. The resulting permeability reductions of 25 to 95% in cores and sandpacks indicate increased sweep efficiency.

Premature breakthrough of the injected fluids often results in much of the oil in a subterranean reservoir being by-passed (see Figure 1). Thief zones can result from reservoir heterogeneity and/or unfavorable fluid properties. In the case of steam, and steam-CO_2 co-injection ($\underline{1}$), gravity segregation of the low-density steam and the high-density reservoir fluids results in steam override near the top of the reservoir (see Figure 2). Cores taken from reservoirs at the termination of a steam drive clearly show depleted oil saturations in the upper reservoir zones.

An effective method to increase the efficiency of a fluid drive is to plug the thief zone(s), direct fluids to areas of higher oil saturation and thus improve the ratio of oil produced to fluid injected. Emulsions have been shown to be effective permeability modification agents, but little work has been reported on their use.

0097-6156/88/0373-0405$07.00/0
© 1988 American Chemical Society

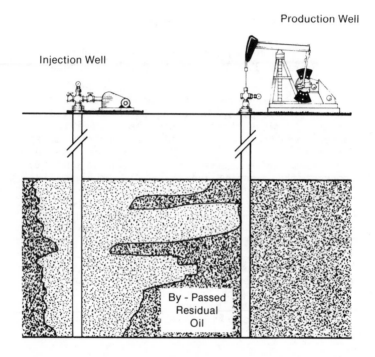

Figure 1.   Premature breakthrough during waterflood.

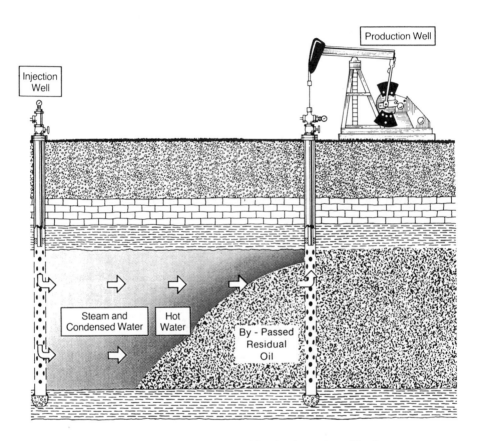

Figure 2.   Gravity override during steamflood.

Literature Background

Review of the literature resulted in several references relating to the use of emulsions as agents for causing permeability reduction. McAuliffe (2) demonstrated that injection of externally produced oil-in-water emulsions at 24°C effectively reduces the water permeabilities of sandstone cores. These laboratory findings were later substantiated by a field test of emulsion injection followed by waterflooding in the Midway-Sunset Field (3).

Several waterflood recovery process patents assigned to Texaco, Inc., describe laboratory core studies in which blocking emulsions were produced in situ by injection of surfactant mixtures (4-6). The conditions under which these experiments were performed (relatively high salinity, the presence of divalent ions, crushed limestone cores) allow us to infer the creation of water-in-oil emulsions. Two later patents claim the procedure also applies to profile improvement for steamflooding (7-8). Cooke (9), in laboratory experiments, has observed that a viscous oil-external emulsion may be responsible for the large increase in pressure gradient that is observed immediately behind the displacement front during alkaline waterflooding under saline conditions. A recent waterflood patent assigned to Mobil Oil Corp. relates to the creation of a plugging emulsion within a high salinity stratified reservoir (10). The surfactant/water/oil emulsion described in this patent is likely oil external.

Emulsion Properties. In most cases, the type of an emulsion (oil-in-water or water-in-oil) can be predicted by the appearance of the emulsion. In general, oil-in-water emulsions will appear to be chocolate or brown and will dilute easily with water, although creaming will occur eventually if agitation ceases. The oil droplets in macroemulsions normally are between 1 and 50 microns in diameter and are easily visible with an optical microscope. Emulsions produced via the natural surfactants that occur in some crude oils usually have more uniform droplet sizes and are quite stable (11). For blocking (permeability reduction) purposes, the droplet size distribution is an important property of the emulsion. If the droplets are too small, they may tend to slowly solubilize into the continuous phase or not block at all, and if they are too large, creaming and coalescence may become problems. The viscosity of oil-in-water emulsions remains low enough to be pumped easily. The electrical conductivity of oil-in-water emulsions tends to be that of the aqueous phase.

Water-in-oil emulsions, generally appear to be black, do not dilute with water, and have electrical conductivity lower than that of the brine. The viscosities may be very high, and the shear behavior is thixotropic.

Emulsion Formation. Oil-in-water emulsions can be produced by agitating oil with an aqueous solution of emulsifier (agent-in-water method) or by utilizing the naturally occurring surfactants already present in some oils (agent-in-oil method). Either method is suitable for creation of the emulsion above ground and injection of that emulsion into the reservoir.

In situ formation of oil-in-water emulsions adds the requirement that the emulsification proceed spontaneously or at least with very little energy input due to mixing. Most such systems are associated with the agent-in-oil procedure, and spontaneous emulsification to oil-in-water emulsions sometimes occurs when aqueous caustic is mixed with petroleum oils containing naphthenic acids. Some researchers propose that mass transfer of the naturally occurring surfactants across the interface is the mechanism that causes this phenomena (12,13,14).

Emulsification with caustic is possible with oils that have a fairly high total acid number (TAN). Below about 1.5 mg KOH/gm oil, the oils either will not emulsify or will form water-in-oil emulsions. The rate of emulsification with caustic is much faster than emulsification with surfactant mixtures, which is a characteristic property for emulsions generated via the agent-in-oil procedure (15).

Blockage Mechanisms. McAuliffe's (2) concept of the mechanism by which an oil-in-water emulsion can cause a permeability reduction is shown in Figure 3. In this case, the oil droplet is large enough to cause blockage by lodging within the pore throat. In this situation, the flow of a dilute, stable emulsion in a porous media is similar to a filtration process. If the pressure gradient across the drop becomes great enough, the drop may be forced on through the pore throat. Another process for reducing permeability has been observed by Soo and Radke (16). When emulsions are injected into a porous media micro-model, drops not only block pores of throat sizes smaller than their own, but they are also observed to adhere to pore walls and to be trapped in crevices (16).

It is important to observe that a reduction in permeability from emulsion plugging may not necessitate that the median droplet size equal or exceed the median pore throat diameter. Competition from an ensemble of smaller droplets "crowding" into a single pore throat would have the same effect in blocking a pore throat as would one large droplet, as shown in Figure 3. Another important (but speculative) mechanism of emulsion plugging to consider is the decrease in relative permeability of the fluid phase due to the presence of an additional competing emulsion phase. Here again, emulsion droplets smaller than the median pore throat size in the porous structure would possibly play a role in the overall blocking mechanism.

Finally, permeability reductions attributed in the literature to the formation of water-in-oil emulsions are evidently caused by the high viscosity of those emulsions or to the formation of an oil film (lamella) across the pore throat (9).

Experimental

Study of Emulsification. The emulsions used for this experimental study were the oil-in-water type. Water external emulsions are easily injected, stable at the low salinities encountered in steamflooded reservoirs, and expected to produce more stable blockage at elevated temperatures (17).

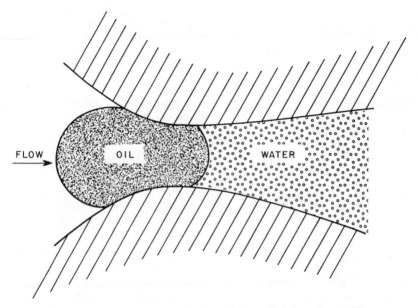

Figure 3.   Pore throat with oil droplet blocking flow.

The crude oils selected for this study are an 18.4° API California crude from the Wilmington field, an 11.5° API California crude from the Kern River field, and a 33° API mid-continent crude from the Delaware-Childers field in Oklahoma.

Emulsions were tested by mixing an emulsifier (either caustic or commercial surfactants dissolved in water) and the crude oil, then heating the sealed container to 110°C in an oven. After heating, the sample was removed from the oven and placed in a mechanical shaker for 15 to 20 minutes, then returned to the oven. This procedure was repeated three times before the sample was left in the oven for observation.

Emulsification With Caustic. Emulsification of the Wilmington oil with caustic at ambient and elevated temperatures proceeds almost spontaneously. The stability of 50% oil-in-water emulsions produced with Wilmington oil and caustic is given in Table I. These oil-in-water emulsions are quite stable at 110°C for long periods (weeks) of time. The optimal concentration of sodium hydroxide occurs at 0.42% NaOH where a uniform single-layer oil-in-water emulsion is produced. At higher NaOH concentrations, the increasing ionic strength of the solution results in formation of upper layer water-in-oil emulsions.

Oils obtained from the Kern River field also emulsify readily with caustic at ambient temperatures; however, heating to 43°C causes the emulsification to be more rapid. Water and sediment constitute 50% of the produced crude oil. Treated oil (generator feed) is also available in the field which contains only 2% water and, consequently, a higher TAN (total acid number) than the produced crude oil (see Table II). The treated Kern River oil forms an extremely stable 75% oil-in-water emulsion at an optimal 0.50% NaOH concentration.

Figure 4 reveals that the interfacial tension between Kern River crude oil and caustic becomes vanishingly small near the optimal (as determined by the bottle tests) caustic concentration. Preparing the NaOH solution with softened brine results in a qualitatively more stable emulsion than use of unsoftened brine.

The total acid numbers (TAN) and the experimentally determined optimal NaOH concentrations for Kern River, Wilmington, and other viscous, asphaltic crude oils are given in Table III.

Emulsification With Surfactant Mixture. Oils with low TAN such as the Delaware-Childers crude cannot be emulsified with caustic. Attempts were made to produce an oil-in-water emulsion which is stable at 110°C by using petroleum sulfonates of different average equivalent weights as the emulsifying agent. The Delaware-Childers oil was mixed with the aqueous emulsifier at a ratio of one part oil to 6 parts water and tested according to the procedure described previously. If all of the oil emulsified, this would correspond to a 14.3% oil-in-water emulsion. Results of emulsification tests performed with petroleum sulfonates at 3.75% active concentration and Delaware-Childers oil are given in Table IV. The largest volume of stable oil-in-water emulsion was obtained for the mixture of 3.75% Petrostep 420®, a medium equivalent weight petroleum sulfonate, and 2.5% SE-463, an alkylaryl polyether sulfonate, in 4.25% NaCl. The optimal system consisting of 3.75% Petrostep 420®

TABLE I. Emulsification of Wilmington Crude Oil with Caustic

NaOH(%)	Vol (ml.) Oil	Vol (ml.) Aqueous	Apparent emulsion type Upper	Apparent emulsion type Lower	Room temperature Upper	Room temperature Lower	4 hrs. at 110°C Upper	4 hrs. at 110°C Lower	8 days at 110°C Upper	8 days at 110°C Lower
0.00	3.0	3.0	w/o	w	5.0	1.0	3.3	2.7	3.1	2.9
0.03	3.0	3.0	o/w	o/w	5.5	0.5	5.3	0.7	3.6	2.4
0.08	3.0	3.0	o/w	o/w	4.5	1.5	4.1	1.9	3.5	2.5
0.17	3.0	3.0	o/w	o/w	5.1	0.9	4.7	1.3	3.8	2.2
0.25	3.0	3.0	o/w	o/w	4.8	1.2	4.4	1.6	0.4	5.6
0.42	3.0	3.0	o/w	o/w	5.7	0.3	0.0	6.0	0.1	5.9
0.58	3.0	3.0	w/o	o/w	2.7	3.3	2.8	3.2	2.8	3.2
0.75	3.0	3.0	w/o	o/w	3.5	2.5	3.2	2.8	1.7	4.3
1.00	3.0	3.0	w/o	o/w	3.5	2.5	3.2	2.8	3.0	3.0

W/O - water-in-oil
O/W - oil-in-water

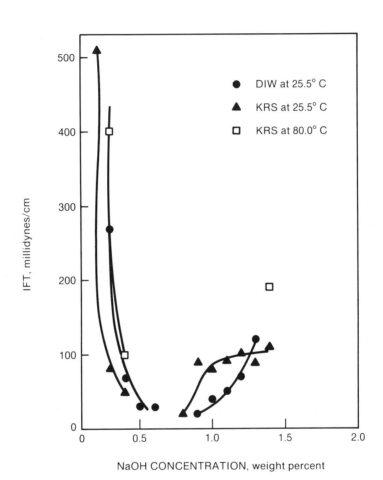

Figure 4. Interfacial tension - Kern River oil/caustic. DIW = deionized water; KRS = softened Kern River brine.

TABLE II. Kern River Oil

	Oil Produced	Oil Treated
Specific Gravity @ 60°/60° F	0.9892	0.9868
A.P.I. Gravity, degrees	11.5	11.9
Kinematic Viscosity @ 130° F, centistoke	–	1374.5
Bottom Sediment and Water, Vol. %	50.0	2.0
Total Acid Number	1.331	2.364

TABLE III. Emulsification of Asphaltic Crude Oils with Caustic

Oil	Gravity (°API)	TAN (mg KOH/gm oil)	Water	Optimal NaOH Concentration (%)
Wilmington 5G	18.4	2.04	deionized	0.42
Wilmington B66099	11.9	2.31	deionized	0.18
Midway-Sunset B76067	14.4	4.15	deionized	0.42
Hasley Canyon B77023	11.1	0.68	deionized	None
Kern River 85GL392	11.5	1.33	reservoir brine	0.83
Kern River 85GL1417	11.9	2.36	reservoir brine	0.50
Freeman	16	2.80	deionized	0.25
Chaffee	19.5	1.40	deionized	0.25

TABLE IV. Static Emulsification Tests Performed with Surfactants
and Delaware-Childers Oil[*]

Petroleum Sulfonate	Average eq. wt.	Ethoxylated sulfonate	NaCl (%)	Volume, oil-in-water emulsion 110°C (16 hours elapsed) (ml)
Witco-TRS 40® (3.75%)	335	None	0 2.12 4.25	5.9 6.0 5.9
Witco-TRS 40® (3.75%)	335	[**]SE-463 (2.5%)	0 2.12 4.25	6.0 6.0 6.0
Petrostep-420® (3.75%)	420	None	0 2.12 4.25	5.9 6.2 0
Petrostep-420® (3.75%)	420	SE-463 (2.5%)	0 2.12 4.25	6.0 6.1 6.9 (optimum)
Witco-TRS 18® (3.75%)	495	None	0 2.12 4.25	0 0 0
Witco-TRS 18® (3.75%)	495	SE-463 (2.5%)	0 2.12 4.25	6.5 0 0

[*](7 ml Samples)
[**]supplied by GAF Chemical Co.

and 2.5% SE-463 was particularly stable, and emulsions containing up to 33% Delaware-Childers oil were easily prepared.

The emulsion produced with surfactant mixtures (and also those produced via caustic) would eventually separate into two distinct layers. At low salinities, the lower layer consists of a stable oil-in-water emulsion. At higher salinities inversion occurs, and the lower layer separates as a clear liquid with no oil droplets-- the upper layer then becoming a water-in-oil-emulsion.

Particle Size. After determining, with bottle tests, which systems easily produced stable oil-in-water emulsions, the droplet size distributions for the oil-in-water emulsions were determined with a Model TA II Coulter Counter. The quantitative results obtained with the Coulter Counter were verified by qualitative observations with an optical microscope. The droplet size distributions for several oil-in-water emulsions are given in Figure 5. A qualitative correlation between droplet size and emulsion stability was observed. The smaller the median droplet size, the more stable was the emulsion. The pore size distribution for a 300-md Berea sandstone core is given for comparison.

Emulsification Properties of Crude Oils. Although impossible to predict with certainty which crude oil systems will easily emulsify to form stable O/W emulsions, some general rules were observed to apply:

1. Oils with high acid numbers emulsify with caustic, some spontaneously.

2. Heavy oils in general emulsify with cationic surfactants.

3. Heavy oils often form stable emulsions.

4. Light oils emulsify with anionic surfactants; however, substantial mechanical energy is usually required.

5. Light oil emulsions usually cream faster than emulsions made with heavy oils.

6. Oil-in-water emulsions are usually more stable in low salinity reservoirs.

Coreflood Test Procedure. Laboratory coreflood experiments were performed to test the effectiveness of emulsion blocking in improving sweep efficiency at elevated temperatures. The emulsions, prepared as previously described, were diluted before injection into the cores. The emulsion reservoir was stirred slowly to prevent the dispersed oil droplets from creaming. Creaming was more of a

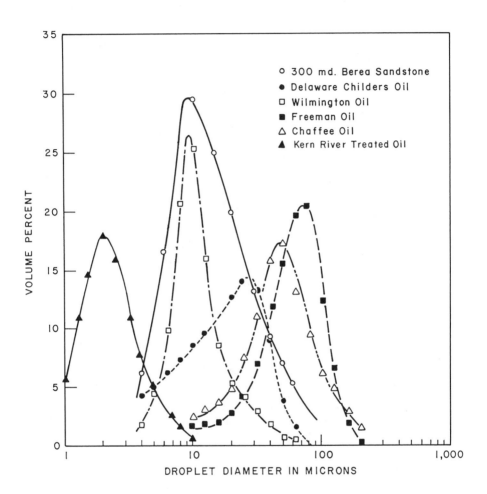

Figure 5. Droplet size distribution for emulsions.

problem with the light oil emulsion than with the heavy oil
emulsions.
    Berea sandstone cores (25.4 cm by 3.8 cm) used in the
experiments with Wilmington and Delaware-Childers crude oils were
fired at 427°C. After firing, the cores were saturated with brine,
mounted in a Hassler type core holder, and placed in a temperature-
controlled oven.   After initial absolute permeability was
determined, the cores were left either oil-free or saturated with
oil and waterflooded to residual oil saturation.
    The core experiments with Kern River oil were performed using
sandpacks (25.4 cm by 3.7 cm) made from unconsolidated field core.
The core material was packed into Teflon sleeves with Teflon end
caps and then placed into a Hassler type core holder.   Before
packing, the sand was cleaned by Soxhlet extraction with toluene and
was not fired. Otherwise, the procedure was similar to that for
saturating the consolidated Berea sandstone cores.
    Fluid injection, pressure monitoring, and temperature were
controlled by an HP85 microcomputer system. Injections were done at
constant flow rate with a Constametric III metering pump, from which
the filters were removed.

Coreflood Tests With Oil-Free Cores. The coreflood experiments were
at first performed at ambient temperature and then extended to hot
water conditions at 110°C as an approach to saturated steam
conditions. Pilot experiments with the light mid-continent crude
were extended to the heavier California crude oils, with steamfloods
at saturated steam conditions to test the steam stability of
"emulsion blocks" created with the heavier oil. The data for these
coreflood tests are summarized in Table V.
    Figures 6 and 7 illustrate the effects on effective permeability
to water from injecting the 0.5% oil-in-water emulsions created from
Delaware-Childers crude oil.   At 25°C, a 68% reduction in
permeability occurred after injecting 9.5 PV of emulsion. At 90°C,
10 PV of emulsion resulted in a 95.2% reduction in permeability,
with most of the reduction occurring within 2 PV.
    Figures 8, 9, and 10 show the permeability reductions that
resulted from injecting 0.5% emulsions produced from the Wilmington
crude oil and caustic.   The temperatures are, respectively, 25°,
90°, and 110°C. The reductions in permeability were from 84 to 88%,
with most of the reduction occurring within 1 PV of emulsion
injection.
    These experiments were conducted at constant flow rate.
Blocking effects at constant pressure (more similar to field
conditions) would probably show more dramatic effect. In all of the
experiments with injected emulsions, the effective permeability to
water was decreased far more than an equivalent amount of residual
oil would have reduced the permeability. The emulsion droplets are
more efficient in reducing the effective permeability of a core than
is the same amount of oil that is not emulsified.
    Similar results were obtained when injecting an externally
produced emulsion into a core which was subsequently steamflooded.
The results (Table V) show that the reduction in permeability caused
by the emulsion (created with Wilmington oil) was stable at
steamflood conditions.   This experiment was conducted with a 25-in.
core and saturated steam at 160°C.   Before emulsion injection,

TABLE V. Injection of Externally Produced Emulsion into Oil-Free Cores

Temp. (°C)	Emulsion Concentration (%)	Emulsion Injected (PV)	Permeability Absolute (md)	Final (md)	Reduction (%)	Flow rate (cc/min)
Delaware-Childers Oil (Light)						
25	0.5	9.5	219	71	68.0	3.4
90	0.5	10.0	148	7	95.2	3.4
Wilmington Oil (Medium)						
25	0.5	8.9	266	31	88.3	3.4
90	0.5	9.9	74	12	84.3	3.4
110	0.5	8.5	187	26	86.1	3.4
160	0.5	6.5	90	21	76.7	5.0
Kern River Oil (Heavy)						
150	2.5	1.0	1624	406	75.0	2.0

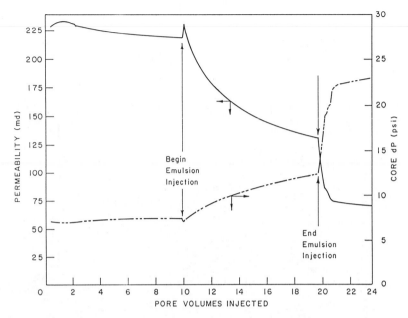

Figure 6. Permeability reduction - 25 °C Delaware-Childers oil.

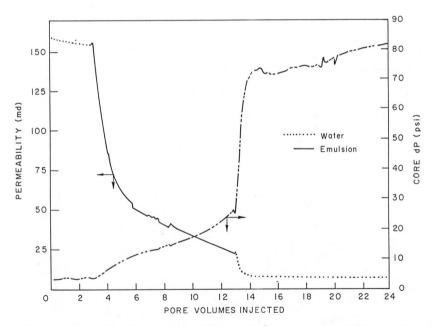

Figure 7. Permeability reduction - 90 °C Delaware-Childers oil.

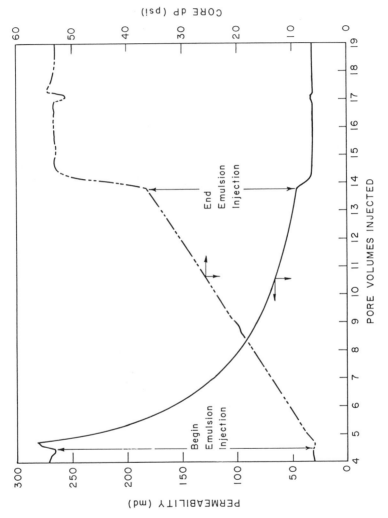

Figure 8. Permeability reduction – 25 °C Wilmington oil.

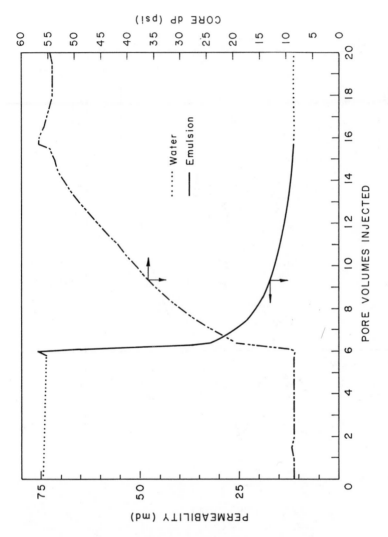

Figure 9.   Permeability reduction – 90 °C Wilmington oil.

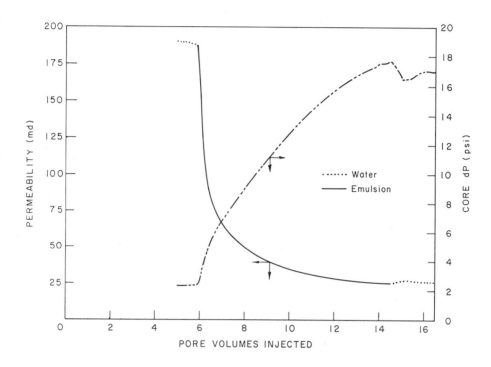

Figure 10.   Permeability reduction - 110 °C Wilmington oil.

permeabilities were measured before and after steam to make sure the steamflood itself did not cause a permeability reduction.   A 77% reduction in permeability from emulsion injection was observed under these steamflood conditions providing strong evidence for the development and use of this type of blocking procedure in the field.

Injecting 0.5 PV of 2.5% Kern River oil-in-water emulsion into a sandpack made from Kern River sand reduced the permeability from 1624 to 397 md, a reduction of 76%.   This is a significant result since injection of 0.5 pore volume of a steam-swept zone is economically viable should a field test be performed.   An additional 0.5 PV of the emulsion was injected, lowering the permeability to 226 md.   The stability of the block under steamflooding conditions was tested by injecting saturated steam at 150°C.   After steam injection, the permeability was 406 md--still a 75% reduction in effective permeability (see Table VI).

TABLE VI.   Injection of 2.5% emulsion into Kern River sand

Absolute permeability (md)	1624
Permeability after 0.5 PV emulsion (md)	397
Permeability after 1.0 PV emulsion (md)	226
Permeability after 150°C steam (md)	406

Emulsion Injection Into Cores Containing Residual Oil.   These experiments were performed because of uncertainty about the effect of residual oil on an "emulsion block." In the case of residual oil remaining in the core, the effective permeability to water is much lower at the beginning of emulsion injection than with an oil-free core. The results for Delaware-Childers oil are summarized in Table VII.

In one experiment, emulsion injection was begun after waterflood, and in the others emulsion injection was begun after tertiary recovery. The number of pore volumes of emulsion injected was 10, 7, and 8, respectively. The reductions in effective permeability, 52, 33, and 56%, were significant, but not as high as those when using oil-free cores.

For the tests listed in Table VIII, the emulsions were injected after high-permeability cores were saturated with Wilmington oil and then steamflooded. In the first test, emulsion injection lowered the permeability by 91%; however, subsequent injection of steam at 194°C resulted in destruction of most of that emulsion block.   In

TABLE VII.   Injection of Externally Produced Emulsions into
Cores Containing Residual Oil

Delaware-Childers Oil - 25°C

Emulsion Concentration (%)	Emulsion Injected (PV)	Permeability				Flow rate (cc/min)
		Absolute (md)	Before emulsion (md)	Final (md)	Reduction (%)	
0.5	10.0	285	33	16	52	3.4
0.5	7.0	324	26	18	33	3.4
0.5	8.0	297	84	38	56	3.3

the second test, emulsion injection lowered the permeability by 85%, and that emulsion block was stable to the subsequent injection of steam at 177°C.

TABLE VIII. Injection of externally produced emulsions into cores containing residual oil. Wilmington Oil - after steam

Emulsion Injected (PV)	Absolute (md)	Before Emulsion (md)	Final (md)	Reduction (%)	Flow Rate (cc/min)
8.1	1436	862	79	91	3.4
8.2	823	744	108	85	3.4

In Situ Emulsification. Coreflood experiments designed to cause permeability reductions by in situ creation of oil-in-water emulsions have been less successful than injection of externally produced emulsions, but still show significant reductions in permeability. The data are summarized in Table IX.

The first two tests were performed in the usual manner of saturating the core with Wilmington oil and then waterflooding to residual oil saturation, resulting in oil saturation of 45 and 49%, respectively, before caustic injection. After injection of caustic slugs, the effective permeabilities were reduced 27 and 25%, respectively.

In the third test, the core was not saturated with oil before waterflooding, and the oil saturation was only 34%, resulting in higher initial permeability. Under this condition, the reduction in effective permeability increased to 43%. In all three tests, oil-in-water emulsions were produced from the core which had droplet size distributions appropriate to cause blockage of pore throats. These three tests illustrate that it is difficult to simulate in a one-dimensional model the conditions which exist in an actual reservoir after a steamflood, but that it is possible to create "emulsion blocks" in situ under appropriate conditions.

The injection of emulsifier caustic into the Kern River sandpack (test 4) caused a significant permeability reduction, but not until injection of 1 PV of water. The stability of this "block" to steam was not tested. The photograph of the sandpack in Figure 11 reveals that brown, oil-in-water emulsions were formed in situ during the experiment.

Injection of surfactant slugs alone did not spontaneously produce macroemulsions which would significantly lower the permeability of the cores. Sometimes subsequent injection of an inert gas produced an oil-in-water emulsion that was part of the liquid phase of the foam.

Possible Extension to Field Situation. Use of emulsions for profile improvement may have merit in a field situation, but this has not been demonstrated. Before the process can be applied to a field

TABLE IX. In Situ Emulsion Formation

| Emulsifier Injected (PV) | Permeability | | | | | |
	Absolute (md)	Before emulsion (md)	Final (md)	Reduction (%)	Emulsifier	Flow rate (cc/min)
		Wilmington Oil - 50°C				
0.45	325	45	33	27	0.55% NaOH	3.4
0.47	257	24	18	25	1.06% NaOH	3.4
0.83	1400	88	50	43	1.06% NaOH	3.4
		Kern River Oil - 76°C				
0.49	1775	74	24	67	1.25% NaOH	2.0

Figure 11. Emulsion formation in Kern River sandpack.

situation,   additional,   field   specific   experiments   need   to   be
performed.
   The   emulsification   properties   of   the   crude   oil   must   be
determined.   Some   crude   oils   can   be   emulsified   with   surfactant
mixtures, others with caustic.   Some crudes, such as Hasley Canyon
(Table   III),   are   difficult   to   emulsify.   Experiments   can   be
performed to determine if in situ emulsification is feasible, or if
an   emulsion   must   be   injected.   If   in   situ   emulsification   is
feasible, loss of chemicals to reservoir rock is a problem to be
addressed.   If   in   situ   emulsification   is   employed   in   conjunction
with steam, it must be determined if chemicals are most effective
when   injected   with   the   flowing   steam   or   when   chemical/steam
injections are alternated.   Relative permeabilities of the injected
fluids should be determined.   All of this information is needed to
calculate the economics of scale-up to a specific field situation.
   Although not included in this paper, a field experiment has been
designed for the Kern River Unit located at the edge of the Kern
River field, Kern County, California.   The ideal emulsification
properties of the crude oil have already been discussed in this
paper.   This field is an example of reservoir properties that should
be considered ideal when selecting a reservoir for future testing of
the process: heterogeneous production zones, areas of low and high
oil saturations, low salinity, low clay content, and crude oil that
is easily emulsified.

Summary

This research has shown that emulsions which are stable at elevated
temperatures and survive dilution with fresh water can be formed.
They have drop sizes appropriate for blocking pores in a porous
medium at elevated temperatures and in the presence of saturated
steam.   Emulsion blocking also occurs in the presence of residual
oil.   It was also demonstrated that injection of a small slug of
emulsion made from oil and water available in a specific field
caused a significant reduction in the permeability of core material
from that field.

Acknowledgments

The authors wish to acknowledge support for this research by the
U.S. Department of Energy (Cooperative Agreement DE-FC22-83FE60149)
and Energy Associates.

Literature Cited

1.   Frauenfeld,   T.   W.   J.,   et   al.   The   Effect   of   an   Initial   Gas
     Content on Thermal EOR as Applied to Oil Sands.   Proc. 56th
     Annual California Regional Meeting, paper SPE 15086, Oakland,
     California, 1986.
2.   McAuliffe,   C.   D.   Oil-in-Water   Emulsions   and   Their   Flow
     Properties in Porous Media, J. Pet. Tech., June 1973, p. 727-33.
3.   McAuliffe, C. D.   Crude Oil-in-Water Emulsions to Improve Fluid
     Flow in an Oil Reservoir, J. Pet. Tech., June 1973, p. 721-26.

4. Varnon, J. E., et al. High Conformance Enhanced Oil Recovery Process, U. S. Patent 4,161,218, July 17, 1979.
5. Schievelbein, V. H. High Conformance Oil Recovery Process, U. S. Patent 4,161,983, July 24, 1979.
6. Schievelbein, V. H., et al. High Conformance Enhanced Oil Recovery Process, U. S. Patent 4,161,218, July 17, 1979.
7. Schievelbein, V. H. High Conformance Oil Recovery Process, U.S. Patent 4,184,549, Jan. 22, 1980.
8. Wu, C. H., et al. High Vertical and Horizontal Conformance Thermal Oil Recovery Processes, U.S. Patent 4,175,618, Nov. 27, 1979.
9. Cooke, C. E., Williams R. E., and Kolodzie P. A. Oil Recovery by Alkaline Waterflooding, J. Pet. Tech., December 1974, p. 1365-74.
10. Hurd, B. G. Oil Recovery Process for Stratified High Salinity Reservoirs, U. S. Patent 4,458,760, July 10, 1984.
11. Wasan, D. T., et al. The Mechanism of Oil Bank Formation, Coalescence in Porous Media and Emulsion Stability, Report prepared for the U. S. Department of Energy under Contract No. AER-76-14904.
12. Mayer, E. H., et al. Alkaline Waterflooding - Its Theory, Application and Status, proc. Second European EOR Symposium, Paris, 1982.
13. Cash, R. L., et al. Spontaneous Emulsification - a Possible Mechanism for Enhanced Oil Recovery. Proc. 50th SPE Annual Fall Meeting, paper SPE 5562, Dallas, Texas, 1975.
14. Miller, C.A. and Neogi, P. Thermodynamics of Microemulsions: Combined Effects of Dispersion Entropy of Drops and Bending Energy of Surfactant Films, AIChE Journal, v. 26, No. 2, 1980, p. 212-20.
15. Becher, P. Emulsions: Theory and Practice. Reinhold Publishing Corporation, New York, (1983).
16. Soo, H. and Radke, C. J. The Flow of Dilute, Stable Emulsions in Porous Media, Ind. Eng. Chem. Fundam., v. 23, No. 3, 1984, p. 342-47.
17. French, T. R., J. S. Broz, P. B. Lorenz, and Bertus, K. M. Use of Emulsions for Mobility Control During Steamflooding. Proc. 56th SPE Calif. Regional Meeting. Paper SPE 15052, 1986.

RECEIVED January 5, 1988

Chapter 22

Injectivity and
Surfactant-Based Mobility Control

Field Tests

Duane H. Smith

Enhanced Oil Recovery Group, Morgantown Energy Technology Center,
U.S. Department of Energy, Morgantown, WV 26507–0880

Several field tests of injectivity and of various
types of surfactant-based mobility control for gas
flooding have been reported in the literature. These
are briefly reviewed for clues to possible future
directions for the technology and as a guide to the
research and development needed for achieving technical success and commercialization of surfactant-
based mobility control for gas flooding.

Completion of several successful field tests, usually by several different operators and often performed intermittently over a period of
many years, is the final step before commercialization of a new
enhanced oil recovery technology. For surfactant-based, gas-flood
sweep and mobility control, several field tests covering different
injection schemes and various control techniques have been performed.
These include a test in the very large SACROC CO_2 project[1]; various
injection tests at Siggins Field, Illinois[2]; the test of CO_2/
surfactant solution coinjection at Rock Creek in West Virginia[3];
and the use of foam as a diverting agent in the Wilmington Field[4].
Other field tests are planned[1]. The completed projects and their
conclusions are summarized and discussed here as a guide to needs for
further research and development and for clues to possible future
directions for surfactant-based reservoir sweep control for gas
flooding.

Field Tests

SACROC(1). The SACROC CO_2 flood is the largest and perhaps most
important flood of its type. In 1984 Chevron performed a test of
surfactant-based mobility control in this field. Although the test
has not been fully described in the open literature, it has been
widely discussed in the oil industry[1]. The surfactant used was
Alipal CD-128, a mixture of alkyl ethoxysulfates having an average
of about 2.5 ethoxy groups and 8 to 10 carbons in the alkyl chains.
This commercial surfactant has long been sold for other uses, but has
been studied by Holm and others for gas mobility control[5-7]. The

surfactant was used as an in situ dispersing agent, that is, surfactant and CO_2 were injected sequentially in the hope that in situ
shear would be sufficient to disperse CO_2 in the aqueous phase.
     Although no significant adverse effects occurred, the test pro-
duced essentially no benefits either. Because of the dual porosity
of the reservoir (i.e., open fractures as well as intergranular
pores), the test has been widely accepted as demonstrating that
"foams" are unsuited for correcting gross rock heterogeneities such
as fractures(1). Unfortunately, no conclusion has been advanced in
the open literature as a guide to future tests.

Siggins Field Injectivity Tests(2). An extensive series of injection
tests was performed in the Siggins Field in Illinois over a 3-year
period from 1964 to 1967, following the early laboratory work of
Bernard and Holm(8). The field had been discovered in 1906, extensive water flooding had been started in the 1940's and 1950's, and
the water-to-oil ratio (WOR) had reached 25:1. The tests were performed in a typical fine-grained, Pennsylvanian-aged sandstone whose
thickness ranged from about 30-50 ft, at a depth of 400 ft. The
sandstone had an average porosity of 19% and an average permeability
to air of 56 mD. The saturation of the oil, whose viscosity was 8 cP
at the reservoir temperature of 65°F, was 26% of the pore volume.
     Fluids were injected into a central well, surrounded by 6 pro-
duction wells at distances from about 200-360 ft (Figure 1). Injection equipment consisted of an air compressor, a pump for the
25 drums of 60% active surfactant that were used, a pump for dilution
of the concentrated surfactant solution with plant water, a foam
generator (2-ft long, 3-in. diameter pipe packed with ceramic particles) that optionally could be bypassed, and a sight glass for visual
observation of the fluids before they entered the well.
     An ethoxylated alkyl sulfate with small amounts of amide stabi-
lizers was used at concentrations (in the aqueous phase) of 0.5, 1.0,
and 1.5%. The air-to-aqueous phase ratio (at reservoir pressure) was
10:1. Two types of tests were performed: (1) alternate injections
of surfactant solution and air, as in WAG (water alternating with
gas) cycles; and (2) simultaneous injections of air and surfactant
solution, using the foam generator. Diagnostic chemical tracer
injections were used before and after the tests.
     The tests were designed to achieve five specific objectives:
     1. Check for mechanical problems associated with generation of
foam around the wellbore.
     2. Measure the effects of surfactant use on the water injection
profile.
     3. Determine the effects of surfactant use on gas and water
injectivity.
     4. Examine the influence of foam on channeling of gas and of
water between the injection well and producing wells.
     5. Detect transport of surfactant from the injection well to
the producing wells.
     The following results were observed:
     1. No mechanical problems were noted either during the alter-
nating injection of surfactant solution and air or during their
simultaneous injection as foam.

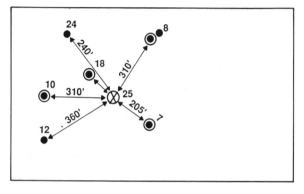

Figure 1. Well pattern for the Siggins Field, Illinois, foam
injection test. (Redrawn with permission from Ref. 2.
Copyright 1970 SPE.)

2. At a concentration as low as 0.1% the surfactant substantially improved the injection profile, increasing the vertical
uniformity of the distribution of water into the pay zone. This
uniformity improvement continued during all subsequent tests
(described below).
3. Sequential injection of 0.1% surfactant solution and air had
little effect on injection rates. After three cycles of alternating
0.5% solution and air followed by two cycles of 1.0% solution and
air, water and air injectivity each decreased about 20%.
4. Simultaneous injection (using the foam generator) of
0.2 pore volume of air and 1.0% surfactant solution reduced the
mobility of water about 30%, but had little effect on the injectivity
of air. Simultaneous injection for 0.6 total pore volume of solution
further reduced the water mobility to 35% of its original value.
These mobility reductions decreased the WOR in nearby wells to 12,
during the period in which the WOR increased to 28 in the rest of the
field.
      The injection of 0.6 pore volume of surfactant solution also
reduced the mobility of air by 50%, and stopped the severe channeling
of air to one of the production wells (No. 7).
5. No surfactant or chemical tracer was detected at any of the
wells in the pilot area. However, about 2 years after the tests
began, surfactant was detected in an abandoned well about 1,500 ft
from the injection well.

Rock Creek(3). The Rock Creek Field, located in Roane County in
southwest West Virginia and containing the Big Injun sandstone, was
discovered in 1906. Primary production (by solution gas drive) produced about 10% of the original oil in place (OOIP), and six different secondary projects were tried in the field. The first and only
successful method was low-pressure gas recycling, which began in 1935
and was extended in 1972. In between these dates three separate
water floods and a steam flood were tried, all without commercial
success. The characteristics of the oil (paraffinic, 43 API gravity,
3.2 cP viscosity) and the reservoir (21.7% porosity, 21.5 mD permeability, 73°F temperature, 24.6-ft thickness, 1,975-ft depth, 34% oil
saturation) suggested the use of a miscible CO_2 flood. Injection of
CO_2 in a pilot of two adjoining 10-acre, 5-spot patterns and 13 surrounding backup water-injection wells began in 1979. The original
design called for continuous injection of CO_2 (following the injection of water to boost the pressure to 1,000 psi). But field problems and the unavailability of CO_2 resulted in unplanned WAG cycles,
water being injected when CO_2 could not be.
      During the latter stages of the CO_2 flood, two new injection
wells (PI-7 and PI-8) and one production well (OB-1) were placed in
operation in an external corner of one of the five spots (Figure 2)
as a "miniflood" test of a conventional CO_2 process. A substantial
amount (11% OOIP) of oil was recovered from this new producer,
demonstrating that the WAG injections up to that time had left a considerable amount of oil. This area (PI-6, PI-7, PI-8, and OB-1) was
selected for a test of the ability of "CO_2-foam" flooding to produce
"quinary" oil, following all of the other methods (primary, low-
pressure gas, water, miscible CO_2) that had been applied.

Laboratory tests with Alipal CD-128 indicated that a steady-state reciprocal apparent viscosity of 0.2-0.4 cP^{-1} (about equal to that of the oil) could be obtained and that surfactant retention by the reservoir rock would not be excessive (0.00056 lb surfactant/ft^3 of rock "permanent" adsorption). Eventually, the following injection sequence was planned: (1) 15 kg of NH_4SCN tracer in 371 bbl of water, followed by a water "spacer" between the tracer and the subsequent surfactant solution; (2) 2,764 bbl of 0.1% surfactant, intended to satisfy the adsorption requirements of the rock; (3) coinjection at 1,000 psi (6.9 MPa) of 5,335 bbl (total) of CO_2 and of 0.05% surfactant in fresh water; (4) displacement, by water, of the dispersion formed during step (3).

Unfortunately, the original location for which the test was planned (with its several closely spaced injection and observation wells) had to be abandoned when it was discovered that one of the injectors was losing large amounts of fluid from the formation. Hence, another site was chosen.

As actually performed, the test had one injection well (PI-2), one observation well (OB-2) 75 ft from it, and a production well (the No. 4 center pilot producer). (See Figure 2.) There was essentially no means to monitor the results of the test if -- as seems to have occurred -- most of the injected fluids flowed in one or more other directions.

The following behavior was observed during the test:
1.   During the injection of 0.1% surfactant solution (before CO_2 injection began) loss of injectivity occurred. This problem could not be duplicated in the laboratory, went away when the solute was switched to chemical tracer, and did not return during the coinjection of surfactant solution and CO_2.
2.   None of the thiocyanate tracer was ever detected at the observation well. (No attempt was made to detect it elsewhere.)
3.   Production of CO_2 at the observation well began about 7 months after its coinjection began. No direct channeling of CO_2 to the well was observed.
4.   Surfactant was detected in six samples taken from the observation well about 10 months after injection of the 0.1% solution began, but the concentration of produced surfactant was not determined.
5.   No sign of an oil bank was detected at the observation well. Before the test began, the average oil saturation in this well had been 19% in the lower 20 ft of the formation and just 4% in the upper 10 ft.
6.   The rate of coinjection of CO_2 and surfactant solution (70-80 vol % CO_2) was approximately 73 bbl/d, compared to the average water injection rate of about 120 bbl/d. The injection rate for CO_2 was about three to five times the water injection rate.

The operation of the test as a quinary flood, the low oil saturation (average 14%) at the observation well before the test began, and the vertical distribution of that oil (virtually all in the lower two-thirds of the formation) made it very difficult for the test to recover oil. With the forced relocation of the test from its originally planned multiwell site, virtually all chance of success came to depend on flow past the single observation well and the ability to accurately measure solute concentrations in the aqueous phase at that well.

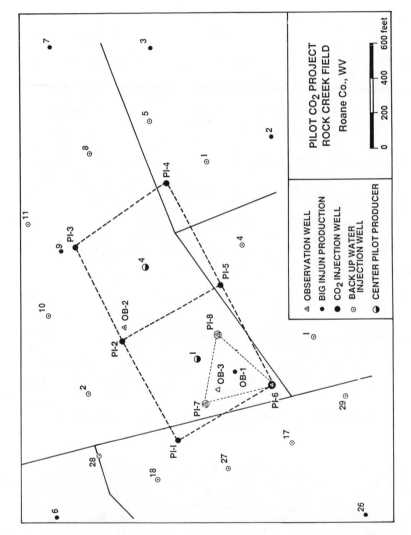

Figure 2. Well patterns for the CO_2 WAG and subsequent surfactant-based CO_2 mobility control test at Rock Creek, West Virginia.

The experience from Rock Creek and Siggins Field suggests that tests may be substantially more informative if maximum use of pressure tests and multiple tracers is made to determine reservoir flow patterns before a site is selected and a test is designed, instead of being used as part of the test after a site has been selected.

Long Beach CO_2 Diversion Test(4). The injection of an 85:15 mixture of CO_2 and N_2 into the Tar Zone of Fault Block V of the Wilmington, California, field began in February 1982, after the field had been water flooded for 20 years. The field pressure (1,100 psi) was well below the miscibility pressure (> 3,000 psi) of the heavy (14 API gravity) oil. In this 320 acre WAG project gas was injected in 2 wells and water in 8, with the gas injection rotated among the 10 injection wells. One of these 10 wells was selected as the injector for a test of the surfactant Alipal CD-128 as a diverting agent.

Figure 3 is a schematic of the sand zones near the injection well. The three zones of unconsolidated sand had varying thicknesses and permeabilities, but the vertical effective pays were estimated to be 70 ft, 45 ft, and 45 ft, respectively, in the S, T, and F1 zones. Permeabilities ranged from 100 to 1,000 mD, being greater in the S zone.

At the time of the surfactant test, the WAG process had increased oil production, but there were two main problems: (1) excessive gas production in some wells, and (2) poor distribution of fluids among the three zones. In the diversion test injection of surfactant solution was alternated with the injection of gas. Both fluids were expected to go preferentially into the highly water-saturated S zone that was taking the largest volumes of fluids, creating an in-depth dispersion that would reduce the mobility of both gas and water in that zone, and diverting these fluids into the two lower, less-swept zones.

Several different laboratory floods were performed with Wilmington crude before the field test began, to supplement previous laboratory experiments with the chosen surfactant(5-7). In these studies significant increases in oil recovery were obtained with the use of surfactant (e.g., an increase from 23% OOIP without surfactant, to 35% with).

The operator decided that the project would be successful if one or more of three goals were met:

1. A decrease of flow of both fluids into the S zone, accompanied by increase of flow into the other two zones, especially the intermediate T zone.

2. A decrease in the injectivity of gas into the well.

3. A decrease in the production of gas from neighboring wells.

The injection procedure was composed of eight cycles, each cycle consisting of about 2,600 bbl of 1% surfactant in formation brine followed by nine times that volume of CO_2-N_2 gas. These volumes were equivalent to radially uniform distributions of about 40 ft for the aqueous solution and 110 ft of fluid-fluid dispersion. Gas and water injection profile surveys made before, during, and after these cycles indicated that the fluids entered a 70-ft vertical section (the S zone) of the formation.

The test results fulfilled the first two objectives. But because of the way in which the test was conducted, it could not be determined if any effect was observed for the third:

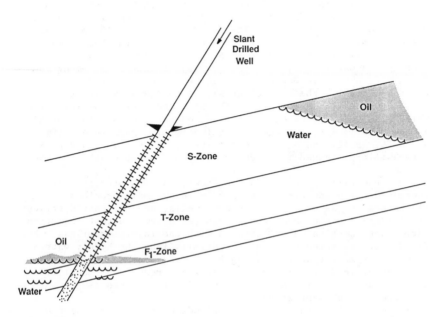

Figure 3.   Schematic of the injection well and pay zones of the
           Wilmington Field immiscible CO_2 diversion test.

1. The entry of gas into the S zone was reduced from 99% to as little as 57%, with the T zone receiving up to 43% of the gas.
2. At the same injection pressure, the gas injection rate following injection of the surfactant was reduced by as much as two-thirds from the pretest rate. Pressure falloff tests showed that in-depth reductions of permeability had been achieved.
3. No reduction of gas production occurred from surrounding wells, but it could not be determined whether the produced gas came from the test well or from other injection wells in the field. In retrospect, use of a gas tracer in the test injection well might have answered this question.

Originally it was planned to continue the final period of gas injection for a much longer period of time to determine how long the flow diversion would continue without further surfactant injection. In the Siggins Field test performed many years before, continued injection of gas had caused little collapse of foam(2).

Instead, the test well was returned to water injection in its rotation as part of the regular WAG production cycle. After about three times as much water had been injected as surfactant solution had been used, the injectivity and injection profile returned to their pretest behavior and the benefit of the surfactant was lost.

Future Field Tests(1). A second surfactant test is being planned for the SACROC Field, with the objective of improving the areal sweep. This test will use a different, proprietary surfactant from the 1984 test, and will undoubtedly include other changes as well. Planning for possible field tests is also underway at other companies, although many of these future tests are still in their very early stages. The urgency of these tests may depend heavily on the near-term behavior of the gas floods now in progress.

Conclusions

It should be noted that the field tests were made with only one type of surfactant, and without benefit of many recent research advances in such areas as high-pressure phase behavior and surfactant design, mechanisms of dispersion formation and disappearance, and mechanisms of dispersion flow through porous media. Furthermore, the design and successful performance of field tests pose many technological challenges in addition to those encountered in the prerequisite experimental and theoretical research.

Although the number of field tests is still limited, the extant tests do include a variety of rock types (sandstone, unconsolidated sand, carbonates), dips (updip and none), and oils (light and heavy). Overall, the success rate has been more or less typical of field tests of new oil recovery technologies. Based on these tests, the following conclusions may be reached:
1. No mobility-control test produced any irreversible reservoir damage or ultimate loss of oil recovery. Future tests should not be avoided because of fears of this type.
2. Field flow patterns often differ greatly from what would be expected for uniform fields. Hence, maximum possible characterization of flow before the test site is selected may greatly improve the value of the test.

3. During the test it is important that mechanical breakdowns be avoided, that the test not be interrupted by regular field production cycles, and that maximum use of tracers, pressure falloff measurements, observation wells, and other measurement techniques be employed.

4. Reasonable injectivities were maintained with either alternate injection of surfactant solution and gas, or with coinjection. In tests where vertical injection profiles were measured, the use of surfactant substantially improved the profile(2,4). When injectivity reductions occurred, these indicated desirable reductions in mobilities and in the loss of fluids to high-permeable zones(2,4). In the one case where injectivity problems initially occurred, these happened before any dispersion was formed, and later disappeared(3).

5. In-depth dispersion placement and permeability reduction was achieved in the immiscible-CO_2 diversion test(4). In the Siggins Field test, the severe channeling of gas to a nearby producer was blocked(2).

6. Dispersions have proved to be much more stable to gas injection than to water injection(2,4). The maximum ratio of water to surfactant solution that can be used in a three-phase cycle of surfactant solution, gas, and water is uncertain, but is probably no more than three(4).

Therefore, methods of stabilizing dispersions against water injection (e.g., use of other surfactants, lamellar liquid crystals, use of very dilute solutions of water-soluble polymers with surfactants) might aid commercialization by reducing surfactant costs.

7. More field tests will be needed, especially to incorporate research advances in such areas as materials design; phase behavior and dispersion morphology; mechanisms of dispersion formation, flow, and breakdown; and simulation of dispersion-based sweep control.

Acknowledgments

The author thanks L. W. Holm and R. J. Watts for reading the manuscript, and L. W. Holm for Figure 3.

Literature Cited

1. Soc. Petrol. Engrs. Forum Series "Monitoring Performance of Full-Scale CO_2 Projects," Durango, Colorado, August 17-21, 1987.
2. Holm, L. W. J. Petrol. Technol. 1970, 22, 1499.
3. Heller, J. P.; Boone, D. A.; Watts, R. J. Proc. Soc. Petrol. Engrs. East. Reg. Mtg., Morgantown, WV, November 6-8, 1985, SPE 14519.
4. Holm, L. W.; Garrison, W. H. Proc. SPE/DOE Fifth Symp. Enhanced Oil Recovery, Tulsa, April 20-23, 1987, p. 497, SPE/DOE 14963.
5. Bernard, G. G.; Holm, L. W. U.S. Patent 4,113,011, 1978.
6. Bernard, G. G.; Holm, L. W.; Harvey, C. P. Soc. Petrol. Engrs. J. 1980, 20, 481.
7. Patton, J. T. Enhanced Oil Recovery by CO_2 Foam Flooding, Final Report, U.S. DOE, DE8200905, 1982.
8. Bernard, G. G.; Holm, L. W. Soc. Petrol. Engrs. J. 1964, 4, 267.

RECEIVED January 29, 1988

Author Index

Affiliation Index

Subject Index

Production by Meg Marshall
Indexing by Deborah H. Steiner
Jacket design by Carla L. Clemens

Elements typeset by Hot Type Ltd., Washington, DC
Printed and bound by Maple Press, York, PA

Recent ACS Books

Biotechnology and Materials Science: Chemistry for the Future
Edited by Mary L. Good
160 pp; clothbound; ISBN 0–8412–1472–7

Chemical Demonstrations: A Sourcebook for Teachers
By Lee R. Summerlin and James L. Ealy, Jr.
Volume 1, Second Edition, 192 pp; spiral bound; ISBN 0–8412–1481–6

Chemical Demonstrations: A Sourcebook for Teachers
By Lee R. Summerlin, Christie L. Borgford, and Julie B. Ealy
Volume 2, Second Edition, 229 pp; spiral bound; ISBN 0–8412–1535–9

Practical Statistics for the Physical Sciences
By Larry L. Havlicek
ACS Professional Reference Book; 198 pp; clothbound; ISBN 0–8412–1453–0

The Basics of Technical Communicating
By B. Edward Cain
ACS Professional Reference Book; 198 pp; clothbound; ISBN 0–8412–1451–4

The ACS Style Guide: A Manual for Authors and Editors
Edited by Janet S. Dodd
264 pp; clothbound; ISBN 0–8412–0917–0

Personal Computers for Scientists: A Byte at a Time
By Glenn I. Ouchi
276 pp; clothbound; ISBN 0–8412–1000–4

Writing the Laboratory Notebook
By Howard M. Kanare
146 pp; clothbound; ISBN 0–8412–0906–5

Principles of Environmental Sampling
Edited by Lawrence H. Keith
458 pp; clothbound; ISBN 0–8412–1173–6

Phosphorus Chemistry in Everyday Living
By Arthur D. F. Toy and Edward N. Walsh
362 pp; clothbound; ISBN 0–8412–1002–0

Chemistry and Crime: From Sherlock Holmes to Today's Courtroom
Edited by Samuel M. Gerber
135 pp; clothbound; ISBN 0–8412–0784–4

For further information and a free catalog of ACS books, contact:
American Chemical Society
Distribution Office, Department 225
1155 16th Street, NW, Washington, DC 20036
Telephone 800–227–5558